# Race and Ethnicity in Society

## The Changing Landscape

### THIRD EDITION

**ELIZABETH HIGGINBOTHAM**
University of Delaware

**MARGARET L. ANDERSEN**
University of Delaware

WADSWORTH
CENGAGE Learning

Australia • Brazil • Japan • Korea • Mexico • Singapore • Spain • United Kingdom • United States

WADSWORTH
CENGAGE Learning™

**Race and Ethnicity in Society: The Changing Landscape, Third Edition**

Elizabeth Higginbotham and Margaret L. Andersen

Publisher/Executive Editor: Linda Schreiber-Ganster

Acquisitions Editor: Erin Mitchell

Assistant Editor: Linda Stewart

Editorial Assistant: Mallory Ortberg

Media Editor: Melanie Cregger

Marketing Manager: Andrew Keay

Marketing Assistant: Dimitri Hagnere

Marketing Communications Manager: Laura Locallio

Content Project Management: PreMediaGlobal

Creative Director: Rob Hugel

Art Director: Caryl Gorska

Print Buyer: Karen Hunt

Rights Acquisitions Account Manager, Text and Image: Dean Dauphinais

Production Service: PreMediaGlobal

Copy Editor: PreMediaGlobal

Cover Designer: Reizebos Holzbaur/Brie Hattey

Cover Image: Panoramic Images

Compositor: PreMediaGlobal

For product information and technology assistance, contact us at
**Cengage Learning Customer & Sales Support, 1-800-354-9706**

For permission to use material from this text or product, submit all requests online at **www.cengage.com/permissions**
Further permissions questions can be e-mailed to
**permissionrequest@cengage.com**

Library of Congress Control Number: 2010942353

ISBN-13: 978-1-111-51953-7

ISBN-10: 1-111-51953-6

**Wadsworth**
20 Davis Drive
Belmont, CA 94002-3098
USA

Cengage Learning is a leading provider of customized learning solutions with office locations around the globe, including Singapore, the United Kingdom, Australia, Mexico, Brazil, and Japan. Locate your local office at **www.cengage.com/global.**

Cengage Learning products are represented in Canada by Nelson Education, Ltd.

To learn more about Wadsworth, visit **www.cengage.com/ Wadsworth**

Purchase any of our products at your local college store or at our preferred online store **www.cengagebrain.com.**

Printed in the United States of America
1 2 3 4 5 6 7 15 14 13 12 11

# Table of Contents

# Preface

The study of race and ethnicity is changing, just as the character of race and ethnic relations is changing. Among other things, with the extraordinary election of President Barack Obama, the nation has its first African American President. Obama's election and his presidency signal tremendous changes in American race relations. It took a wide coalition of groups to elect him, an outcome that would have been unimaginable not that many years ago. As a result, in the aftermath of the election, many declared the United States to be a "post-racial" society—as if all of the injustices and inequalities of race were now gone because of this one man. But one can hardly call the United States "post-racial" when stark realities of racial inequality still persist, even though they are sometimes hidden from view. So, Obama's presidency does signal a new reality—one where racism in some forms is declining. But, at the same time, racial inequality—and its sister, ethnic inequality—are continuing realities of American life.

The truth is that the United States is now a very multiracial society. New patterns of immigration since the late 1960s have brought new populations to the United States—populations that are changing the racial and ethnic composition of the nation. Whereas the focus of study about race and ethnicity has long assumed a "Black/White" framework, there is now more attention to the different racial and ethnic groups that make up the United States. This is especially obvious in the rather heated national discussions about immigration and immigration policy.

This anthology is intended to introduce students to the study of race by engaging them in the major topics and themes now framing the study of race in the United States. Because society is changing in its racial and ethnic makeup, current scholarship on race is also changing, focusing more on the social construction of race, racial and ethnic identities, and the containing patterns of racial exclusion. Reflecting changes in the scholarship on race, this book departs from the traditional "smorgasbord approach" to studying race—an approach that documents the different histories of an array of groups, but where ethnicity is seen as only applying to Whites of European ancestry. Within this model, ethnicity has been perceived as mostly "white," race as "color." Such an approach can no longer capture the complexities and dynamism of race and ethnicity in the United States.

In this society, race and ethnicity are still mechanisms for sorting people, what the American Sociological Association has called "stratifying practices". Disparities in jobs, housing, health, education, and other facets of life can be largely attributed to race and ethnicity, along with their relationship to other social factors such as social class, gender, age, nationality, and so on. Yet, many people claim that the United States is now a colorblind society where race no longer matters. The current generation of college students has come of age not just in a more diverse nation, but at a time when race *appears* to have lost some of its significance in organizing relationships and social institutions. Popular culture makes it appear that interracial relationships are common and that race no longer matters in shaping social relations. The dominant ideology is one of "color blindness," as if recognizing race is the same as being racist.

But, race does matter—and it matters a lot—in education, in workplaces, in communities, in health care systems, in courts and policing, in everyday interactions, and in our identities, complex as those might be because of the multiple social locations in which people's identities emerge. In fact, race and ethnicity are present throughout day-to-day life; they are embedded in social institutions and in social relationships. And they have real and recurring social and historical consequences. Neither race nor ethnicity just go away by denying that they are there.

As the nation becomes more racially and ethnically diverse, studying race has become more complicated. We offer this anthology to those teaching courses on race and ethnicity who want material and readings for their students that reflect this complexity, while also keeping the book's contents accessible to undergraduate readers. The book is intended primarily for courses on the sociology of race and ethnicity, although it can be used for courses in other departments and interdisciplinary programs where courses on race are being taught (such as education, political science, ethnic studies, and some humanities departments).

We have organized the third edition to reflect the different themes that underlie the study of race and ethnicity—consciously doing so around themes instead of around particular group experiences. The major themes of this book include:

- showing the diversity of experiences that now constitute "race" in the United States;
- teaching students the significance of race as a socially constructed system of social relations;
- showing the connection between different racial identities and the social structure of race;
- understanding how racism works as a belief system rooted in societal institutions;

- providing a social structural analysis of racial inequality;

- providing a historical perspective on how the racial order has emerged and how it is maintained;

- examining how people have contested the dominant racial order;

- exploring current strategies for building a just multiracial society.

## NEW TO THE THIRD EDITION

We have made substantial changes in the third edition, reflecting both changes in society since the publication of the second edition and the newest developments in the scholarship on race and ethnicity. The third edition of *Race and Ethnicity in Society: The Changing Landscape* has also benefited tremendously from the comments of reviewers and those who used the first two editions of the book. As a result, we have **reorganized part of the Table of Contents into four major parts and thirteen different topical sections** (described further below). Instead of putting all social institutions into one very large part, as we did in the second edition, we have broken this part into different sections, each on an important social institution; thus, Part III on "Race and Social Institutions" has five separate sections within it (on work; families and communities; housing and education; health care and the environment; and, the criminal justice system).

We have included **two entirely new sections** in this edition—one on "Media and Popular Culture" (Section III) and one on "Race, Class, and Gender Inequality" (Section VII). The section on media and popular culture is included because of the tremendous influence of the media—and its representations of race and ethnicity—on today's students. Articles in this section explore how race and ethnicity are represented in diverse forms of media, including hip hop, beauty culture, sports mascots, and even Halloween costuming.

In the new section on race, class, and gender inequality (Section VII, "Exploring Intersections: Race, Class and Gender Inequality") we examine how race and ethnicity are manifested differently, given their intersection with class and gender. This section provides a conceptual framework for thinking about overlapping forms of inequality and does so by also looking at various myths about race and masculinity (in particular, the myth of the absent father), as well as how middle-class Latinas negotiate some of their work environments.

There are **33 new articles** in the third edition of the book, selected to keep the anthology current and to reflect the very good work that people are doing as they think about race and ethnicity in society. The new articles, like those we have re-used, have been selected for their importance, but also for their *accessibility to undergraduate readers*. The new articles are on such topics as the impact of the current recession on people of color (Melvin L. Oliver and Thomas M. Shapiro, "Sub-Prime as a Black Catastrophe"); the significance of President Obama's election (Michael Omi and Taeku Lee, "Barack Like Me: Our First Asian American President" and Thomas F. Pettigrew, "Post-Racism? Putting

Obama's Victory in Perspective"); racial-ethnic profiling of Arab Americans (Susan M. Akram and Kevin R. Johnson, "Race, Civil Rights, and Immigration Law After September 11, 2001: The Targeting of Arabs and Muslims"); the new social and geographic locations of immigrants (Charles Hirschman and Douglas S. Massey, "Places and Peoples: The New American Mosaic"); white identities (Tim Wise, "White Like Me: Reflections on Race from a Privileged Son"); re-segregation of the schools (Gary Orfield and Chungmei Lee, "Historic Reversals, Accelerating Resegregation and the Need for New Integration Strategies"); and, diversity in corporate life (Frank Dobbins, Alexandra Kalev, and Erin Kelly, "Diversity Management in Corporate America"), to name a few.

We have added a new feature, included at the end of each of the thirteen introduction called "**Face the Facts.**" These are graphic depictions of basic information relevant to the section topics. For example, in Part IV on "Race and Identity in U.S. Society," we include a chart on changes in the racial identities of the U.S. population, taken from the U.S. Census. As another example, in the section on race, class, and gender inequality, we include data on median income by race and gender. Each section includes this feature so that instructors can help students interpret basic demographic and other data pertinent to the study of race and ethnicity. This feature will be especially helpful for instructors who want to use the anthology as a stand-alone text. For others, this can be a good companion reader to any textbook or to other books.

## ORGANIZATION OF THIS BOOK

This book is organized in four major thematic parts, subdivided into thirteen different sections. Part I ("The Social Basis of Race and Ethnicity") establishes the analytical frameworks that are now being used to think about race in society. The section examines the social construction of race and ethnicity as concepts and experiences. Together, the articles show the grounding of racial and ethnic categories in specific historical and social contexts and their fluidity over time.

Part II ("Continuity and Change: How We Got Here and What It Means") explores both the historical patterns of inclusion and exclusion that have established racial and ethnic inequality, while also explaining some of the contemporary changes that are shaping contemporary racial and ethnic relations. Critical to this section is the concept of citizenship—that is, who belongs, both in the formal and informal sense—and how belonging is shaped by social policies and social beliefs. Thus, immigration is a central topic in this section, but we also use this section to explore overlapping forms of inequality, reminding students that race and ethnicity are neither static nor uni-dimensional. Rather, racial, class, and gender stratification are overlapping and intersecting realities; the social location of groups is configured through these intersecting social factors.

Part III ("Race and Social Institutions") examines the major institutional structures in contemporary society and investigates patterns of racial inequality within these institutions. Persistent inequality in the labor market and in patterns of community, residential, and educational segregation continue to shape the life chances of different groups. Indeed, the well-being of different groups is patterned by institutional racism, which we examine in the context of these different institutional forms.

Finally, Part IV ("Building a Just Society") concludes the book by looking at both large-scale contexts of change, such as those reflected in the movement to elect the first African American president. But here we also ask students to think about the contexts in which they might find themselves working and living—beyond their college experience. Together, the articles in this section show the different ways that change can occur and the efforts needed to build a more just society.

Within these four major thematic parts are sections that organize an understanding of the changing landscape of race and ethnicity. In Part I, we include four sections beginning with Section I ("The Social Construction of Race and Ethnicity'). This section lays a basic foundation for understanding race and ethnicity as social, not biological or genetic, facts. Section II ("What Do You Think?: Prejudice, Racism, and Stereotyping") examines the most immediately experienced dimensions of race: beliefs and ideology. We think that students learn a lot about race by examining the manifestations of a racially-stratified society in people's beliefs. This section also includes material on the current ideology of colorblind thinking and how it perpetuates racial inequality.

Section III ("Representing Race and Ethnicity: The Media and Popular Culture") is a new section that includes empirical analysis of one of the greatest influences on students' thinking about race and ethnicity: the media and popular culture. Articles in this section ask readers to think critically about the media and popular culture that they consume and the racial representations that are rampant but often taken-for-granted. The articles specifically examine diverse forms and subjects in popular culture, such as hip hop, dominant images of beauty and race, sports icons, and holiday costuming. We think that having students examine these popular culture forms early in the book will make them more critical observers of the cultures in which they likely participate.

Section IV ("Who Are You?: Race and Identity") examines racial identities—a topic that we think is especially interesting to students. In a racially diverse and changing society, many people now form racial identities that cross racial and ethnic boundaries. This section also includes a new article on white identity, to bring the current interest in whiteness into this book.

In Part II we examine patterns of exclusion, how they have been established, how they define citizenship and belonging, and how racial exclusions intersect with class and gender inequities. This part includes three sections, beginning with Section V ("Who Belongs?: Race, Rights and Citizenship"). This section retains the focus on conceptualizing citizenship that was present in the first edition, but strengthens the analysis of citizenship and state policy by including new pieces on state policy and American Indians, as well as new material on the suspicions cast on some new immigrant groups as the result of the politics of fear subsequent to 9/11.

Section VI ("The Changing Face of America: Immigration") includes articles that examine different dimensions of immigration experiences. There is a review of immigration policy here with a discussion of the implications of trends in policy for the future of the treatment of immigrants, as well as articles that explore generational change within immigrant families. The section also addresses comparisons with current and past immigration patterns, as well as discussing the global context of immigrant movement.

Section VII ("Exploring Intersections: Race, Class, Gender, and Inequality") is a new section that includes material on the connections between racial inequality, social class inequality, and gender inequality. It is framed by articles that introduce students to intersectional thinking, followed by work that shows how race, class, and gender influence the lives of people in different social contexts—specifically, fatherhood for African American men, the challenges young African American women face in the inner city, and professional work for Latinas.

Part III ("Race and Social Institutions") includes five sections, each focusing on racial stratification within major institutions of U.S. society. Section VIII ("Race and the Workplace") details how racial oppression in the United States was originally grounded in economic exploitation. This section shows how racial inequality shapes employment patterns and employment options for people of color. We also see here how patterns of segregation and inequality are found in the work of immigrants, but at the same time how people of color negotiate the work environments in which they are employed.

Section IX ("Shaping Lives and Love: Race, Families, and Communities") shows that racism is not an abstract force. It harms people, their families, and the places they live even while people actively resist psychic assaults and work to build families and communities for social support. Racial segregation disrupts not only families and communities but also relationships that people might otherwise have.

Section X ("How We Live and Learn: Segregation, Housing, and Education") focuses on racial segregation in housing and its co-dependent, education. As long as racial groups do not live near each other nor are educated with each other, inequalities result. This section also documents the increasing segregation that is a fact of life in contemporary American society.

Section XI ("Do We Care?: Race, Health Care and the Environment") shows the consequences for group health, as well as the health of the planet, because the patterns of racial inequality that mark society influence them. Without attending to the disparities in health, we do not develop a healthy society, especially when racial-ethnic communities are also put at greater risk of toxic environments.

Section XII ("Criminal Injustice?: Courts, Crime, and the Law") studies the institutions that are ravaging poor communities of color—the criminal justice system. How just is a society where racial inequality marks every phase of the criminal justice system? The articles contained in this section examine the injustices that have marked the experiences of people of color, including immigrants, and raise serious questions about the administration of justice in contemporary society.

Finally, Section XIII ("Moving Forward: Analysis and Social Action") concludes the book by asking students to think about social changes that likely will affect the future of race and ethnicity in U.S. society. Together, the articles in this section show how change occurs at the individual, institutional, and societal level—that is, collective action toward a more racially just society.

## PEDAGOGICAL FEATURES

In addition, we have included *new introductions to each of the thirteen sections* of the book. The introductions not only frame each section, but also provide discussion of major concepts needed to interpret the articles and, in many cases, some historical framing of the issues covered in the section.

We have included *brief introductions for each article*, intended to help students understand the major contribution of the article and its connection to the rest of the book. We have included *Discussion Questions* for each article in this book, with the goal of helping students grasp the major points of each argument. These questions can also be used for student paper assignments, research exercises, and class discussion.

There are *student exercises* included at the end of each section. These will provide opportunities for student discussion and both in and out of class projects. They have been revised to reflect the new additions and revisions to the book.

*Face the Facts* is a new feature, included at the conclusion of each section introduction. This feature includes graphs, drawn from census and other public data, that help students visualize some of the contemporary evidence for racial inequalities. This feature will also help students interpret data shown in graphic form so that they can be better educated about the realities of race and ethnicity in America.

## ACKNOWLEDGMENTS

We have benefited from the support and encouragement of many people who have either discussed the contents of the book with us or provided clerical and computer assistance or other forms of help that enabled us to complete/work on this book even in the midst of many other commitments. We thank Maxine Baca Zinn, Bonnie Thornton Dill, Dianna Dilorenzo, Ben Fleury-Steiner, Valerie Hans, Sarah Hedrick, Linda Keen, Carole Marks, Nancy Quillen, Richard Rosenfeld, Joan Stock, Judy Watson and Marilyn Whittington for all the advice, help, and support they provide us. Special thanks go to Katie Grunert for her research assistance. We also appreciate the support provided by the University of Delaware and the Center for the History of Business, Technology, and

Society at the Hagley Museum and Library. The editors at Cengage/Wadsworth have also been enthusiastic about this project, so we thank Erin Mitchell and Chris Caldeira for their enthusiasm and guidance. The suggestions of those who carefully reviewed the first drafts of this book were extremely valuable and have helped make this a stronger anthology, thus we thank Julius H. Bailey, University of Redlands; Marianne Cutler, East Stroudsburg University; Karen Hayden, Merrimack College; Karen Tejada, State University of New York— University at Albany; Lisa Wade, Occidental College; Pamela Williams-Paez, College of the Canyons; and George Wilson.

Elizabeth Higginbotham
Margaret L. Andersen

# About the Editors

**Elizabeth Higginbotham** (B.A., City College of the City University of New York; M.A., Ph.D., Brandeis University) is Professor of Sociology, Black American Studies, and Women's Studies at the University of Delaware. She is the author of *Too Much to Ask: Black Women in the Era of Integration* (University of North Carolina Press, 2001) and co-editor of *Women and Work: Exploring Race, Ethnicity, and Class* (Sage Publications, 1997; with Mary Romero). She has also authored many articles in journals and anthologies on the work experiences of African American women, women in higher education, and curriculum transformation. While teaching at the University of Memphis, she received the Superior Performance in University Research Award for 1991–92 and 1992–93. Along with colleagues Bonnie Thornton Dill and Lynn Weber, she is a recipient of the American Sociological Association Jessie Bernard Award (1993) and Distinguished Contributions to Teaching Award (1993) for the work of the Center for Research on Women at the University of Memphis. She also received the 2003–2004 Robin M. Williams Jr. Award from the Eastern Sociological Society, given annually to one distinguished sociologist. Elected in 2006, she serves a term as Vice President of the Eastern Sociological Society from 2007 to 2008. She is currently in her second term on the Delaware Humanities Council.

**Margaret L. Andersen** (B.A., Georgia State University; M.A., Ph.D. University of Massachusetts, Amherst) is the Edward F. and Elizabeth Goodman Rosenberg Professor of Sociology at the University of Delaware where she also holds appointment in Black American Studies and Women's Studies. She is the author of *On Land and On Sea: A Century of Women in the Rosenfeld Collection; Thinking about Women; Race, Class, and Gender: An Anthology* (co-edited with Patricia Hill Collins); *Sociology: Understanding a Diverse Society* (with Howard F. Taylor); *Sociology: The Essentials* (with Howard F. Taylor); and *Understanding Society: An Introductory Reader* (co-edited with Kim Logio and Howard F. Taylor). Professor Andersen was the

Vice President of the American Sociological Association (ASA) for 2008–09 and received the ASA's prestigious Jessie Bernard Award (2006). She also received the 2004 Sociologists for Women in Society Feminist Lecturer Award and the 2007–2008 Eastern Sociological Society Robin F. Williams Lecturer Award. She has received two teaching awards at the University of Delaware: the University Excellence in Teaching Award and the College of Arts and Sciences Outstanding Teacher Award. She currently chairs the National Advisory Board for the Stanford University Center for Comparative Studies in Race and Ethnicity and is the former editor of *Gender & Society*.

# PART I

# The Social Basis of Race and Ethnicity

# Section I

# The Social Construction of Race and Ethnicity

### Elizabeth Higginbotham
### and Margaret L. Andersen

W hen studying racial inequality, one of the first questions to answer is, "What is race?" Most people assume that they can tell someone's race by just looking. This is because many people think of race in terms of skin color or other biological features that seemingly distinguish different groups. But is this a reasonable understanding of the concept?

Race has been used historically to differentiate groups, but why not use eye color or height or some characteristic other than color to categorize people into so-called races? The actual meaning of race lies not in people's physical characteristics, but in the historical treatment of different groups and the significance that society gives to what is believed to differentiate so-called racial groups. In other words, *what is important about race is not biological difference, but the different ways groups of people have been treated in society*.

This means that *race is a social construction*. This is one of the most important lessons from analyzing race in society. The idea of race has developed within the context of the social institutions and historical practices in which groups defined as races have been exploited, controlled, and—in some cases—enslaved. Imagine this scenario: A social and economic system is created in which some people are forced into slavery based on their presumed inferiority. The dominant group then creates a belief system that tries to explain this exploitation. The solution is creating "race" as an idea that can be used to justify the treatment and abuse of people perceived as different.

Sociologists define a **race** as a group that is treated as distinct in society because of certain perceived characteristics that have been defined as signifying superiority or inferiority. Note that in this definition, perception, belief, and social treatment are the key elements defining race, not the actual physical characteristics of human groups. Furthermore, so-called races are created within a system of social dominance. *Race is thus a social construction, not an attribute of certain groups of people.*

Over time, as Howard Taylor explains in "Defining Race," race has become a complex and multifaceted social phenomenon. It is fundamentally rooted in social definitions: how people see each other, how they define their own identity, and how they are situated within a social order—an order that has been structured along lines of inequality. Taylor's essay explores the different meanings of race in society.

Still, the complex social reality of race does not stop people from thinking of it as biologically rooted. But scientists who have mapped human genetic makeup have concluded that there is no "race gene." Yes, there are certainly physical differences between people, some of which make people identifiable, but at the level of genetic composition, as Joseph L. Graves, Jr. ("The Race Myth") points out, there are far more similarities among people than there are differences, even when you take into account such things as eye color, skin color, and other physical characteristics. Scientists have concluded that genetic variation among human beings is indeed very small, and thus, that there is no biological basis for race.

The idea of race is, in fact, a relatively recent development in modern history. Abby Ferber ("Planting the Seed: The Invention of Race") takes us back to the establishment of race as a concept. She explains that the idea of race emerged through the work of quasi-scientists in the eighteenth century—a time when White Europeans sought to explain and rationalize the exploitation of African, Asian and indigenous people. She shows how the process of developing systems of racial classification is tangled up with the history of racism. This is important for understanding how racist thinking has emerged and how it is tied to the exploitation of people of color worldwide. Her work also shows how racism arose as science was emerging. Her essay will also make you think twice about using the term *Caucasian*, commonly used to refer to White people, once you learn the term's racist origins.

Karen Brodkin ("How Did Jews Become White Folks?") illustrates the social construction of race in a different context. She writes about the long history of **anti-Semitism** (defined as the hatred of Jewish people) and tells us how at different times in world history anti-Semitic thinking constructed Jews, not as a religious ethnic group, but as a separate race. Thus, the Nazi regime of Germany in the 1930s and 1940s constructed Jewish people as an inferior race— systematically murdering millions of Jewish people as a result. Brodkin's essay links anti-Semitism to other forms of racism in that anti-Semitism was supported by *scientific racism* (the use of quasi-science to support racism). As Brodkin shows, even in the United States, racism defined the only "real Americans" as native-born White people. However, as Brodkin shows, Jewish Americans have been upwardly mobile. Why? Because over time they became seen as "White," and conditions in the United States after World War II opened up opportunities to those defined as "White."

Brodkin's essay also brings out another important point: that ethnic groups can become racialized. **Ethnic groups** are those that share a common culture and that have a shared identity. Ethnicity can thus stem from religion, national origin, or other shared characteristics. Important to this definition of an ethnic group is not just the shared culture, but the sense of group belonging. Thus, Jewish people are an ethnic group, as are Italian Americans, Cajuns, and Irish-American Catholics. Ethnicity can exist even within a so-called racial group (such as Jamaicans, Haitians, or Cape Verdeans among Black Americans). Some European ethnic groups have historically been racialized, as Brodkin's essay about Jews shows. This has typically occurred during periods of high rates of immigration when anti-immigrant sentiment swelled and groups, such as the Italians, were defined by dominant groups as racially inferior. The fact that we do not now consider Jews or the Italians to be a race shows how powerful the social construction of race can change over time.

Although concepts of race change over time, race is still "real"—but real in a social and historical sense. Michael Omi and Howard Winant ("Racial Formation") introduce the concept of **racial formation** to refer to the social and historical process by which racial categories are created. Historically in the United States, one's racial membership was determined by law—though, inter-estingly enough, the meaning of race varied from state to state. In Louisiana, for example, you were defined as Black if you had one Black ancestor out of thirty-two; in Virginia, it was one in sixteen; in Alabama, you were Black by law if you had *any* Black ancestry. By just crossing state borders, a person would legally change his or her race!

Racial formation means that social structures, not biology, define race. Because a group's perceived racial membership has been the basis for group oppression, this understanding of race shows how systems of authority and governance construct concepts of race. There are, of course, sometimes observ-able differences between individuals, but it is what these differences have come to mean in history and society that matters.

Together, the articles in this section show us that racism is not just a matter of individual attitudes and prejudices, although those surely are important, as we will see in the next part.

## FACE THE FACTS: THE U.S. CENSUS COUNTS RACE

**Reproduction of Questions on Race and Hispanic Origin From Census 2000**

**NOTE: Please answer BOTH Questions 5 and 6.**

5. **Is this person Spanish/ Hispanic/ Latino?** *Mark **X** the "No" box if **not** Spanish /Hispanic /Latino.*

   □ **No**, not Spanish/Hispanic/Latino          □ Yes, Puerto Rican
   □ Yes, Mexican, Mexican Am., Chicano          □ Yes, Cuban
   □ Yes, other Spanish,/Hispanic/Latino—*Print group*

   [ | | | | | | | | | | | | | | | | | | | | ]

6. **What is this person's race?** ***Mark X one or more races*** *to indicate what this person considers himself/herself to be.*

   □ White
   □ Black, African Am., or Negro
   □ American Indian or Alaska Native—*Print name of enrolled or principal tribe.*

   [ | | | | | | | | | | | | | | | | | | | | ]

   □ Asian Indian          □ Japanese          □ Native Hawaiian
   □ Chinese               □ Korean            □ Guamamian or Chamarro
   □ Filipino              □ Vietnamese        □ Samoan
   □ Other Asian—*Print race.*                 □ *Other Pacific Islander—Print race*

   [ | | | | | | | | | | | | | | | | | | | | ]

   □ *Some other race—Print race*

   [ | | | | | | | | | | | | | | | | | | | | ]

SOURCE: U.S. Census Bureau. 2000. www.census.gov

***Think about it:*** This is how race and ethnic identity are tabulated by the U.S. Census. What would you put? What does this form of questioning tell you about the government's role in the social construction of race and ethnicity?

# 1

# Defining Race

HOWARD F. TAYLOR

*This article shows the complex and multiple dimensions around which the concept of race is socially constructed. Taylor shows that race is a social status, one that takes on meaning within the context of social identities and social relationships.*

Race is with us every minute of every day. Despite our constant protestations against it and its realities, it, like gender, permeates every fiber of our very existence. Race is causally and intimately related to hundreds of forces and has hundreds of consequences. Our race determines, beyond chance, how long we will live. In general, racial minorities in America have a lower life expectancy than Whites. American minorities have less access to medical care; and, as a consequence, minorities are more burdened with chronic and life-threatening illnesses American minorities, particularly Blacks, Hispanics, and American Indians, have considerably lower annual incomes relative to Whites, even Whites who have the same level of education as their minority counterparts. Thus Blacks, Hispanics, American Indians, and also Asians have greater odds of being poor than Whites. Blacks, Hispanics and American Indians are on average promoted less often in the workplace than are Whites with the same or even less education. All three racial minorities routinely suffer housing discrimination and are less likely to be offered low-interest housing mortgages in an urban area than are Whites who live in the same urban area. Surveys find that members of all three minority groups feel more alienated from the institutions of society than do Whites of the same socioeconomic status. Finally, Blacks and Hispanics are more likely to be arrested for crime, more likely to be held without bail, more likely to be sentenced, and more likely to receive longer sentences than are Whites from the same part of the city or town, of the same or similar social class, and who have exactly the same record of prior arrests.

So race matters. It matters a lot. What, then, is "race?" What are the definitions of race that are used in American society? How can one tell who is of which race? If the definitions of race in American society are inexact, as they are, then can one define race in some way so as to be able to research its effects? How can we actually measure one's race?

Race is multiply defined in this society; that is, there is no one single definition, but several definitions. All these definitions apply simultaneously, and no

SOURCE: Howard F. Taylor, "Defining Race." Reprinted by permission of the author.

one definition takes precedence over another. These definitions do, however, have one thing in common: They are all creations of society. They have been put into place by humans and by their social interactions and their societal institutions. Let us have a look at these definitions.

## RACE AS A BIOLOGICAL CATEGORY

Most people grew up thinking that the real definition of "race" is that it is a strictly biological category. This is only partly correct. We were taught to place people into "race" categories such as Caucasian (White), Negroid (Black; Negro; African American), Mongolian (Asian), and so on, on the basis of secondary physical characteristics such as skin color, hair texture, lip form, nose form, and so on. This definition and classification, put forth in the late nineteenth and early twentieth centuries by some physical anthropologists and now considered vastly outdated, has had considerable influence on how people think about the matter of race right up to the present time. It gave rise to the idea that racial classification was "scientific" since it was based on physical traits and upon thinking at the time in the field of physical anthropology. Even some sociology texts as recently as the 1940s defined race in this manner, and furthermore even defined race as a "subspecies" of humankind (Young, 1942)!

It is now generally agreed upon by researchers in both the physical as well as the social and behavioral sciences that only a small part of the definition of race in this society is based on these secondary physical characteristics. Such characteristics do play a part in how people themselves define race, to be sure, but only a part. This is because the human variability between what are regarded in society as "racial groups" occurs mostly within racial groups rather than between them. Thus, the skin color of African Americans varies from extremely light/white (even blond and blue-eyed) all the way to very dark brown. There are White people who have dark skin and very curly hair and full lips yet they are still White, and are so regarded in their immediate communities. Similarly, there are African Americans with white skin, thin lips, and straight hair. So skin color (and lip form and hair texture) is not a very good indicator of race.

The population geneticist Richard Lewontin (1996) notes that even for strictly physical traits involving body chemistry, blood type, and other strictly physical traits, almost all the variability on such traits is within racial categories; almost no variability in such traits exists between races. Lewontin estimates that the overlap in physical traits between any two groups designated as "races" in American society is more than 99 percent. That applies as well to genetics: Any two racial groups compared are 99 percent similar genetically. Clearly, then, in physical characteristics, the races are far more similar than they are different. It is in this sense that one hears that race is "not a true scientific concept." In the sense of thousands of physical traits and characteristics, this is certainly true.

Does this mean that race is no longer important, that race no longer matters? Certainly not. We have already noted that race is a fundamental and firmly ingrained part of human existence, not only in America, but in most other societies as well. How then is it that a concept—with virtually no "scientific" validity (at least, in the physical science sense)—has come to be so important and intrusive in human existence? The answer lies in the realization that the definition of race in America, and in most other societies as well, is largely social.

## RACE AS A SOCIAL CONSTRUCTION

To say that race is a social construction means that it and its definition grow out of the process of human interaction. This means that race is what interacting humans define it to be. In this sense, how you are perceived in a community of peers in part defines your race: If your friends and associates see you as African American, then in that one respect you are indeed African American.

As another example of social construction, if race-like divisions among people in this society were made only on the basis of who had red hair and who did not, then people would come to think of races as being defined by hair color—a physical characteristic. Over time, this definition would come to be upheld by society's institutions—by the courts, by the educational system, by the federal government, and so on. It might well come to be thought of as a truly "scientific" classification, since it is based on a physical characteristic (hair color instead of skin color), and as everyone knows, redheads are different from everyone else—they have fiery tempers and argue a lot, don't they? Thus, racial stereotypes would soon be applied to redheads!

Social construction means that people learn, through socialization and interaction processes, to attribute certain characteristics to people who are classified into a racial category. These are just what racial stereotypes are: attributions that are for the most part not true and yet stubbornly persist over time. These stereotypes are social constructions. They are generally based on only a small truth (if on any truth at all) and are then thought of in society as applying to all members, and to "typical" members, of some racial category.

Stereotypes are generally negative. We have all heard the common stereotypes: Blacks are inherently musical, possess "natural" rhythm, are loud, and crime-prone; Asians are sneaky and overly conforming; Hispanics are naturally violence-prone and carry knives; American Indians are quiet, subservient, and underachievers. These traits are seen by society as inherent to any member of the particular group. In fact they are seen as essential to the group identified by the stereotype. Sociologists call this process essentialization: Such negative stereotypical traits are regarded by society as essential (inherent) to the character of any person identified by the stereotype. Negative stereotypes, thus negative essentializations, are applied far more to minorities than to Whites, and they thus help define what it means to be minority in society, even though Whites are sometimes ridiculed as having a few negatively stereotyped traits ("White boys can't jump," "blondes are dumb," and so forth).

## RACE AS AN ETHNIC GROUP

A group is an ethnic group if its members are united by a common culture. Culture includes language, religion, tools, music, habits, socialization practices, and many other elements. Racial groups are ethnic groups. They generally share a common culture—not perfectly, of course, but to an extent great enough so as to be able to identify some set of cultural elements held in common. Thus African Americans are not only regarded as a racial group but they are also an ethnic group. They possess many elements of a common culture (music, art, linguistic similarities, a sense of "we" feeling and identity, and so on). There are even multiple ethnic groups among African Americans such as Cape Verdeans, Haitian Americans, Gullah Islanders, and so forth. Jews are also an ethnic group—a group sharing a common or nearly common religion (religious culture), but Jews are not regarded as a race.

Sometimes an ethnic group in a society comes to be thought of in that society as a race. The best example of this is Hitler's definition of Jews as a race in Nazi Germany in the 1930s and early 1940s—the "Jewish race." This led to heavy negative stereotyping of Jews in Germany and other countries and eventually to the killing of millions of Jews by means of starvation and gas chambers during the Holocaust. What had been an ethnic group came to be defined in Nazi society as a "race," though we now know that Jewish people are not a race.

When any group—whether an ethnic group or a social class—comes to be thought of as a race and then is actually defined by society as a race, this means the group has become racialized. Thus Hitler racialized Jews. By so doing, it was easier to then stereotype Jews and to regard them as a separate and inferior category of people. It was the process of racialization that allowed Hitler to do this and to convince many German citizens that Jews were bad and needed to be totally eliminated. He almost succeeded. This underscores the point that races tend to be defined in a society by social processes, less so than by physical characteristics. This shows again that race is a social construction.

## RACE AS A SOCIAL CLASS OR PRESTIGE RANK

In parts of Brazil, such as the province of Bahia, the higher one's social class status and wealth, the more likely one is to be formally listed in the country's census as "White" (Surratt and Inciardi 1998; Patterson 1982). This is true even if the person in question is dark-skinned and clearly of largely African ancestry. Thus while skin color is certainly important to defining who is of what race ("color") in Brazil, wealth and social standing are just as important. A favorite expression in Brazil is: "Money whitens" (o dinheiro embranquece)! People with mixed physical characteristics may be labeled "White" if they appear well-dressed and occupy a prestigious professional occupation. Similarly, a poor, light-skinned person may be labeled "Black" to indicate low social standing.

The use of color as a label for people in Brazil is far more complex than is the case in the United States. The simple distinctions of "Black," "White," "Yellow," and "Red" used in the United States are seen in Brazil as utterly ridiculous. There, a large number of labels are used to identify different colors of individuals—reddish brown; reddish dark brown; reddish brown with a hint of yellow; and so on. A fairly recent survey in Brazil revealed 143 color labels used for the population, including gray, pink, dirty white, and cinnamon (Ellison 1995). These race-like labels for individuals are used in conjunction with their wealth and social standing. This shows that how your color is perceived and thus labeled within a society can be significantly determined by your social class or prestige rank within that society.

## RACE AS A "RACIAL FORMATION" OF SOCIETY'S INSTITUTIONS

What a race is, and who is of what race, can be defined by one or more of society's institutions, such as the government, the educational institution, the state and federal legal system, and the criminal justice system. An insightful theory in this regard is Orm and Winant's (1994) racial formation theory. This theory argues that over time, society's powerful institutions define what race is and who is to be classified within what racial category.

Such has been the case in the United States. During slavery, and after the Emancipation Proclamation of 1863, Black people were defined in the law as "Negro" if they had only a small fraction of Black ancestry ("Black blood"), even so small an amount as one great-great-great grandparent who was Black. This is the origin of the one-drop rule, namely, one was "Negro" if he or she had but one drop of Negro blood. The amount of ancestry ("blood") required varied somewhat from state to state, but it was a small amount. The negative stigma of Blackness needed to be only minuscule for one to be totally labeled. (In some states, one Black great-great-great-great grandparent was sufficient to render one a "total" Negro!) The point is that by law it was the government, via the law, that determined how race was to be defined. The individual was not allowed to define her or his own race; it was up to the government, and the government carried the strong sanction of the law. This is what is meant by racial formation. The "one-drop rule" exists even to this day in the public mind.

A clear illustration of how the government determines race is how one is designated as American Indian (Native American Indian) in the United States. According to the U.S. government, one cannot simply decide to call oneself "Indian." Through what is called the federal acknowledgement process, the U.S. government declares one to be an American Indian, but one must go through an elaborate bureaucratic process, involving paper forms and legal documents, and then be certified as a member of some Indian tribe. The U.S. government maintains a list of Indian tribes that it considers to be "legitimate," and the individual must be able to prove membership in one of those pre-designated tribes. Moreover, the tribes themselves must be certified as legitimate—and experts estimate that about 569 American Indian groups have been so designated. This means that the government, through this type of racial formation, has more say in whether or not American Indian people are American Indians than do the people themselves! Obviously, not all Native American Indians agree with this procedure.

## RACE AS SELF-DEFINED

Finally, a person's race may be defined in part simply by what that person calls herself or himself. When asked "What race are you?" (or even simply "What are you"?), what then does the person say? A person with one Black parent and one White parent may well say "I am Black" (an implicit use of the one-drop rule); or they may say "I am White" (which may conflict with how they appear to others); or they may say "I am bi-racial" or "I am of mixed race." Or, still further, the person may refuse to be classified by race, arguing that race is a false classification in the first place. (If such a person has light skin and as a result chooses to live as a White person—severing all ties with Black relatives and also the Black community— that person by tradition could be accused by the Black community of "passing," that is, passing for White. Designating someone as "passing" is less common now than it was a decade or two ago, but it is still done.) On some present-day college campuses, a light-skinned Black person who appears to be passing for White and who denies being Black is (only somewhat) jokingly called an "incogNegro" (or simply, "incog")—a pun on "incognito!"

A person is, in principle, free to designate himself or herself in any way they might wish. The problem is that all the other five definitions that we have given also, simultaneously, come into play. What if the person with one Black parent and one White parent has brown skin and looks to everyone else like a Black person, that is, the person is dark enough of skin in order to be called "Black" among her or his peers? This would show intervention of both the biological as well as the social constructionist definitions, and perhaps the racial formation definition as well. (Golf star Tiger Woods is of African American and Asian parentage and has racially designated himself—though somewhat jokingly—as Cablinasian. What would you call him?)

The point is that there is more to race than one's personal definition even as applied to one's own self. It may sound odd to quarrel with a person about what they choose to call themselves racially. But such is the reality of race in the United States—and most other countries as well: It is not totally up to the person alone!

We are reminded in this context of the story of a dark-skinned Black man in the southern U.S. who in the early 1950s boarded a bus and proceeded to sit in the front row. The laws and norms of segregation in much of the South [at that time] prevented Black people from sitting up front on a bus. Consequently, the bus driver turned to the man and said: "Boy, you can't sit there!" The man replied,

"Well, why not?" "Because," said the bus driver, "you are a Negro, and Negroes cannot sit up here in the front of the bus!" "Oh, well in that case," said the man, "I can stay right here. I have resigned from the Negro race!"

One is left to contemplate what race would be like in this country if one could easily "resign" from a race!

## CONCLUSION

"Race" in the United States is not defined by one single definition, but simultaneously by several definitions. Six definitions of race were explored here. One's race is defined by a combination of the following: by one's physical appearance, such as by skin color (the biological definition); by social construction (any definition arising out of the process of human interaction, such as how those around you define you); as an ethnic group (Hitler defined the Jews as a race); as a social class rank (as in Brazil); as racial formation (such as the U.S. government's definition of who is American Indian); and finally, by one's own self-definition. No one definition is dominant over another in U.S. society. Each definition shows the significance of society in defining race.

## REFERENCES

Ellison, K. 1995. "Brazil's Blacks, Building Pride, Invoke the Legend of Rebel Warrior Zumbi." *Miami Herald* (November 19): 1A.

Lewontin, Richard. 1966. *Human Diversity*. New York: W. H. Freeman.

Omi, Michael, and Howard Winant. 1994. *Racial Formation in the United States*, 2nd ed. New York: Routledge.

Patterson, Orlando. 1982. *Slavery and Social Death: A Comparative Study*. Cambridge: Harvard University Press.

Surratt, Hilary L., and James A. Inciardi. 1998. "Unraveling the Concept of Race in Brazil: Issues for the Rio de Janeiro Cooperative Agreement Site." *Journal of Psychoactive Drugs* 30 (July–September): 255–260.

Young, Kimball. 1942. *Sociology: A Study of Society and Culture*. New York: American Book Co.

## DISCUSSION QUESTIONS

1.  What are the different ways race can be defined in society? How does this challenge the understanding of race as simply a fixed, biological category?

2.  How do social beliefs shape our understanding of race?

3.  How does Taylor's essay indicate that race matters, even if it is a social construction?

# 2

## The Race Myth

JOSEPH L. GRAVES, JR.

*The idea that race is a biological or genetic fact is wrong, as Graves here demonstrates. There is no such thing as a "race gene," and, as Graves shows, there is vastly more genetic similarity among human beings than there are differences across so-called races. This article underscores, from current biological research, that race is not a thing so much as it is an idea that human beings have created.*

Sometimes, the scientific investigation of one problem presents solutions to another. In 1986, scientists proposed a major undertaking: to sequence the entire human genome, to draw a map of the 100,000 or so genes that make humans distinct from apes or dolphins or squirrels. We knew that each of those genes could be found in a specific place on one of the twenty-three pairs of human chromosomes, but for the most part, we didn't know exactly where.

Remember those twisting ladders of DNA molecules from high-school biology? The DNA of each gene is made up of letters, which are pairs of molecule combinations. There are three billion letters in the human genome. At the Technology Center of Silicon Valley (now called the Tech Museum for Innovation), scientists built a model of what this looks like. Think of a spiral staircase and each step as a telephone directory. Now, wind a second staircase around the first to make a double spiral. Every phonebook is a gene, and the contents are the letters. If you have 100,000 phonebooks, and the information in them isn't listed alphabetically, and you aren't sure which phonebook goes where along the spirals, you've got a big project on your hands.

Producing all three billion letters of the human genome might be the greatest achievement in the history of biology, possibly of all science. The human genome map could tell us about the origin of human disease, the general function of our bodies, and possibly what it means to be human. A number of theoretical and technical problems had to be solved to sequence the genome, not the least of which was whose genome should be read. The scientists directly involved with the genome project immediately began to wrestle with the problem of human genetic variation and the concept of race.

SOURCE: Joseph L. Graves, Jr. 2004. *The Race Myth: Why We Pretend Race Exists in America.* New York: Dutton.

## FOR EVERY DISCOVERY, A CONTROVERSY

Celera Genomics was the only major private corporation in the quest to map the human genome. In February 2001, Celera's CEO, Craig Venter, touched off a minor firestorm when he commented that "race is not a scientific concept."[1] He knew that it wasn't possible to distinguish people who were ethnically African American, Chinese, Hispanic, or white at the genome level. Celera's sequencing of the human genome showed that the average pair of human beings who are not close relatives differ by 2.1 million genetic letters out of those 3 billion, yet only a few thousand of those differences account for the biological differences between individuals.

Venter argued that we all are essentially identical twins at the level of the genome. Celera used DNA extracted from five volunteers, three women and two men, who were ethnically African American, Chinese American, Hispanic, and Europeon American. Their results showed that at the DNA level you could clearly tell the females from the males (due to the genetic differences in the X and Y chromosomes), but you could not identify the race of the individual from the DNA.

Venter's comment should not have been controversial. The Celera study only confirmed at the molecular level something population geneticists and physical anthropologists had recognized for well over fifty years: the nonexistence of biological races in the human species. Still, some prominent biologists felt compelled to attack Venter and defend the race concept in biology. Among them was James Crow (yes, that is his name), who in early 2002 defended the legitimacy of identifying races in humans. In the publication of the National Academy of Arts and Sciences, *Daedalus*, he commented: "Whenever an institution or society singles out individuals who are exceptional or outstanding in some way, racial differences will become more apparent. That fact may be uncomfortable, but there is no way around it."[2]

In support of this notion, rather than citing scientific evidence, he gave as an example a social phenomenon: the overrepresentation of African Americans in track and field, and their underrepresentation in physics and engineering, relative to Asian Americans. His exact comment was: "A stopwatch is colorblind." It's amazing when geneticists of James Crow's stature demonstrate this blind spot when it comes to human variation and the concept of race. It still happens because the social construction of race is deeply ingrained in the thinking of most American intellectuals, including biologists and medical practitioners.

... The fact is that no biological races exists in modern humans. In the next few pages, I'll explain how a race is defined and why there isn't enough variation in humans for our differences to qualify as races.

## WHAT DOES IT TAKE TO BE A RACE?

To qualify officially as a biological race (or subspecies or variety), an animal or plant has to meet one of two requirements:

1. It can have its own distinct genetic lineage, meaning that it evolved in enough isolation that it never (or rarely) mated with individuals outside its borders, or

2. The genetic distance between one population and another has to be significantly greater than the genetic variability that exists within the populations themselves.

The first requirement is pretty straightforward, but the second takes some explaining. Think of it as a formula, and we're looking for two percentages to plug into it: one for distance and one for variability. To do that, we have to understand how geneticists measure variety, and we have to define genetic distance and genetic variability as painlessly as possible.

## VARIETY IS THE SPICE OF GENETICS

We're all pretty familiar with DNA by now: It's the DNA molecules along our chromosomes that determine what makes us the same as other animals, what makes us different from them, and what makes us different from each other.

Along each chromosome, there are specific parts—somewhere between twenty-five thousand and forty thousand of them—that control our traits. At each of these parts there are two chemical messages that code for a trait, like brown eyes or the ability to roll your tongue. Some messages are dominant over others, so that if you get a message from Mom that says "brown" and a message from Dad that says "blue," your eyes are going to be brown. Many more messages have a mixing effect, producing physical features that fall somewhere in the middle of both parents, so that, say, your nose might be bigger than your mother's but smaller than your father's.

The range of possible combinations of these messages is pretty mindboggling: One egg cell or one sperm cell can have 8,388,608 possible combinations of chromosomes, so for a couple producing a child, the number of potential chromosome arrangements for that child would be 8,388,608! This means that there is a tremendous amount of diversity that can be produced by even one pair of parents, yet on average the physical traits of the offspring still resemble a mixture of both of the parents.

No one's saying that these differences amount to races, of course, otherwise, you'd be a different race from your parents. The large number of combinations of traits just means that, any way you look at it, human genetics is complex.

# MEASURING VARIATION: WE'RE NOT ALL THAT DIFFERENT

There are ways to compare the complexity of one group of humans with another, no matter how you want to define *group*. A group could be all the people who were born and raised on Maui, or a number of people who have Down's syndrome. With all the possible variety, where do we start measuring variation?

Let's look again at those messages in our DNA. About 33 percent of the spots on the chromosomes where they live allow a lot of possibilities for a particular trait—people can have brown, blue, green, hazel, or golden eyes, for instance. In the remaining 67 percent of the spots, only one type of message is allowed, so that nearly every human being will have identical messages for a particular trait at that spot (wrong messages in these spots can result in genetic disease). So, about a third of our messages are responsible for all of the variety we see in people worldwide.

That seems like a lot, but it's not all. From studying these messages for traits, scientists now know that two individuals from anywhere in the world can potentially share 86 percent of the traits out of that 33 percent. Doing the math, that leaves only 4.62 percent of our genetic makeup responsible for all our individuality. Put another way: the traits of an Irish businessperson, an African-American lawyer, and the prime minister of India are quite likely to be 95.38 percent identical. Geography does play a part, so that if two people are from the same continent, you can reduce the variability by another 10 percent, and if they're from the same village, you can reduce it by yet another 4 percent. So, variety is measured by looking at the percentage of our DNA that makes us unique. What's genetic distance?

# GETTING WITHIN SHOUTING DISTANCE

Genetic distance is a statistic calculated by examining different groups of people. It is a measurement of how frequently the genetic messages for traits occur in populations. People who share larger proportions of the same messages across their genome are genetically close, and people who do not share the same messages are genetically more distant. For instance, let's examine the message that causes sickle cell anemia, a disease that people think of as associated with race. This message occurs in large numbers in people who live in tropical areas because, it turns out, if you have one sickle cell message and one normal message, you have a better chance of surviving malaria, a typical disease of the tropics. So, the sickle cell anemia message is in high frequencies in populations in western Africa, the Middle East, the Persian Gulf, the Mediterranean, and India. In the case of this message, someone from Ghana is genetically closer to someone from Syria than to someone from Kenya, because Kenyans (who live at high altitudes where there isn't much malaria) don't have a high frequency of sickle cell messages. This clearly shows that sickle cell anemia cannot be associated with any particular race.

This points to an important fact: Geographical distance does not necessarily equal genetic distance. In fact, assuming that two people are genetically different because they look like they came from different parts of the world can be really dangerous for their health. Why? Because things like people's blood type or their ability to accept transplanted organs are dictated by how genetically close they are, not necessarily by where their ancestors came from geographically.

Because people equate race with external physical characteristics, they assume more often than not that a person is more likely to find an organ donor among the members of their own supposed racial group. This misconception in biomedical research or clinical practice that insists on sticking to these false racial categories causes many errors and lost lives.

## LOCATION, LOCATION, LOCATION?

If geographic distance doesn't necessarily equal genetic distance, does that make geographic distance irrelevant? No. There is still a strong correlation between geographic distance and genetic distance. After all, it's geographic isolation that is the biggest factor in the development of new traits. It's just a mistake to assume that one type of distance equals the other.

So, whether we're looking at geographically close or distant populations of humans, the question that really matters is still this: If all the genetic information from two populations that we think of as races is examined, how similar are they? And are their differences big enough to qualify them as separate races?

## SIZE MATTERS: THE STANDARD FOR MEASURING
## GENETIC VARIATION

To figure this out, we would have to look at a lot more than a couple of indicators like a sickle cell message or blood types. We'd have to evaluate the genetic information from enough people across a big enough number of separated populations for the statistics to be meaningful. Fortunately for you and me, it's been done for us already.

This analysis has used a number of approaches, examining the big three sources: protein variation, nuclear DNA, and mitochondrial DNA. All of these techniques agree that anatomically modern humans are a young species—too young to have developed any significant genetic distances between populations before mass mobility on a global scale started blurring what few differences there were. The result is that there is no unambiguous way to describe biological races within our species.

It's not for lack of a standard: Biologists have studied genetic variation in a wide variety of organisms other than humans for over fifty years, and have described many geographical races or subspecies. The races we have identified outside of humans usually show about 20 percent total genetic distance between their populations, as is the case in, say, the various species of fruit flies.

We do not see anywhere near that much genetic variation in modern humans. The genetic distances in humans are statistically about ten times lower (2 percent) than the 20 percent average in other organisms, even when comparing the most geographically separated populations within modern humans. There is greater genetic variability found within one tribe of western African chimpanzees than exists in the entire human species! In fact, there has never been any degree of natural selection in modern humans equivalent even to the levels used to create the differences in the breeds of domestic pigeons or dogs. In order to support the existence of biological races in modern humans, you would have to use a very different sort of reasoning than has been applied to all other species of animals—and that would be bad science.

## ONE REQUIREMENT BITES THE DUST

Okay, now we have the first number to plug into our formula for race: Genetic distance between populations of humans is about 2 percent. We still need the other number, for genetic variation within those populations.

The same types of studies that showed us that human genetic distances average only 2 percent have also shown that there is about 8.6 times more genetic variation within the classically defined racial groups than between them. Why? Because there is about 8.6 times more genetic variation between any given individual on the planet and another individual than there is between the populations they belong to. In other words, the variability that makes one African-American person different from another is greater than the variability between African Americans and Swedes or Tibetans or Amazonian tribes.

Remember we said that modern humans share 86 percent of their genetic variations, so that less than 5 percent of our genome is responsible for our individuality. Now, it's time to look back at the formula for genetic distance versus genetic variability. For any group of humans to be a race, their genetic distance from another group, around 2 percent, would have to be greater than their unique genetic variability, around 5 percent. Two is not bigger that five.

It's a fact that if you add the 10 percent that accounts for shared variations between populations on the same continent, plus the 4 percent shared by local populations, you end up with about 1 percent genetic variability. Since this is lower than the 2 percent statistic for genetic distance, this would technically meet our definition of a race. There are a couple of problems with that conclusion. For one, we would be accepting percentages of variability in humans that are far below the percentages that are applied to all other organisms on the planet as evidence of race. For another, treating populations with 1 percent variability as distinct races would result in identifying over 1,000 races of humans. Following the argument to its logical conclusion, since families are even more genetically similar than geographic neighbors, we could say that every individual family is a race, or every cultural group that tries to keep itself separate (such as the Amish) is a race. In other words, the distinction becomes meaningless.

That's really the point. The majority of genetic variation in humans occurs between individuals, without regard to membership in socially constructed race. And none of the unique variations we see approach the minimum levels used to identify races in other species.

## HOMAGE TO ALEX HALEY: OUR COMMON ROOTS

Our formula didn't work, so we know that populations of humans just don't meet the required distance-to-variability ratio for being separate races. Identifying a race by physical characteristics such as skin color or eye shape is as invalid as saying that all people who are tall or who have straight hair or who are pigeon-toed constitute separate races. Let's look now at the other requirement for race identification, genetic lineage.

There is no evidence that any group of humans now in existence—geographically or in socially defined races—has an evolutionary or genetic line of descent that is distinct from other groups.

Every person is descended from two parents, four grandparents, and eight great-grandparents. With each generation back into our ancestry, the number of lineal ancestors doubles (if all the ancestors are unrelated). If we go back twenty-four generations into our past, about fifteen hundred years ago, we each would have had 33,554,432 ancestors! At that time, the world population was around 206,000,000 people. That means that if we all really had separate, unrelated ancestors, every single one of us would be descended from about 16 percent of the world's population in the year 502 A.D. Obviously the math doesn't work, which strongly suggests that not all of our ancestors came from independent families and that many individuals must have contributed to many family lineages.

In a place like the United States, where the population has been intermarrying for a long time, and has been engaged in interracial sex and conception outside of marriage, it is fairly easy to show that the socially described races do not exist as separate lines of genetic descent. Still, we can and should ask if there ever was a time when the world's populations were truly independent.

In this sense, an independent lineage would be a population that was isolated from mixing genetically with other humans—one that never mated and bred outside its own group. In the animal kingdom, kangaroos are an example of this: The geographic isolation of Australia and its surrounding islands was so complete that this particular species of marsupial developed there and nowhere else. Since we know that race is an intermediate step toward becoming a species, what we're looking for genetically is the human equivalent of a marsupial on its way to becoming a kangaroo....

## ALL ROADS LEAD TO …

More reliable methods of studying human populations can be used, such as *neighbor joining* techniques, which allow relative genetic distances between groups to be shown. The data from these show that humans vary in small increments from group to group, rather than differing drastically depending on where they originated.

Humans have been on the move since we first left sub-Saharan Africa 80,000 years ago, combining our DNA at the same time as we were adapting genetically to our local conditions. Distinct genetic lineages cannot be traced within or between human populations because we've been mixing with each other since we first evolved. We're the only species of the *Homo* genus to have survived into modernity—an analogy would be if the only members of *Canis* to survive were domestic dogs (no more coyotes, jackals, or wolves). Genetically, we're not even separate breeds: We're all mutts.

## AREN'T PHYSICAL DIFFERENCES PROOF
## THAT RACES EXIST?

We kind of look like separate breeds, though. Observable physical features are what people point to first to justify their belief in different races. After everything you've read so far, you know that's a dead horse but, just to make the point, let's beat it a little longer.

What determines physical traits in populations? Genes do this in combination with environmental factors. So, if genetic variation can't be divvied up evenly into races, neither can physical traits. Skin color, hair type, body stature, blood groups, or tendencies to get certain diseases do not alone or in combination define the racial groups that have been socially constructed in North America.

It's absolutely true that these physical traits vary among geographical populations. What most people don't realize is the way that they vary. For example, Sri Lankans of the Indian subcontinent, Nigerians, and aboriginal Australians share a dark skin tone, but differ in hair type, facial features, and genetic predisposition for disease. If you try to use characteristics such as height, body proportions, skull measurements, hair type, and skin color to create a tree showing how human populations are related, you get a tree that doesn't match the measured genetic relatedness and known evolutionary history of our species.

A tree like that would say, "all short, extremely dark-skinned people with thick curly hair are the same race," and would link Papuans from New Guinea and aboriginal Australians most closely to sub-Saharan Africans. We know, however, from genetic analyses, that Papuans and aboriginal Australians are the group most genetically distant from sub-Saharan Africans. And that makes total sense, because sub-Saharan Africans and aboriginal Australians are two of the most geographically separated human groups, so intermarriage has been minimal. Australia, after all, was cut off from human migration for most of its history.

We also know that within sub-Saharan African populations, everything from skin color to skull types to total genetic diversity is more variable than in any other of the world's populations. In other words, a person from the Congo and a person from Mali are more likely to be different genetically from each other than either is from a person from Belgium. Yet, if everyone from this region got up and moved to the United States, we'd call them all African Americans and see them as members of the same race.

Physical traits fail to define races because local populations produce traits that adapt to climate and other environmental factors wherever those factors occur. This means that, however genetically or geographically distant they are, tropical populations will have physical traits that match tropical conditions, like the sickle cell message. Kenyans and Peruvians will have greater lung capacities and red blood cell counts from living at high altitudes. These features are completely independent of the other genetic aspects of their physical makeup, and cannot be used to determine membership in a socially defined race.

## TAKING IT BACK HOME

America has perfected the concept of socially defined races. On first inspection, it might have seemed that these races had biological legitimacy. The English colonists, the native peoples they called *Indians*, and the West Africans they brought as slaves all came from different places along the range of genetic diversity. It is entirely possible that the social construction of race in America would have proceeded differently if the full spectrum of the world's populations had immigrated to America along with western Europeans and Africans. Certainly, things would have been different if the prior history of the world had allowed these groups to come together under conditions fostering social equality.

But, as we know, cultural rather than biological traits were used to define our races. The rules of cultural evolution differ from those of biological evolution, but sadly scientists, doctors, philosophers, and law makers have for the most part, not yet acknowledged the difference. We must stop masking the real social issues with racist ideologies in order to build a truly just society.

## NOTES

1. Fox, Maggie. "First Look at Human Genome Shows How Little There Is." Washington (Reuters), February 11, 2001.
2. Crow, James F. "Unequal by Nature: A Geneticist's Perspective on Human Differences." *Daedalus*, Winter 2002: 85.

# DISCUSSION QUESTIONS

1.  What does Graves mean when he says that the majority of genetic variation in humans occurs between individuals, not between races?

2.  Over the years, many have spent a great deal of time and effort arguing that there is a biological basis to race, but most have been proven wrong, and now we know that there is no genetic basis for race. Why do you think there has been so much more attention given to presumed biological bases for race, compared to social bases for race?

# 3

## Planting the Seed:

### The Invention of Race

ABBY L. FERBER

*The history of the concept of race is deeply linked to racist thinking, particularly as it developed in some of the quasi-scientific notions developed in the sixteenth and seventeenth centuries. Ferber shows how racist thought is a relatively recent historical development, one that stems from the exploitation of human groups by others. In this history of racist thought, many have tried to use science (illegitimately) to try to justify such exploitation.*

My students are always surprised to learn that race is a relatively recent invention. In their minds, race and racial antagonisms have taken on a universal character; they have always existed, and probably always will, in some form or another. Yet this fatalism belies the reality—that race is indeed a modern concept and, as such, does not have to be a life sentence.

Winthrop Jordan has suggested that ideas of racial inferiority, specifically that blacks were savage and primitive, played an essential role in rationalizing slavery.[1] There was no conception of race as a physical category until the eighteenth century.[2] There was, however, a strong association between blackness and evil, sin, and death, long grounded in European thought. The term "race" is believed to have originated in the Middle Ages in the romance languages, first used to refer to the breeding of animals. Race did not appear in the English language until the sixteenth century and was used as a technical term to define human groups in the seventeenth century. By the end of the eighteenth century, as emphasis upon the observation and classification of human differences grew, "race" became the most commonly employed concept for differentiating human groups according to Northern European standards. Audrey Smedley argues that because "race" has its roots in the breeding of animal stock, unlike other terms used to categorize humans, it came to imply an innate or inbred quality, believed to be permanent and unchanging.[3]

Until the nineteenth century, the Bible was consulted and depended upon for explanations of human variation, and two schools of thought emerged. The first asserted that there was a single creation of humanity, monogenesis, while the

SOURCE: From Abby L. Ferber, *White Man Falling: Race, Gender, and White Supremacy* (Lanham, MD: Rowman & Littlefield, 1998), pp. 27–43. Reprinted by permission of Rowman & Littlefield Publishers, Inc.

second asserted that various human groups were created separately, polygenesis. Polygenesis and ideas about racial inferiority, however, gained few believers, even in the late 1700s when the slave trade was under attack, because few were willing to support doctrines that conflicted with the Bible.[4]

While European Americans remained dedicated to a biblical view of race, the rise of scientific racism in the middle of the eighteenth century shaped debate about the nature and origins of races.[5] The Enlightenment emphasized the scientific practices of observing, collecting evidence, measuring bodies, and developing classificatory schemata. In the early stages of science, the most prevalent activity was the collection, examination, and arrangement of data into categories. Carolus Linnaeus, a prominent naturalist in the eighteenth century, developed the first authoritative racial division of humans in his *Natural System*, published in 1735.[6] Considered the founder of scientific taxonomy, he attempted to classify all living things, plant and animal, positioning humans within the matrix of the natural world. As Cornel West demonstrates, from the very beginning, racial classification has always involved hierarchy and the linkage of physical features with character and cultural traits.[7] For example, in the descriptions of his racial classifications, Linnaeus defines Europeans as "gentle, acute, inventive ... governed by customs," while Africans are "crafty, indolent, negligent ... governed by caprice."[8] Like most scientists of his time, however, Linnaeus considered all humans part of the same species, the product of a single creation.

Linnaeus was followed by Georges Louis Leclerc, Comte de Buffon, who is credited with introducing the term "race" into the scientific lexicon. Buffon also believed in monogenesis and in his 1749 publication *Natural History*, suggested that human variations were the result of differences in environment and climate. Whiteness, of course, was assumed to be the real color of humanity. Buffon suggested that blacks became dark-skinned because of the hot tropical sun and that if they moved to Europe, their skin would eventually lighten over time. Buffon cited interfertility as proof that human races were not separate species, establishing this as the criterion for distinguishing a species.

Buffon and Johann Friedrich Blumenbach are considered early founders of modern anthropology. Blumenbach advanced his own systematic racial classification in his 1775 study *On the Natural Varieties of Mankind*, designating five human races: Caucasian, Mongolian, Ethiopian, American, and Malay. While he still considered races to be the product of one creation, he ranked them on a scale according to their distance from the "civilized" Europeans.[9] He introduced the term "Caucasian," chosen because he believed that the Caucasus region in Russia produced the world's most beautiful women. This assertion typifies the widespread reliance upon aesthetic judgments in ranking races....

The science of racial classifications relied upon ideals of Greek beauty, as well as culture, as a standard by which to measure races. Race became central to the definition of Western culture, which became synonomous with "civilization."[10]...

The history of racial categorizations is intertwined with the history of racism. Science sought to justify *a priori* racist assumptions and consequently rationalized and greatly expanded the arsenal of racist ideology. Since the eighteenth century, racist beliefs have been built upon scientific racial categorizations and the linking of social and cultural traits to supposed genetic racial differences. While some social critics have suggested that contemporary racism has replaced biology with a concept of culture, the [1994] publication of *The Bell Curve*[11] attests to the staying power of these genetic notions of race. Today, as in the past, racism weaves together notions of biology and culture, and culture is assumed to be determined by some racial essence.

Science defined race as a concept believed to be hereditary and unalterable. The authority of science contributed to the quick and widespread acceptance of these ideas and prevented their interrogation. Equally important, the study of race and the production of racist theory also helped establish scientific authority and aided discipline building. While the history of the scientific concept of race argues that race is an inherent essence, it reveals, on the contrary, that race is a social construct. Young points out that "the different Victorian scientific accounts of race each in their turn quickly became deeply problematic; but what was much more consistent, more powerful and long-lived, was the cultural construction of race."[12]

Because race is not grounded in genetics or nature, the project of defining races always involves drawing and maintaining boundaries between those races. This was no easy task. It is important to pay attention to the construction of those borders: how was it decided, in actual policy, who was considered white and who was considered black? What about those who did not easily fit into either of those categories? What were the dangers of mixing? How could these dangers be avoided? These issues preoccupied policy makers, popular culture, and the public at large....

Throughout the second half of the nineteenth century, discussion of race and racial purity grew increasingly popular in both academic and mainstream circles as Americans developed distinctive beliefs and theories about race for the first time. As scientific beliefs about race were increasingly accepted by the general public, support for the one-drop rule became increasingly universal. Popular opinion grew to support the belief that no matter how white one appeared, if one had a single drop of black blood, no matter how distant, one was black....

Throughout the history of racial classification in the West, miscegenation and interracial sexuality have occupied a place of central importance. The science of racial differences has always displayed a preoccupation with the risks of interracial sexuality. Popular and legal discourses on race have been preoccupied with maintaining racial boundaries, frequently with great violence. This [essay] suggests that racial classification, the maintenance of racial boundaries, and racism are inexorably linked. The construction of biological races and the belief in maintaining the hierarchy and separation of races has led to widespread fears of integration and interracial sexuality....

The history of racial classification, and beliefs about race and interracial sexuality, can be characterized as inherently white supremacist. White supremacy has been the law and prevailing worldview throughout U.S. history, and the ideology of what is today labeled the white supremacist movement is firmly rooted in this tradition. Accounts that label the contemporary white supremacist movement as fringe and extremist often have the consequence of rendering this history invisible. Understanding this history, however, is essential to understanding and combating both contemporary white supremacist and mainstream racism.

## NOTES

1. Jordan, Winthrop. 1969. *White over black*. Chapel Hill: University of North Carolina Press.
2. Banton, Michael, and Jonathan Harwood. 1975. *The race concept*. New York: Praeger; Mencke, John G. 1979. *Mulattoes and race mixture: American attitudes and images, 1865–1918*. Ann Arbor, Mich.: University Microfilms Research Press.
3. Smedley, Audrey. 1993. *Race in North America: Origin and evolution of a worldview*. Boulder, Colo.: Westview Press.
4. Banton and Harwood 1975, 19.
5. Banton and Harwood 1975, 24.
6. West, Cornel. 1982. *Prophesy deliverance! An Afro-American revolutionary Christianity*. Philadelphia: Westminster Press.
7. West 1982.
8. West 1982, 56.
9. Smedley 1993, 166.
10. Young, Robert J. C. 1995. *Colonial desire: Hybridity in theory, culture and race*. New York: Routledge.
11. Hernstein, Richard J., and Charles Murray. 1994. *The bell curve: Intelligence and class structure in American life*. New York: Free Press.
12. Young, 1995, p. 94.

## DISCUSSION QUESTIONS

1. What does Ferber mean when she writes that "the history of racial categorizations is intertwined with the history of racism"?
2. What role have science and religion played in the social construction of racism?

# 4

# How Did Jews Become White Folks?

KAREN BRODKIN

*Brodkin's discussion of how Jewish people have come to be defined as "White" shows how race can be socially constructed in particular social and historical contexts. As Jewish immigrants to the United States became upwardly mobile, racial ideologies changed to redefine Jewish people in terms different than those they first encountered.*

> The American nation was founded and developed by the Nordic race, but if a few more million members of the Alpine, Mediterranean and Semitic races are poured among us, the result must inevitably be a hybrid race of people as worthless and futile as the good-for-nothing mongrels of Central America and Southeastern Europe.
> —Kenneth Roberts,
> "Why Europe Leaves Home"

The late nineteenth century and early decades of the twentieth saw a steady stream of warnings by scientists, policymakers, and the popular press that "mongrelization" of the Nordic or Anglo-Saxon race—the real Americans—by inferior European races (as well as by inferior non-European ones) was destroying the fabric of the nation.

I continue to be surprised when I read books that indicate that America once regarded its immigrant European workers as something other than white, as biologically different. My parents are not surprised, they expect anti-Semitism to be part of the fabric of daily life, much as I expect racism to be part of it. They came of age in the Jewish world of the 1920s and 1930s, at the peak of anti-Semitism in America. They are rightly proud of their upward mobility and think of themselves as pulling themselves up by their own bootstraps. I grew up during the 1950s in the Euro-ethnic New York suburb of Valley Stream, where Jews were simply one kind of white folks and where ethnicity meant little more to my generation than food and family heritage. Part of my ethnic heritage was the belief that Jews were smart and that our success was due to our own efforts and abilities, reinforced by a culture that valued sticking together, hard work, education, and deferred gratification.

I am willing to affirm all those abilities and ideals and their contribution to Jews' upward mobility, but I also argue that they were still far from sufficient to account for Jewish success. I say this because the belief in a Jewish version of Horatio Alger has become a point of entry for some mainstream Jewish organizations to adopt a racist attitude against African Americans especially and to oppose affirmative action for people of color. Instead I want to suggest that Jewish success is a product not only of ability but also of the removal of powerful social barriers to its realization.

It is certainly true that the United States has a history of anti-Semitism and of beliefs that Jews are members of an inferior race. But Jews were hardly alone. American anti-Semitism was part of a broader pattern of late-nineteenth-century racism against all southern and eastern European immigrants, as well as against Asian immigrants, not to mention African Americans, Native Americans, and Mexicans. These views justified all sorts of discriminatory treatment, including closing the doors, between 1882 and 1927, to immigration from Europe and Asia. This picture changed radically after World War II. Suddenly, the same folks who had promoted nativism and xenophobia were eager to believe that the Euro-origin people whom they had deported, reviled as members of inferior races, and prevented from immigrating only a few years earlier, were now model middle-class white suburban citizens.

It was not educational epiphany that made those in power change their hearts, their minds, and our race. Instead, it was the biggest and best affirmative action program in the history of our nation, and it was for Euromales. That is not how it was billed, but it is the way it worked out in practice. I tell this story to show the institutional nature of racism and the centrality of state policies to creating and changing races. Here, those policies reconfigured the category of whiteness to include European immigrants. There are similarities and differences in the ways each of the European immigrant groups became "whitened." I tell the story in a way that links anti-Semitism to other varieties of anti-European racism because this highlights what Jews shared with other Euro-immigrants.

## EURORACES

The U.S. "discovery" that Europe was divided into inferior and superior races began with the racialization of the Irish in the mid-nineteenth century and flowered in response to the great waves of immigration from southern and eastern Europe that began in the late nineteenth century. Before that time, European immigrants—including Jews—had been largely assimilated into the white population. However, the 23 million European immigrants who came to work in U.S. cities in the waves of migration after 1880 were too many and too concentrated to absorb. Since immigrants and their children made up more than

70 percent of the population of most of the country's largest cities, by the 1890s urban American had taken on a distinctly southern and eastern European immigrant flavor. Like the Irish in Boston and New York, their urban concentrations in dilapidated neighborhoods put them cheek by jowl next to the rising elites and the middle class with whom they shared public space and to whom their working-class ethnic communities were particularly visible.

The Red Scare of 1919 clearly linked anti-immigrant with anti-working-class sentiment—to the extent that the Seattle general strike by largely native-born workers was blamed on foreign agitators. The Red Scare was fueled by an economic depression, a massive postwar wave of strikes, the Russian Revolution, and another influx of postwar immigration. Strikers in the steel and garment industries in New York and New England were mainly new immigrants....

Not surprisingly, the belief in European races took root most deeply among the wealthy, U.S.-born Protestant elite, who feared a hostile and seemingly inassimilable working class. By the end of the nineteenth century, Senator Henry Cabot Lodge pressed Congress to cut off immigration to the United States; Theodore Roosevelt raised the alarm of "race suicide" and took Anglo-Saxon women to task for allowing "native" stock to be outbred by inferior immigrants. In the early twentieth century, these fears gained a great deal of social legitimacy thanks to the efforts of an influential network of aristocrats and scientists who developed theories of eugenics—breeding for a "better" humanity—and scientific racism.

Key to these efforts was Madison Grant's influential *The Passing of the Great Race*, published in 1916. Grant popularized notions developed by William Z. Ripley and Daniel Brinton that there existed three or four major European races, ranging from the superior Nordics of northwestern Europe to the inferior southern and eastern races of the Alpines, Mediterraneans, and worst of all, Jews, who seemed to be everywhere in his native New York City. Grant's nightmare was race-mixing among Europeans. For him, "the cross between any of the three European races and a Jew is a Jew." He didn't have good things to say about Alpine or Mediterranean "races" either. For Grant, race and class were interwoven: the upper class was racially pure Nordic; the lower classes came from the lower races.

Far from being on the fringe, Grant's views were well within the popular mainstream. Here is the *New York Times* describing the Jewish Lower East Side of a century ago:

> The neighborhood where these people live is absolutely impassable for wheeled vehicles other than their pushcarts. If a truck driver tries to get through where their pushcarts are standing they apply to him all kinds of vile and indecent epithets. The driver is fortunate if he gets out of the street without being hit with a stone or having a putrid fish or piece of meat thrown in his face. This neighborhood, peopled almost entirely by the people who claim to have been driven from Poland and Russia, is the eyesore of New York and perhaps the filthiest place on the western continent. It is impossible for a Christian to live there because he will be

driven out, either by blows or the dirt and stench. Cleanliness is an unknown quantity to these people. They cannot be lifted up to a higher plane because they do not want to be. If the cholera should ever get among these people, they would scatter its germs as a sower does grain.[1]

Such views were well within the mainstream of the early twentieth-century scientific community....

By the 1920s, scientific racism sanctified the notion that real Americans were white and that real whites came from northwest Europe. Racism by white workers in the West fueled laws excluding and expelling the Chinese in 1882. Widespread racism led to closing the immigration door to virtually all Asians and most Europeans between 1924 and 1927, and to deportation of Mexicans during the Great Depression.

Racism in general, and anti-Semitism in particular, flourished in higher education. Jews were the first of the Euro-immigrant groups to enter college in significant numbers, so it was not surprising that they faced the brunt of discrimination there. The Protestant elite complained that Jews were unwashed, uncouth, unrefined, loud, and pushy. Harvard University President A. Lawrence Lowell, who was also a vice president of the Immigration Restriction League, was open about his opposition to Jews at Harvard. The Seven Sister schools had a reputation for "flagrant discrimination." M. Carey Thomas, Bryn Mawr president, may have been some kind of feminist, but she was also an admirer of scientific racism and an advocate of immigration restriction....

Jews are justifiably proud of the academic skills that gained them access to the most elite schools of the nation despite the prejudices of their gatekeepers. However, it is well to remember that they had no serious competition from their Protestant classmates. This is because college was not about academic pursuits. It was about social connection—through its clubs, sports and other activities, as well as in the friendships one was expected to forge with other children of elites. From this, the real purpose of the college experience, Jews remained largely excluded.

This elite social mission had begun to come under fire and was challenged by a newer professional training mission at about the time Jews began entering college. Pressures for change were beginning to transform the curriculum and to reorient college from a gentleman's bastion to a training ground for the middle-class professionals needed by an industrial economy.... Occupational training was precisely what had drawn Jews to college. In a setting where disparagement of intellectual pursuits and the gentleman C were badges of distinction, it certainly wasn't hard for Jews to excel. Jews took seriously what their affluent Protestant classmates disparaged, and, from the perspective of nativist elites, took unfair advantage of a loophole to get where they were not wanted.

Patterns set by these elite schools to close those "loopholes" influenced the standards of other schools, made anti-Semitism acceptable, and "made the aura of exclusivity a desirable commodity for the college-seeking clientele."[2] Fear that colleges "might soon be overrun by Jews" were publicly expressed at a 1918 meeting of the Association of New England Deans. In 1919 Columbia

University took steps to decrease the number of its Jewish students by a set of practices that soon came to be widely adopted. They developed a psychological test based on the World War I army intelligence tests to measure "innate ability—and middle-class home environment"; and they redesigned the admission application to ask for religion, father's name and birth place, a photo, and personal interview. Other techniques for excluding Jews, like a fixed class size, a chapel requirement, and preference for children of alumni, were less obvious....

Columbia's quota against Jews was well known in my parents' community. My father is very proud of having beaten it and been admitted to Columbia Dental School on the basis of his skill at carving a soap ball. Although he became a teacher instead because the tuition was too high, he took me to the dentist every week of my childhood and prolonged the agony by discussing the finer points of tooth-filling and dental care. My father also almost failed the speech test required for his teaching license because he didn't speak "standard," i.e., nonimmigrant, nonaccented English. For my parents and most of their friends, English was the language they had learned when they went to school, since their home and neighborhood language was Yiddish. They saw the speech test as designed to keep all ethnics, not just Jews, out of teaching.

There is an ironic twist to this story. My mother always urged me to speak well, like her friend Ruth Saronson, who was a speech teacher. Ruth remained my model for perfect diction until I went away to college. When I talked to her on one of my visits home, I heard the New York accent of my version of "standard English," compared to the Boston academic version.

My parents believe that Jewish success, like their own, was due to hard work and a high value placed on education. They attended Brooklyn College during the Depression. My mother worked days and went to school at night; my father went during the day. Both their families encouraged them. More accurately, their families expected it. Everyone they knew was in the same boat, and their world was made up of Jews who were advancing just as they were. The picture for New York—where most Jews lived—seems to back them up. In 1920, Jews made up 80 percent of the students at New York's City College, 90 percent of Hunter College, and before World War I, 40 percent of private Columbia University. By 1934, Jews made up almost 24 percent of all law students nationally and 56 percent of those in New York City. Still, more Jews became public school teachers, like my parents and their friends, than doctors or lawyers....

How we interpret Jewish social mobility in this milieu depends on whom we compare them to. Compared with other immigrants, Jews were upwardly mobile. But compared with nonimmigrant whites, that mobility was very limited and circumscribed. The existence of anti-immigrant, racist, and anti-Semitic barriers kept the Jewish middle class confined to a small number of occupations. Jews were excluded from mainstream corporate management and corporately employed professions, except in the garment and movie industries, in which they were pioneers. Jews were almost totally excluded from university faculties (the few who made it had powerful patrons). Eastern European Jews were concentrated in small businesses, and in professions where they served a largely Jewish clientele....

Although Jews, as the Euro-ethnic vanguard in college, became well established in public school teaching—as well as visible in law, medicine, pharmacy, and librarianship before the postwar boom—these professions should be understood in the context of their times. In the 1930s they lacked the corporate context they have today, and Jews in these professions were certainly not corporation-based. Most lawyers, doctors, dentists, and pharmacists were solo practitioners, depended upon other Jews for their clientele, and were considerably less affluent than their counterparts today.

Compared to Jewish progress after World War II, Jews' pre-war mobility was also very limited. It was the children of Jewish businessmen, but not those of Jewish workers, who flocked to college. Indeed, in 1905 New York, the children of Jewish workers had as little schooling as the children of other immigrant workers. My family was quite the model in this respect. My grandparents did not go to college, but they did have a modicum of small business success. My father's family owned a pharmacy. Although my mother's father was a skilled garment worker, her mother's family was large and always had one or another grocery or deli in which my grandmother participated. It was the relatively privileged children of upwardly mobile Jewish immigrants like my grandparents who began to push on the doors to higher education even before my parents were born.

Especially in New York City—which had almost one and a quarter million Jews by 1910 and retained the highest concentration of the nation's 4 million Jews in 1924—Jews built a small-business-based middle class and began to develop a second-generation professional class in the interwar years. Still, despite the high percentages of Jews in eastern colleges, most Jews were not middle class, and fewer than 3 percent were professionals—compared to somewhere between two-thirds and three-quarters in the postwar generation.

My parents' generation believed that Jews overcame anti-Semitic barriers because Jews are special. My answer is that the Jews who were upwardly mobile were special among Jews (and were also well placed to write the story). My generation might well respond to our parents' story of pulling themselves up by their own bootstraps with, "But think what you might have been without the racism and with some affirmative action!" And that is precisely what the post–World War II boom, that decline of systematic, public, anti-Euro racism and anti-Semitism, and governmental affirmative action extended to white males.

## WHITENING EURO-ETHNICS

By the time I was an adolescent, Jews were just as white as the next white person. Until I was eight, I was a Jew in a world of Jews. Everyone on Avenue Z in Sheepshead Bay was Jewish. I spent my days playing and going to school on three blocks of Avenue Z, and visiting my grandparents in the nearby Jewish neighborhoods of Brighton Beach and Coney Island. There were plenty of Italians in my neighborhood, but they lived around the corner. They were a kind of Jew, but on the margins of my social horizons. Portuguese were even more distant, at the end of the bus ride, at Sheepshead Bay. The *shul*, or temple, was on Avenue Z, and I begged my father to take me like all the other fathers took their kids, but religion wasn't part of my family's Judaism. Just how Jewish my neighborhood was hit me in first grade, when I was one of two kids to go to school on Rosh Hashanah. My teacher was shocked—she was Jewish too—and I was embarrassed to tears when she sent me home. I was never again sent to school on Jewish holidays. We left that world in 1949 when we moved to Valley Stream, Long Island, which was Protestant and Republican and even had farms until Irish, Italian, and Jewish ex-urbanites like us gave it a more suburban and Democratic flavor.

Neither religion nor ethnicity separated us at school or in the neighborhood—except temporarily. During my elementary school years, I remember a fair number of dirt-bomb (a good suburban weapon) wars on the block. Periodically, one of the Catholic boys would accuse me or my brother of killing his god, to which we'd reply, "Did not," and start lobbing dirt bombs. Sometimes he'd get his friends from Catholic school and I'd get mine from public school kids on the block, some of whom were Catholic. Hostilities didn't last for more than a couple of hours and punctuated an otherwise friendly relationship. They ended by our junior high years, when other things became more important. Jews, Catholics and Protestants, Italians, Irish, Poles, "English" (I don't remember hearing WASP as a kid), were mixed up on the block and in school. We thought of ourselves as middle class and very enlightened because our ethnic backgrounds seemed so irrelevant to high school culture. We didn't see race (we thought), and racism was not part of our peer consciousness. Nor were the immigrant or working-class histories of our families.

As with most chicken-and-egg problems, it is hard to know which came first. Did Jews and other Euro-ethnics become white because they became middle class? That is, did money whiten? Or did being incorporated into an expanded version of whiteness open up the economic doors to middle-class status? Clearly, both tendencies were at work.

Some of the changes set in motion during the war against fascism led to a more inclusive version of whiteness. Anti-Semitism and anti-European racism lost respectability. The 1940 Census no longer distinguished native whites of native parentage from those, like my parents, of immigrant parentage, so Euro-immigrants and their children were more securely white by submersion in an expanded notion of whiteness.

Theories of nurture and culture replaced theories of nature and biology. Instead of dirty and dangerous races that would destroy American democracy, immigrants became ethnic groups whose children had successfully assimilated into the mainstream and risen to the middle class. In this new myth, Euro-ethnic suburbs like mine became the measure of American democracy's victory over racism and Jewish mobility became a new Horatio Alger story. In time and with hard work, every ethnic group would get a piece of the pie, and the United States would be a nation with equal opportunity for all its people to become part of a prosperous middle-class majority. And it seemed that Euro-ethnic immigrants and their children were delighted to join middle America.

## NOTES

1. *The New York Times,* July 30, 1893; cited in Allon Schoener, 1967. *Portal to America: The Lower East Side 1870–1925.* New York: Holt, Rinehart, and Winston, pp. 57–58.
2. Synott, Marcia Graham. 1986. "Anti-Semitism and American Universities: Did Quotas Follow the Jews?" In *Anti-Semitism in American History*, edited by David A. Gerber. Urbana, IL: University of Illinois Press, p. 250.

## DISCUSSION QUESTIONS

1. How does Brodkin see anti-Semitism as linked to other forms of racism? How is this revealed by historical events?
2. What does Brodkin mean by arguing the "Jews have become white folks?" How does her own family history reveal this process?

# 5

# Racial Formation

MICHAEL OMI AND HOWARD WINANT

*The concept of racial formation, first developed by Omi and Winant, has be-
come central to the sociological study of race. It refers to the social and historical
processes by which groups come to be defined in racial terms and it specifically
locates those processes in state-based institutions, such as the law.*

In 1982–83, Susie Guillory Phipps unsuccessfully sued the Louisiana Bureau
of Vital Records to change her racial classification from black to white. The
descendant of an 18th-century white planter and a black slave, Phipps was designated "black" in her birth certificate in accordance with a 1970 state law which
declared anyone with at least 1/32nd "Negro blood" to be black.

The Phipps case raised intriguing questions about the concept of race, its
meaning in contemporary society, and its use (and abuse) in public policy. Assistant Attorney General Ron Davis defended the law by pointing out that some
type of racial classification was necessary to comply with federal recordkeeping
requirements and to facilitate programs for the prevention of genetic diseases.
Phipps's attorney, Brian Begue, argued that the assignment of racial categories
on birth certificates was unconstitutional and that the 1/32nd designation was
inaccurate. He called on a retired Tulane University professor who cited research
indicating that most Louisiana whites have at least 1/20th "Negro" ancestry.

In the end, Phipps lost. The court upheld the state's right to classify and
quantify racial identity....

Phipps's problematic racial identity, and her effort to resolve it through state
action, is in many ways a parable of America's unsolved racial dilemma. It illustrates the difficulties of defining race and assigning individuals or groups to racial
categories. It shows how the racial legacies of the past—slavery and bigotry—
continue to shape the present. It reveals both the deep involvement of the state
in the organization and interpretation of race, and the inadequacy of state institutions to carry out these functions. It demonstrates how deeply Americans both
as individuals and as a civilization are shaped, and indeed haunted, by race.

Having lived her whole life thinking that she was white, Phipps suddenly
discovers that by legal definition she is not. In U.S. society, such an event is indeed catastrophic. But if she is not white, of what race is she? The state claims

SOURCE: From Michel Omi and Howard Winant, *Racial Formation in the United States. From the
1960s to the 1990s*, 2nd ed. (New York: Routledge, 1994), pp 53–61. Reprinted by permission of
Routledge/Taylor & Francis Books, Inc.

that she is black, based on its rules of classification ... and another state agency, the court, upholds this judgment. But despite these classificatory standards which have imposed an either-or logic on racial identity, Phipps will not in fact "change color." Unlike what would have happened during slavery times if one's claim to whiteness was successfully challenged, we can assume that despite the outcome of her legal challenge, Phipps will remain in most of the social relationships she had occupied before the trial. Her socialization, her familial and friendship networks, her cultural orientation, will not change. She will simply have to wrestle with her newly acquired "hybridized" condition. She will have to confront the "Other" within.

The designation of racial categories and the determination of racial identity is no simple task. For centuries, this question has precipitated intense debates and conflicts, particularly in the U.S.—disputes over natural and legal rights, over the distribution of resources, and indeed, over who shall live and who shall die.

A crucial dimension of the Phipps case is that it illustrates the inadequacy of claims that race is a mere matter of variations in human physiognomy, that it is simply a matter of skin color. But if race cannot be understood in this manner, how can it be understood? We cannot fully hope to address this topic—no less than the meaning of race, its role in society, and the forces which shape it—in one [article], nor indeed in one book. Our goal in this [article], however, is far from modest: we wish to offer at least the outlines of a theory of race and racism.

## WHAT IS RACE?

There is a continuous temptation to think of race as an essence, as something fixed, concrete, and objective. And there is also an opposite temptation: to imagine race as a mere illusion, a purely ideological construct which some ideal non-racist social order would eliminate. It is necessary to challenge both these positions, to disrupt and reframe the rigid and bipolar manner in which they are posed and debated, and to transcend the presumably irreconcilable relationship between them.

The effort must be made to understand race as an unstable and "decentered" complex of social meanings constantly being transformed by political struggle. With this in mind, let us propose a definition: race is a concept which signifies and symbolizes social conflicts and interests by referring to different types of human bodies. Although the concept of race invokes biologically based human characteristics (so-called "phenotypes"), selection of these particular human features for purposes of racial signification is always and necessarily a social and historical process. In contrast to the other major distinction of this type, that of gender, there is no biological basis for distinguishing among human groups along the lines of race.... Indeed, the categories employed to differentiate among human groups along racial lines reveal themselves, upon serious examination, to be at best imprecise, and at worst completely arbitrary.

If the concept of race is so nebulous, can we not dispense with it? Can we not "do without" race, at least in the "enlightened" present? This question has been posed often, and with greater frequency in recent years.... An affirmative answer would of course present obvious practical difficulties: it is rather difficult to jettison widely held beliefs, beliefs which moreover are central to everyone's identity and understanding of the social world. So the attempt to banish the concept as an archaism is at best counterintuitive. But a deeper difficulty, we believe, is inherent in the very formulation of this schema, in its way of posing race as a *problem*, a misconception left over from the past, and suitable now only for the dustbin of history.

A more effective starting point is the recognition that, despite its uncertainties and contradictions, the concept of race continues to play a fundamental role in structuring and representing the social world. The task for theory is to explain this situation. It is to avoid both the utopian framework which sees race as an illusion we can somehow "get beyond," and also the essentialist formulation which sees race as something objective and fixed, a biological datum. Thus we should think of race as an element of social structure rather than as an irregularity within it; we should see race as a dimension of human representation rather than an illusion. These perspectives inform the theoretical approach we call racial formation.

## Racial Formation

We define *racial formation* as the sociohistorical process by which racial categories are created, inhabited, transformed, and destroyed. Our attempt to elaborate a theory of racial formation will proceed in two steps. First, we argue that racial formation is a process of historically situated projects in which human bodies and social structures are represented and organized. Next we link racial formation to the evolution of hegemony, the way in which society is organized and ruled. Such an approach, we believe, can facilitate understanding of a whole range of contemporary controversies and dilemmas involving race, including the nature of racism, the relationship of race to other forms of differences, inequalities, and oppression such as sexism and nationalism, and the dilemmas of racial identity today.

From a racial formation perspective, race is a matter of both social structure and cultural representation. Too often, the attempt is made to understand race simply or primarily in terms of only one of these two analytical dimensions.... For example, efforts to explain racial inequality as a purely social structural phenomenon are unable to account for the origins, patterning, and transformation of racial difference.

Conversely, many examinations of racial difference—understood as a matter of cultural attributes à la ethnicity theory, or as a society-wide signification system, à la some poststructuralist accounts—cannot comprehend such structural phenomena as racial stratification in the labor market or patterns of residential segregation.

An alternative approach is to think of racial formation processes as occurring through a linkage between structure and representation. *Racial projects do the ideological "work" of making these links. A racial project is simultaneously an interpretation, representation, or explanation of racial dynamics, and an effort to recognize and redistribute resources along particular racial lines.* Racial projects connect what race means in a particular discursive practice and the ways in which both social structures and everyday experiences are racially *organized*, based upon that meaning. Let us consider this proposition, first in terms of large-scale or macro-level social processes, and then in terms of other dimensions of the racial formation process.

## Racial Formation as a Macro-Level Social Process

*To interpret the meaning of race is to frame it social structurally.* Consider, for example, this statement by Charles Murray on welfare reform:

> My proposal for dealing with the racial issue in social welfare is to repeal every bit of legislation and reverse every court decision that in any way requires, recommends, or awards differential treatment according to race, and thereby put us back onto the track that we left in 1965. We may argue about the appropriate limits of government intervention in trying to enforce the ideal, but at least it should be possible to identify the ideal: Race is not a morally admissible reason for treating one person differently from another. Period....

Here there is a partial but significant analysis of the meaning of race: it is not a morally valid basis upon which to treat people "differently from one another." We may notice someone's race, but we cannot act upon that awareness. We must act in a "color-blind" fashion. This analysis of the meaning of race is immediately linked to a specific conception of the role of race in the social structure: it can play no part in government action, save in "the enforcement of the ideal." No state policy can legitimately require, recommend, or award different status according to race. This example can be classified as a particular type of racial project in the present-day U.S.—a "neoconservative" one.

Conversely, *to recognize the racial dimension in social structure is to interpret the meaning of race.* Consider the following statement by the late Supreme Court Justice Thurgood Marshall on minority "set-aside" programs:

> A profound difference separates governmental actions that themselves are racist, and governmental actions that seek to remedy the effects of prior racism or to prevent neutral government activity from perpetuating the effects of such racism....

Here the focus is on the racial dimensions of social structure—in this case of state activity and policy. The argument is that state actions in the past and present have treated people in very different ways according to their race, and thus the government cannot retreat from its policy responsibilities in this area. It cannot suddenly declare itself "color-blind" without in fact perpetuating the same type of differential, racist treatment.... Thus, race continues to signify difference and structure inequality. Here, racialized social structure is immediately linked to an interpretation of the meaning of race. This example too can be classified as a particular type of racial project in the present-day U.S.—a "liberal" one.

To be sure, such political labels as "neoconservative" or "liberal" cannot fully capture the complexity of racial projects, for these are always multiply determined, politically contested, and deeply shaped by their historical context. Thus encapsulated within the neoconservative example cited here are certain egalitarian commitments which derive from a previous historical context in which they played a very different role, and which are rearticulated in neoconservative racial discourse precisely to oppose a more open-ended, more capacious conception of the meaning of equality. Similarly, in the liberal example, Justice Marshall recognizes that the contemporary state, which was formerly the architect of segregation and the chief enforcer of racial difference, has a tendency to reproduce those patterns of inequality in a new guise. Thus he admonishes it (in dissent, significantly) to fulfill its responsibilities to uphold a robust conception of equality. These particular instances, then, demonstrate how racial projects are always concretely framed, and thus are always contested and unstable. The social structures they uphold or attack, and the representations of race they articulate, are never invented out of the air, but exist in a definite historical context, having descended from previous conflicts. This contestation appears to be permanent in respect to race.

These two examples of contemporary racial projects are drawn from mainstream political debate; they may be characterized as center-right and center-left expressions of contemporary racial politics.... We can, however, expand the discussion of racial formation processes far beyond these familiar examples. In fact, we can identify racial projects in at least three other analytical dimensions: first, the political spectrum can be broadened to include radical projects, on both the left and right, as well as along other political axes. Second, analysis of racial projects can take place not only at the macro-level of racial policy-making, state activity, and collective action, but also at the micro-level of everyday experience. Third, the concept of racial projects can be applied across historical time, to identify racial formation dynamics in the past.

## DISCUSSION QUESTIONS

1. What do Omi and Winant mean by *racial formation?* What role does the law play in such a process? How is this shown in the history of the United States?

2. What difference does it make to conceptualize race as a property of social structures versus as a property (or attribute) of individuals?

# Student Exercises

1.  Using some random method of assignment, your instructor will divide your class (or some other grouping) into two groups, one of which is designated the Blues and the other the Greens. Over a period of a week, the Greens should serve the Blues in any way the Blues ask—such as carrying their books, running errands for them, delivering meals to places they designate, or any other job that the Blues design. (Since this is a course assignment, you should be reasonable in your demands.)

    As the week progresses, observe how the Blues act among themselves and in front of the Greens. Also observe how the Greens act among themselves and in front of the Blues. What attitudes do the two groups develop toward each other and toward themselves? How do they talk publicly about the other group? Do classmates begin to generalize about the assumed characteristics of the two different groups?

    What does the experiment reveal about the *social construction of race?*

2.  Have members of your class describe the ethnic background of their family members. You can describe such things as when and how your family arrived in the United States, ethnic traditions that your families may observe, and whether ethnic pride is a part of your family experience. After hearing from classmates belonging to different ethnic groups, list what you learned about ethnicity from listening to these different experiences. Is ethnicity more significant for some groups than others? Is ethnicity more important to some generations within a family than others? Why? What does this teach you about the *social construction of ethnicity* in society?

# Section II

# What Do You Think? Prejudice, Stereotyping, and Racism

Elizabeth Higginbotham
and Margaret L. Andersen

The beliefs that people hold are a powerful part of the persistence of racial inequality. When you think of beliefs about race, the term *prejudice* most likely comes to mind. Prejudice is an attitude that tends to denigrate individuals and groups who are perceived to be somehow different and undesirable. The social scientific definition of *prejudice* dates back to the 1950s and the work of psychologist Gordon Allport. Allport defined **prejudice** as "a hostile attitude directed toward a person or group simply because the person is presumed to be a member of that group and is perceived as having the negative characteristics associated with the group" (Allport 1954: 7).

Prejudice can be directed at many groups. One can be prejudiced against women or gays or athletes or foreigners—anyone who is perceived as a member of an "out-group"—that is, a group different from one's own. Although prejudice can be positive (as in thinking all women are nurturing), it is generally a negative attitude involving hostile or derogatory feelings as well as false generalizations about people in the out-groups.

Prejudice can also be expressed by any group, whether dominant or subordinate; thus, racial minorities may be prejudiced against other racial minorities or against the dominant group, just as more powerful people may be prejudiced against less powerful people. In other words, prejudice is a prejudgment and is the basis for much racial intolerance.

Prejudice rests on social **stereotypes;** that is, oversimplified beliefs about members of a particular social group. Stereotypes categorize people based on false generalizations along a narrow range of presumed characteristics, such as the belief that all Jewish people are greedy or that all blondes are dumb. Although stereotypes are perpetuated in many ways in society (e.g., in families, where parents teach children about other groups), one of the most influential mediums is popular culture—music, magazines, films, and television, among others. Thus, because men of color are portrayed in the media as criminals, this is the most common way that they are stereotyped. Or Asian American women may be stereotyped as sexy and beguiling—an image repeatedly shown in magazines, videos, and other popular media. Stereotypes in the media are so strong that we re-serve Section III of this book to examine how race and ethnicity are represented in the media and popular culture. For now, understand that stereotypes control how people come to define each other and, as such, are *controlling images* (Collins 1990). Judith Ortiz Cofer's essay ("The Myth of the Latin Woman: I Just Met a Girl Named Maria") in this section shows how harmful and insulting stereotypes can be.

The prejudice that grows out of social stereotypes is an *attitude*. This is distinct from discrimination, which is *behavior*. **Discrimination** is the negative and unequal treatment of members of a social group based on their perceived membership in a particular group. Although the term *discrimination* sometimes has a positive connotation (as in "she has a discriminating attitude"), such behav-ior generally is negative.

Prejudice and discrimination are typically thought of as related—prejudice causes discrimination—but things are not that simple. Many years ago, sociologist Robert Merton (1949) developed a four-square typology showing different ways that prejudice and discrimination are and are not related. Look at the following:

|  | PREJUDICE: | Positive (+) | Negative (−) |
|---|---|---|---|
| DISCRIMINATION: | Positive (+) | Case 1: Bigot (+ +) | Case 2: Non-prejudiced discriminator (− +) |
|  | Negative (−) | Case 3: Prejudiced non-discriminator (+ −) | Case 4: All-weather liberal (− −) |

In Case 1, someone may both be prejudiced and discriminate—the classic bigot. Both prejudice and discrimination are overt, intentional, and hostile. In Case 4, someone may be free of prejudice and not discriminate (the person Merton called the "all-weather liberal"). It is in Cases 2 and 3 that we see that prejudice and discrimination may not have a causal relationship. In Case 2, one may not be prejudiced, but still discriminate, such as a homeowner who holds no racial prejudice but will only buy a home in an all-White neighborhood "to protect their property value." Such persons may say they hold no prejudice, but they might look out for their own interests and discriminate nonetheless, even without malice. In Case 3, someone may be prejudiced, but not discriminate—for example, when the law prohibits discrimination. A landlord may, for example, rent to a Black tenant despite holding prejudice. The point is that both prejudice and discrimination occur in a larger context—that of society as a whole. This societal context predicts the occurrence of discrimination as much as people's individual attitudes.

Prejudice is not just a free-floating attitude, however. It is linked to group positions in society (Blumer 1958; Bobo and Hutchings 1996). As Matthew Desmond and Mustafa Emirbayer point out ("American Racism in the Twenty-First Century"), individualistic thinking is one of the fallacies about race. Racism is not just about thoughts; it is woven into the fabric of our society. Thus, **racism**, different from prejudice, is a principle of social domination in which a group that is seen as inferior or different because of presumed biological or cultural characteristics is oppressed, controlled, and exploited—socially, economically, culturally, politically, or psychologically—by a dominant group (Wilson 1973).

Note the key elements of this definition. First, racism is a *principle of domination;* that is, embedded in this definition is the thought that racism involves one or more groups' subordinate position within a system of racial inequality. And, as Desmond and Emirbayer's article shows, this also means that the effects of the history of racism are *cumulative* and do not disappear easily through legally abolishing segregation or placing a few people of color in positions of influence.

Second, note that the definition of racism also emphasizes the word *presumed.* As we learned in Section I, race is not "real" in the biological or cultural sense, but it develops meaning through society and history. How people are perceived within a system of hierarchy and power is the key to understanding racism. Who gets the power to define different groups, and what are the means by which they do it? Law? Media? Schools? All of the these? Those who shape how people are represented have enormous power to shape people's consciousness about race.

Third, racism involves domination on a number of fronts: social, economic, political, cultural, and psychological. **Institutional racism** is the complex and cumulative pattern of racial advantage and disadvantage built into the structure of a society. Institutional racism, reflected in the prejudice and discrimination seen in a society, comprises more than an attitude or behavior. It is a system of power and privilege that gives the advantage to some groups over others. Thus you might say that prejudice is lodged in people's minds, but racism is lodged in society.

As several of the authors in Section II show, many people who benefit from institutional racism are often blind to the systemic advantage that it gives them. Thus, just being a White person will open some opportunities that might not be as readily available to others—independent of that White person's own attitudes and behavior. The invisibility of racial privilege to dominant groups is referred to as **color-blind racism**, the belief that race should be ignored and that race-conscious practices and policies only foster more racism. When dominant groups think that racism is no longer an issue, despite its ongoing reality, they are not likely to engage in practices or support policies that challenge racism (Brown et al. 2003; Bonilla-Silva 2003). To be color-blind in a society in which race still structures people's relationships, identities, and opportunities is to be blind to the continuing realities of race.

Charles Gallagher examines color-blind privilege in his essay "Color-Blind Privilege: The Social and Political Functions of Erasing the Color Line in Post-Race America." He points out that we live under the appearance of a multiracial society where race has actually become a commodity, something that White people can buy and display, while at the same time not challenging the privilege that underlies racial stratification. Products are mass marketed using multiracial images and they sell across color lines, but such images legitimate color-blindness. Gallagher's research shows that while many White people believe themselves to be color-blind, once you scratch the surface of this belief they are quick to defend the status quo.

Racism and racial dynamics are now being challenged by the increasing presence and identity of people as biracial or multiracial, as Rainier Spencer's essay ("Mixed Race Chic") argues. This change is challenging some of the old, binary ways of thinking—that is, thinking solely in terms of two categories, Black and White. But racism produces hierarchical ways of thinking and hierarchical structures in society. Until these are changed, racism does not go away.

Rebekah Nathan ("My Freshman Year: What a Professor Learned by Becoming a Student") provides a unique experiment that uncovers one of the ways that racism manifests itself on college campuses. Nathan is a faculty member who spent a year posing as a college student and observing the culture of a college campus. As part of her study, she observed patterns of student dining, focusing on the experiences of students of color whom she finds most often eat "alone"—not just meaning by themselves but also in the sense that students of color find themselves living in "White spaces" where they are psychologically and culturally alone. Her experiment, along with the other articles in this section, reveals how racism as a social structure patterns everyday life in a racialized society.

## REFERENCES

Allport, Gordon. 1954. *The Nature of Prejudice*. Reading, MA: Addison-Wesley.

Blumer, Herbert. 1958. "Race Prejudice as a Sense of Group Position." *Pacific Sociological Review* 1 (Spring): 3–7.

Bobo, Lawrence, and Vincent L. Hutchings. 1996. "Perceptions of Racial Group Competition: Extending Blumer's Theory of Group Position to a Multiracial Social Context." *American Sociological Review* 25 (December): 951–972.

Bonilla-Silva, Eduardo. 2003. *Racism without Racists: Colorblind Racism and the Persistence of Racial Inequality in the United States.* Lanham, MD: Rowman & Littlefield.

Brown, Michael, Martin Carnoy, Elliott Currie, Troy Duster, David Oppenheimer, Marjorie M. Schultz, and David Wellman. 2003. *Whitewashing Race: The Myth of a Color-Blind Society.* New York: Oxford University Press.

Collins, Patricia Hill. 1990. *Black Feminist Thought: Knowledge, Consciousness, and the Politics of Empowerment.* Boston: Unwin Hyman.

Merton, Robert. 1949. "Discrimination and the American Creed." In *Discrimination and the National Welfare*, ed. Robert W. MacIver. New York: Harper and Brothers. Pp. 99–126.

Wilson, William Julius. 1973. *Power, Racism, and Privilege: Race Relations in Theoretical and Sociohistorical Perspectives.* New York: Macmillan.

## FACE THE FACTS: POLLING PREJUDICE

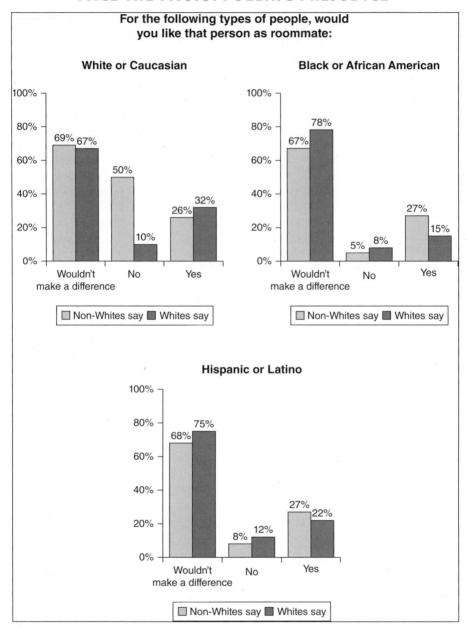

### For the following types of people, would you like that person as roommate:

**White or Caucasian**

69%  67%   50%   10%   26%   32%

Wouldn't make a difference     No     Yes

☐ Non-Whites say  ■ Whites say

**Black or African American**

67%  78%   5%   8%   27%   15%

Wouldn't make a difference     No     Yes

☐ Non-Whites say  ■ Whites say

**Hispanic or Latino**

68%  75%   8%   12%   27%   22%

Wouldn't make a difference     No     Yes

☐ Non-Whites say  ■ Whites say

SOURCE: ROBISON, JENNIFER, 2003. "WILL TEEN TOLERANCE PASS THE ROOMMATE TEST?" *THE GALLUP POLL*. PRINCETON, NJ: THE GALLUP ORGANIZATION.

*Think about it*: The graphs above show the results of a national Internet survey in which people aged 13 to 17 were asked (in an Internet survey): "Many young people have roommates when they are in college or in their first jobs after high school. For the following types of people, would you like that person as a roommate, not like that person as a roommate, or would it make no difference?" The results show white and non-white responses to three different scenarios: White roommates, Black roommates, and Hispanic roommates. Although the large majority of respondents say it would not make a difference, there are some differences both in how Whites and non-Whites respond, depending on the race of the hypothetical roommate. What do these results suggest to you about racial prejudice?

# 6

# American Racism in the Twenty-first Century

MATTHEW DESMOND AND MUSTAFA EMIRBAYER

*These authors explore some of the fallacies that stem from misconceptions about racism. Their article shows that racism has its origins in social structural factors, even though people's understanding of racism in the United States stems from the culture of individualism.*

Pace University is a comprehensive university with over 14,000 students, spread out over six campuses in and around New York City. The student body is diverse, and the university has prided itself on maintaining an accepting and welcoming environment. But in the fall of 2006, several incidents led many students to question how welcome they really were on campus. First, a student discovered a copy of the Koran—the Holy book of Islam—discarded in the toilet of a library bathroom. A few weeks later, a second Koran was given the same maltreatment; this time, the book was desecrated with hateful slurs. Then, a racial slur targeting African Americans was traced in the dew on a car window parked on campus. Shortly after that, swastikas and dozens of racial epithets were found scrolled on a bathroom wall. These events led Muslims and many students of color to fear for their safety. As Ashley Marinaccio, a senior at Pace, confessed, "When one minority is the victim of a hate crime, it certainly provokes fear in other groups, because you cannot help but think, 'Am I next?' I've had discussions with quite a number of people who are worried and do not feel safe because of these incidents."

That these incidents were explicitly hateful and racist in nature, few would dispute. And that they occurred on a college campus is not very unusual. Many of the perpetrators and victims of hate crimes are young adults. In fact, 29% of hate crime offenders are between the ages of 18 and 24, and 11% of documented hate crimes occur in schools and college campuses.

Although the historical period when all people of color, and African Americans in particular, were terrorized by whites—a period that spans a full *two-thirds* of U.S. history—seems far behind us, individuals still carry out overt acts

SOURCE: Matthew Desmond and Mustafa Emirbayer. 2010. *Racial Domination, Racial Progress*, New York: McGraw Hill.

motivated by racial hatred. These acts range in intensity from vandalism to murder.... Hate groups, such as the Ku Klux Klan, neo-Nazis, and skinheads, are still found all around the United States. According to the Southern Poverty Law Center, there are hundreds of active hate groups across the country. These groups are mostly found in the Southern states—Texas, Georgia, and South Carolina have over 40 active groups per state—but California ranks highest in the nation, housing within its borders 53 groups. Members of hate groups cultivate explicitly racist biases against members of other races, ethnicities, or religions and act on those biases through incidents of violence, harassment, and intimidation.

For some people, this is what racism amounts to: intentional acts of humiliation and hate. Although such acts undoubtedly are racist in nature, they are but the tip of the iceberg as to what constitutes racism. To define racism only through extreme groups and their extreme acts is akin to defining weather only through hurricanes. Hurricanes are certainly a type of weather pattern—a most harsh and brutal type—but so too are mild rainfalls, light breezes, and sunny days. Likewise, racism is much broader than hate groups, racial slurs, and defiled Korans. It also comes in much quieter and what one might call everyday-ordinary forms.

## FIVE FALLACIES ABOUT RACISM

There are many misconceptions about the character of racism. Americans are deeply divided over its legacies and inner workings, and much of this division is due to the fact that many Americans understand racism in limited or misguided ways. We have identified five fallacies about racism—logical mistakes, factual or logical errors in reasoning—that are recurrent in many public debates, fallacies one should avoid when thinking about racism.

1. *Individualistic Fallacy.* Here, racism is assumed to belong to the realm of ideas and prejudices. Racism is only the collection of nasty thoughts a "racist individual" has about another group. Someone operating with this fallacy thinks of racism as one thinks of a crime and, therefore, divides the world into two types of people: those guilty of the crime of racism ("racists") and those innocent of the crime ("nonracists"). Crucial to this misconceived notion of racism is intentionality. "Did I intentionally act racist? Did I cross the street because I was scared of the Hispanic man walking toward me, or did I cross for no apparent reasons?" Upon answering "no" to the question of intentionality, one assumes they can classify their actions as "nonracist," despite the character of those actions, and go about their business, as innocent.

This conception of racism simply won't do, for it fails to account for the racism woven into the very fabric of our schools, political institutions, labor markets, and neighborhoods. Conflating racism with prejudice ignores the more systematic and structural forms of racism; it looks for racism within individuals and not institutions. Labeling someone a "racist" shifts our attention from the social surroundings that enforce racial inequalities to the individual with biases. It also lets the accuser off the hook—"He is a racist, I am not"—and treats racism as aberrant and strange, whereas American racism is quite normal.

Furthermore, intentionality is in no way a prerequisite for racism. Racism is often habitual, unintentional, commonplace, polite, implicit, and well-meaning. Thus, racism is not only located in our intentional thoughts and actions; it also thrives in our dispositions and habits, as well as in the social institutions in which we are all embedded.

2. *Legalistic Fallacy.* This fallacy conflates *de jure* legal progress with *de facto* racial progress. *De jure* and *de facto* are Latin expressions meaning, respectively, "based on the law" and "based in fact." Thus, one who operates under the legalistic fallacy assumes that abolishing racist laws (racism in principle) automatically leads to the abolition of racism writ large (racism in practice).

This fallacy begins to crumble after a few moments of critical reflection. After all, we would not make the same mistake when it comes to other criminalized acts: Laws against theft do not mean that one's car will never be stolen. By way of tangible illustration, consider *Brown v. Board of Education*, the landmark 1954 case that abolished *de jure* segregation in schools, making it illegal to enforce racially segregated classrooms. Did that lead to the abolition of *de facto* segregation? Absolutely not. Fifty years later, schools are still drastically segregated and drastically unequal. In fact, some social scientists have documented a nationwide movement of educational resegregation, which has left today's schools even more segregated than those of 1954.

3. *Tokenistic Fallacy.* One who is guilty of the tokenistic fallacy assumes that the presence of people of color in influential positions is evidence of the complete eradication of racial obstacles. This logic runs something like this: "Many people of color, such as Barack Obama, Condoleezza Rice, Colin Powell, Carol Mosely Brown, and Alberto Gonzales, have held high-ranking political posts; therefore, racism does not exist in the political arena. Many people of color, such as Bill Cosby, Oprah Winfrey, Jennifer Lopez, and Lucy Lu, are celebrities and multimillionaires; therefore, there is no racial inequality when it comes to income and wealth distribution. Poor people of color (not society) are to blame for their own poverty."

Although it is true that many people of color have made significant inroads to seats of political and economic power over the course of the last fifty years, a disproportionate number of them remain disadvantaged in these arenas. We cannot, in good conscience, ignore the millions of African Americans living in poverty and, instead, point to Oprah's millions as evidence for economic inequality. Instead, we must explore how Oprah's financial success can coexist with the economic deprivation of millions of black women. We need to explore, in historian Thomas Holt's words, how the "simultaneous idealization of Colin Powell [or, for that matter, Barack Obama] and demonization of blacks as a whole… is replicated in much of our everyday world.[1]"

Besides, throughout the history of America, a handful of nonwhite individuals have excelled financially and politically in the teeth of rampant racial domination. The first black congressman was not elected after the Civil Rights Movement but in 1870! Joseph Rainey, a former slave, served in the House of Representatives for four terms. Madame C. J. Walker is accredited as being the first black millionaire. Born in 1867, she made her fortune inventing hair and beauty products. Few people would feel comfortable pointing to Rainey's or Walker's success as evidence that late-nineteenth-century America was a time of racial harmony and equity. Such tokenistic logic would not be accurate then, and it is not accurate now.

4. *Ahistorical Fallacy*. This fallacy renders history impotent. Thinking hindered by the ahistorical fallacy makes a bold claim: most United States history—namely, the period of time when this country did not extend basic rights to people of color (let alone classify them as fully human)—is inconsequential today. Legacies of slavery and colonialism, the eradication of millions of Native Americans, forced segregation, clandestine sterilizations and harmful science experiments, mass disenfranchisement, race-based exploitation, racist propaganda distributed by the state caricaturing Asians, blacks, and Hispanics, racially motivated abuses of all kinds (sexual, murderous, and dehumanizing)—all of this, purport those operating under the ahistorical fallacy, does not matter for those living in the here-and-now. This idea is so delusional that it is hard to take seriously. Today's society is directed, constructed, and molded by—indeed grafted onto—the past. All that is socially constructed is historically constructed; and since race, as we have seen, is a social construction, it, too, is a historical construction.

A "soft version" of the ahistorical fallacy might admit that events in the "recent past"—such as the time since the Civil Rights Movement or the attacks on September 11, 2001—matter but things in the "distant past"—such as slavery or the colonization of Mexico—have little consequence. But this idea is no less fallacious than the "hard version," since many events in America's "distant past"— especially the enslavement and murder of millions of Africans—are the *most* consequential in shaping present-day society. In this vein, consider the question French historian Marc Bloch poses to us: "But who would dare to say that the understanding of the Protestant or Catholic Reformation, several centuries removed, is not *far more important* for a proper grasp of the world today than a great many other movements of thought or feeling, which are certainly more recent, yet more ephemeral?"[2] We would not dare. (Additionally, any historian would remind us that, since America is just over 200 years old, *all* American history is "recent history.")

5. *Fixed Fallacy*. Those who assume that racism is fixed, that it is immutable, constant across time and space, partake in the fixed fallacy. Since they take racism to be something that does not develop in any way, those who understand racism through the fixed fallacy are often led to ask such questions as "Has racism increased or decreased in the past decades?" And because practitioners of the fixed fallacy usually take as their standard definition of racism only the most heinous forms of racism— racial violence, for example—they confidently conclude that, indeed, things have gotten better.

It is important to trace the career of American racism and to analyze, for example, how racial attitudes or measures of racial inclusion and exclusion have changed over time. Many social scientists have developed sophisticated techniques for doing so. But the question "Have things gotten better or worse?" is legitimate *only* after we account for the morphing attributes of racism. We cannot quantify racism in the same way that we can quantify, say, birthrates. The nature of "birthrate" does not fluctuate over time; thus, it makes sense to ask "Are there more or less births now than there were fifty years ago?" without bothering to analyze if and how a birthrate is different today than it was in previous historical moments.

American racism assumes different forms in different historical moments. Although race relations today are informed by those of the past, we cannot hold to the belief that twenty-first-century racism takes on the exact same form as twentieth-century racism. And we certainly cannot conclude that there is "little or no racism" today because it does not resemble the racism of the 1950s. (Modern-day Christianity looks very different, in nearly every conceivable way, than the Christianity of the early church. But this does not mean that there is "little or no Christianity" today.) So, before we ask "Have things gotten better or worse?" we should ponder the essence of racism today and how it differs from racism experienced by those living in our parents' or grandparents' generation. We should ask, further, to quote Holt once more, "What enables racism to reproduce itself after the historical conditions that initially gave it life have disappeared?"

## RACIAL DOMINATION

We have spent a significant amount of time talking about what racial domination or racism is not. We have yet to spell out what it is. We can delineate two specific manifestations of racial domination: institutional racism and interpersonal racism. **Institutional racism** is systemic white domination of people of color, embedded and operating in corporations, universities, legal systems, political bodies, cultural life, and other social collectives. The word "domination" reminds us that institutional racism is a type of power that encompasses the *symbolic power* to classify one group of people as "normal" and other groups of people as "abnormal," the *political power* to withhold basic rights from people of color and marshal the full power of the state to enforce segregation and inequality, the *social power* to deny people of color full inclusion or membership in associational life, and *economic power* that privileges whites in terms of job placement, advancement, and wealth and property accumulation.

Informed by centuries of racial domination, institutional racism withholds from people of color opportunities, privileges, and rights that many whites enjoy. Examples of institutional racism include the tendency of schools and universities to support curricula that highlight the accomplishments of European Americans, ignoring the accomplishments of non-European Americans; the disproportionate numbers of white people in high-ranking political, economic, and military posts and the ongoing exclusion of people of color from such posts; and the prevalence of law enforcement practices that target people of color, especially African American and Arab Americans, as criminals or terrorists. In all three of these examples, racial domination is carried out at the institutional level, sometimes despite the motives or attitudes of the people working in those institutions. Because institutional racism operates outside the scope of individual intent, many people do not recognize institutional racism as racism when they experience it.

Below the level of institutions, yet informed by the workings of those institutions, we find **interpersonal racism**. This is racial domination manifest in everyday interactions and practices. Interpersonal racism can be *overt*; however, most of the time, interpersonal racism is quite *covert*: it is found in the habitual, commonsensical, and ordinary practices of our lives. Our racist attitudes, as Lillian Smith remarked in *Killers of the Dream*, easily "slip from the conscious mind deep into the muscles."[3] Since we are disposed to a world structured by racial domination, we develop racialized dispositions—some conscious, many more unconscious and bodily—that guide our thoughts and behaviors.

We may talk slowly to an Asian woman at the farmer's market, unconsciously assuming that she speaks poor English. We may inform a Mexican woman at a corporate party that someone has spilled his punch, unconsciously assuming that she is a janitor. We may unknowingly scoot to the other side of the elevator when a large Puerto Rican man steps in, or unthinkingly eye a group of black teenagers wandering the aisles of the store at which we work, or ask to change seats if an Arab American man sits down next to us on an airplane. Many miniature actions such as these have little to do with one's intentional thoughts; they are orchestrated by one's practical sense, one's habitual know-how, and informed by institutional racism.

"Can people of color be racist?" This question is a popular one in the public imagination, and the answer depends on what we mean by racism. Institutional racism is the product of years of white supremacy, and it is designed to produce far-reaching benefits for white people. Institutional racism carries on despite our personal attitudes. Thus, there is no such thing as "black institutional racism" or "reverse institutional racism," since there is no centuries-old socially ingrained and normalized system of domination designed by people of color that denies whites full participation in the rights, privileges, and seats of power of our society. Interpersonal racism, on the other hand, takes place on the ground level and has to do with attitudes and habitual actions. It is certainly true that members of all racial groups can harbor negative attitudes toward members of other groups. An African American may hold ill feelings toward Jews or Koreans. An Asian American may be suspicious of white people. And such prejudiced perceptions are often rampant *within* racial groups as well, as when a Cuban-American feels superior to a Mexican-American, a Japanese-American feels uncomfortable around Chinese-Americans, or dark-skinned African Americans profess to being "more authentically black" than light-skinned African Americans. Indeed, some nonwhite groups have a deep, conflict-ridden history with other nonwhite groups. One thinks here of the Black-Korean conflict, the so-called Black-Brown divide, bitter relations among Latino subgroups, and animus among various American Indian Nations.

People of color, then, can take part in overt and covert forms of interpersonal racism. That said, we must realize that interpersonal racism targeting dominated groups and interpersonal racism targeting the dominant group do not pack the same punch. Two young men, one black, the other white, bump into each other on the street. The black man calls the white man a "honky." In response, the white man calls the black man a "boy." Both racial slurs *are* racial slurs and should be labeled as such, and both reinforce racial divisions. However, unlike "honky," "boy" connects to the larger system of institutional racial domination. The word derives its meaning (and power) from slavery, when enslaved African men were stripped of their masculine honor and treated like children. "Boy" (and many other epithets aimed at blacks) invokes such times—times when murdering, torturing, whipping, and raping enslaved blacks were not illegal acts. Epithets toward white people, including "honky," have no such equivalent. ("Honky" comes from derogatory terms aimed at Bohemian, Hungarian, and Polish immigrants who worked in the Chicago meat-packing plants.) "Boy" also reminds the black man how things stand today: if the confrontation escalates and the police are called, the black man knows that the police officers will probably be white and that he might be harassed or looked on as a threat; if the two men meet in court, the black man knows that the lawyers, judge, and jurors will possibly be mostly (if not all) white; and if the two men are sentenced, the African American man knows, as do many criminologists, that he will get the harsher sentence. "Boy" brings the full weight of institutional racism—systematic, historical, and mighty—down on the African-American man. "Honky," even if delivered with venomous spite, is powerless by comparison.

Moreover, sociologists have shown that, unlike white people, people of color are confronted with interpersonal racism on a regular basis, sometimes daily. For people of color, there is a cumulative character to an individual's racial experiences. These experiences do not take place in isolation. Humiliating or degrading acts always are informed by similar acts that individuals have experienced in the past. To paraphrase sociologist Joe Feagin, the interpersonal events that take place on the street and in other public settings are not simply rare and isolated events; rather, they are recurring events shaped by social and historical racial domination.

## NOTES

1.   Holt, Thomas. 2000. *The Problem of Race in the Twenty-First Century*. Cambridge, MA: Harvard University Press, p. 6.
2.   Marc Bloch. 1953. *The Historian's Craft*. New York: Vintage, p. 41.
3.   Lillian Smith. 1961. *Killers of the Dream*, revised ed. New York: Norton, p. 96.

## DISCUSSION QUESTIONS

1.   What five fallacies do the authors identity about racism, and how does this identification change the usual understanding of racism as compared to individual prejudice?

2.   In what ways do you think the racism of today is different from that of the past?

# 7

# Color-Blind Privilege:

## The Social and Political Functions of Erasing the Color Line in Post Race America

CHARLES A. GALLAGHER

*Popular culture is full of images that portray a multiracial society, but such representations encourage a new form of thinking—color-blind racism. Color-blind racism refers to the dominant belief that race no longer matters in shaping people's experiences—a belief that is contradicted by the reality of race in America.*

The young white male sporting a FUBU (African–American owned apparel company "For Us By Us") shirt and his white friend with the tightly set, perfectly braided cornrows blended seamlessly into the festivities at an all-white bar mitzvah celebration. A black model dressed in yachting attire peddles a New England, yuppie boating look in Nautica advertisements. It is quite unremarkable to observe white, Asian or African-Americans with dyed purple, blond or red hair. White, black and Asian students decorate their bodies with tattoos of Chinese characters and symbols. In cities and suburbs young adults across the color line wear hip-hop clothing and listen to white rapper Eminem and black rapper 50 Cent. It went almost unnoticed when a north Georgia branch of the NAACP installed a white biology professor as its president. Subversive musical talents like Jimi Hendrix, Bob Marley and The Who are now used to sell Apple computers, designer shoes and SUVs. Du-Rag kits, complete with bandana headscarf and elastic headband, are on sale for $2.95 at hip-hop clothing stores and family–centered theme parks like Six Flags. Salsa has replaced ketchup as the best selling condiment in the United States. Companies as diverse as Polo, McDonalds' Tommy Hilfiger, Walt Disney World, Master Card, Skechers sneakers, IBM, Giorgio Armani and Neosporin antibiotic ointment have each crafted advertisements that show an integrated, multiracial cast of characters interacting and consuming their products in a post-race, color-blind world.

Americans are constantly bombarded by depictions of race relations in the media which suggest that discriminatory racial barriers have been dismantled. Social and cultural indicators suggest that America is on the verge, or has already become, a truly color-blind nation. National polling data indicate that a majority of whites now believe discrimination against racial minorities no longer exists.

SOURCE: From *Race, Gender & Class* 10 (2003): 22–37. Reprinted by permission.

A majority of whites believe that blacks have "as good a chance as whites" in procuring housing and employment or achieving middle class status while a 1995 survey of white adults found that a majority of whites (58%) believed that African Americans were "better off" finding jobs than whites (Gallup, 1997; Shipler, 1998). Much of white America now sees a level playing field, while a majority of black Americans sees a field which is still quite uneven.... The color-blind or race-neutral perspective holds that in an environment where institutional racism and discrimination have been replaced by equal opportunity, one's qualifications, not one's color or ethnicity, should be the mechanism by which upward mobility is achieved. Color as a cultural style may be expressed and consumed through music, dress, or vernacular but race as a system which confers privileges and shapes life chances is viewed as an atavistic and inaccurate accounting of U.S. race relations.

Not surprisingly, this view of society blind to color is not equally shared. Whites and blacks differ significantly, however, on their support for affirmative action, the perceived fairness of the criminal justice system, the ability to acquire the "American Dream," and the extent to which whites have benefited from past discrimination (Moore, 1995; Moore & Saad, 1995; Kaiser, 1995). This article examines the social and political functions colorblindness serves for whites in the United States. Drawing on interviews and focus groups with whites from around the country, I argue that colorblind depictions of U.S. race relations serve to maintain white privilege by negating racial inequality. Embracing a colorblind perspective reinforces whites' belief that being white or black or brown has no bearing on an individual's or a group's relative place in the socioeconomic hierarchy.

## DATA AND METHOD

I use data from seventeen focus groups and thirty individual interviews with whites from around the country. Thirteen of the seventeen focus groups were conducted in a college or university setting, five in a liberal arts college in the Rocky Mountains and the remaining eight at a large urban university in the Northeast. Respondents in these focus groups were selected randomly from the student population. Each focus group averaged six respondents ... equally divided between males and females. An overwhelming majority of these respondents were between the ages of eighteen and twenty-two years of age. The remaining four focus groups took place in two rural counties in Georgia and were obtained through contacts from educational and social service providers in each county. One county was almost entirely white (99.54%) and in the other county whites constituted a racial minority. These four focus groups allowed me to tap rural attitudes about race relations in environments where whites had little or consistent contact with racial minorities....

## COLORBLINDNESS AS NORMATIVE IDEOLOGY

The perception among a majority of white Americans that the socio-economic playing field is now level, along with whites' belief that they have purged themselves of overt racist attitudes and behaviors, has made colorblindness the dominant lens through which whites understand contemporary race relations. Colorblindness allows whites to believe that segregation and discrimination are no longer an issue because it is now illegal for individuals to be denied access to housing, public accommodations or jobs because of their race. Indeed, lawsuits alleging institutional racism against companies like Texaco, Denny's, Coke, and Cracker Barrel validate what many whites know at a visceral level is true: firms which deviate from the color-blind norms embedded in classic liberalism will be punished. As a political ideology, the commodification and mass marketing of products that signify color but are intended for consumption across the color line further legitimate colorblindness. Almost every household in the United States has a television that, according to the U.S. Census, is on for seven hours every day (Nielsen 1997). Individuals from any racial background can wear hip-hop clothing, listen to rap music (both purchased at Wal-Mart) and root for their favorite, majority black, professional sports team. Within the context of racial symbols that are bought and sold in the market, colorblindness means that one's race has no bearing on who can purchase a Jaguar, live in an exclusive neighborhood, attend private schools or own a Rolex.

The passive interaction whites have with people of color through the media creates the impression that little, if any, socio-economic difference exists between the races....

Highly visible and successful racial minorities like [former] Secretary of State Colin Powell and ... [former Secretary of State] Condoleezza Rice are further proof to white America that the state's efforts to enforce and promote racial equality have been accomplished.

The new color-blind ideology does not, however, ignore race; it acknowledges race while disregarding racial hierarchy by taking racially coded styles and products and reducing these symbols to commodities or experiences that whites and racial minorities can purchase and share. It is through such acts of shared consumption that race becomes nothing more than an innocuous cultural signifier. Large corporations have made American culture more homogenous through the ubiquitousness of fast food, television, and shopping malls but this trend has also created the illusion that we are all the same through consumption. Most adults eat at national fast food chains like McDonalds' shop at mall anchor stores like Sears and J.C. Penney's and watch major league sports, situation comedies or television drama. Defining race only as cultural symbols that are for sale allows whites to experience and view race as nothing more than a benign cultural marker that has been stripped of all forms of institutional, discriminatory or coercive power. The post-race, color-blind perspective allows whites to imagine that depictions of racial minorities working in high status jobs and consuming the same products, or at least appearing in commercials for products whites desire or consume, is the same as living in a society where color is no longer used to allocate resources or shape group outcomes. By constructing a picture of society where racial harmony is the norm, the color-blind perspective functions to make white privilege invisible while removing from public discussion the need to maintain any social programs that are race-based.

How, then, is colorblindness linked to privilege? Starting with the deeply held belief that America is now a meritocracy, whites are able to imagine that the socio-economic success they enjoy relative to racial minorities is a function of individual hard work, determination, thrift and investments in education. The color-blind perspective removes from personal thought and public discussion any taint or suggestion of white supremacy or white guilt while legitimating the existing social, political and economic arrangements which privilege whites. This perspective insinuates that class and culture, and not institutional racism, are responsible for social inequality. Colorblindness allows whites to define themselves as politically and racially tolerant as they proclaim their adherence to a belief system that does not see or judge individuals by the "color of their skin." This perspective ignores, as Ruth Frankenberg puts it, how whiteness is a "location of structural advantage societies structured in racial dominance" (2001 p. 76).... Colorblindness hides white privilege behind a mask of assumed meritocracy while rendering invisible the institutional arrangements that perpetuate racial inequality. The veneer of equality implied in colorblindness allows whites to present their place in the racialized social structure as one that was earned.

## OPPORTUNITY HAS NO COLOR

Given this norm of colorblindness it was not surprising that respondents in this study believed that using race to promote group interests was a form of (reverse) racism....

Believing and acting as if America is now color-blind allows whites to imagine a society where institutional racism no longer exists and racial barriers to upward mobility have been removed. The use of group identity to challenge the existing racial order by making demands for the amelioration of racial inequities is viewed as racist because such claims violate the belief that we are a nation that recognizes the rights of individuals, not rights demanded by groups....

The logic inherent in the colorblind approach is circular; since race no longer shapes life chances in a color-blind world there is no need to take race into account when discussing differences in outcomes between racial groups. This approach erases America's racial hierarchy by implying that social, economic and political power and mobility is equally shared among all racial groups. Ignoring the extent or ways in which race shapes life chances validates whites' social location in the existing racial hierarchy while legitimating the political and economic arrangements that perpetuate and reproduce racial inequality and privilege.

# REFERENCES

Frankenberg, R. (2001). "The mirage of an unmarked 'whiteness'". In B. B. Rasmussen, E. Klineberg, I. J. Nexica & M. Wray (eds.) *The making and unmaking of whiteness*. Durham: Duke University Press.

Gallup Organization. (1997). "Black/white relations in the U.S." June 10, pp. 1–5.

Kaiser Foundation. (1995). *The four Americas: Government and social policy through the eyes of America's multi-racial and multi-ethnic society*. Menlo Park, CA: Kaiser Family Foundation.

Moore, D. (1995). "'Americans'" most important sources of information: Local news." *The Gallup Poll Monthly*, September, pp. 2–5.

Moore, D. & Saad, L. (1995). "No immediate signs that Simpson trial intensified racial animosity." *The Gallup Poll Monthly*, October, pp. 2–5.

Nielsen, A. C. (1997). *Information please almanac* (Boston: Houghton Mifflin).

Shipler, D. (1998). *A country of strangers: Blacks and whites in America*. New York: Vintage Books.

# DISCUSSION QUESTIONS

1. How does Gallagher see color-blind racism as resulting from White people's privilege? How does privilege influence what White people can understand about racism?

2. In what ways does color-blind racism support the traditional American ideal that any individual can succeed if they only try hard enough?

# 8

# The Myth of the Latin Woman:
## I Just Met a Girl Named María

JUDITH ORTIZ COFER

*Cofer's narrative about her experience in everyday encounters shows the continuing significance of racial and ethnic stereotyping and how it feels when directed at you. She also shows the role of the media in constructing some of these stereotypes, as well as highlighting some of the generational changes that have occurred for Puerto Ricans.*

On a bus trip to London from Oxford University, where I was earning some graduate credits one summer, a young man, obviously fresh from a pub, spotted me and, as if struck by inspiration, went down on his knees in the aisle. With both hands over his heart he broke into an Irish tenor's rendition of "María" from *West Side Story*. My politely amused fellow passengers gave his lovely voice the round of gentle applause it deserved. Though I was not quite as amused, I managed my version of an English smile: no show of teeth, no extreme contortions of the facial muscles—I was at this time of my life practicing reserve and cool. Oh, that British control, how I coveted it. But María had followed me to London, reminding me of a prime fact of my life: you can leave the Island, master the English language, and travel as far as you can, but if you are a Latina, especially one like me who so obviously belongs to Rita Moreno's gene pool, the Island travels with you.

This is sometimes a very good thing—it may win you that extra minute of someone's attention. But with some people, the same things can make *you* an island—not so much a tropical paradise as an Alcatraz, a place nobody wants to visit. As a Puerto Rican girl growing up in the United States and wanting like most children to "belong," I resented the stereotype that my Hispanic appearance called forth from many people I met.

Our family lived in a large urban center in New Jersey during the sixties, where life was designed as a microcosm of my parents' casas on the island. We spoke in Spanish, we ate Puerto Rican food bought at the bodega, and we practiced strict Catholicism complete with Saturday confession and Sunday mass at a church where our parents were accommodated into a one-hour Spanish mass slot, performed by a Chinese priest trained as a missionary for Latin America.

SOURCE: Judith Ortiz-Cofer. 1993. *The Latin Deli: Prose & Poetry*. Athens, GA: University of Georgia Press.

As a girl I was kept under strict surveillance, since virtue and modesty were, by cultural equation, the same as family honor. As a teenager I was instructed on how to behave as a proper señorita. But it was a conflicting message girls got, since the Puerto Rican mothers also encouraged their daughters to look and act like women and to dress in clothes our Anglo friends and their mothers found too "mature" for our age. It was, and is, cultural, yet I often felt humiliated when I appeared at an American friend's party wearing a dress more suitable to a semiformal than to a playroom birthday celebration. At Puerto Rican festivities, neither the music nor the colors we wore could be too loud. I still experience a vague sense of letdown when I'm invited to a "party" and it turns out to be a marathon conversation in hushed tones rather than a fiesta with salsa, laughter, and dancing—the kind of celebration I remember from my childhood.

I remember Career Day in our high school, when teachers told us to come dressed as if for a job interview. It quickly became obvious that to the barrio girls, "dressing up" sometimes meant wearing ornate jewelry and clothing that would be more appropriate (by mainstream standards) for the company Christmas party than as daily office attire. That morning I had agonized in front of my closet, trying to figure out what a "career girl" would wear because, essentially, except for Marlo Thomas on TV, I had no models on which to base my decision. I knew how to dress for school: at the Catholic school I attended we all wore uniforms; I knew how to dress for Sunday mass, and I knew what dresses to wear for parties at my relatives' homes. Though I do not recall the precise details of my Career Day outfit, it must have been a composite of the above choices. But I remember a comment my friend (an Italian-American) made in later years that coalesced my impressions of that day. She said that at the business school she was attending the Puerto Rican girls always stood out for wearing "everything at once." She meant, of course, too much jewelry, too many accessories. On that day at school, we were simply made the negative models by the nuns who were themselves not credible fashion experts to any of us. But it was painfully obvious to me that to the others, in their tailored skirts and silk blouses, we must have seemed "hopeless" and "vulgar." Though I now know that most adolescents feel out of step much of the time, I also know that for the Puerto Rican girls of my generation that sense was intensified. The way our teachers and classmates looked at us that day in school was just a taste of the culture clash that awaited us in the real world, where prospective employers and men on the street would often misinterpret our tight skirts and jingling bracelets as a come-on.

Mixed cultural signals have perpetuated certain stereotypes—for example, that of the Hispanic woman as the "Hot Tamale" or sexual firebrand. It is a one-dimensional view that the media have found easy to promote. In their special vocabulary, advertisers have designated "sizzling" and "smoldering" as the adjectives of choice for describing not only the foods but also the women of Latin America. From conversations in my house I recall hearing about the harassment that Puerto Rican women endured in factories where the "boss men" talked to them as if sexual innuendo was all they understood and, worse, often gave them the choice of submitting to advances or being fired.

It is custom, however, not chromosomes, that leads us to choose scarlet over pale pink. As young girls, we were influenced in our decisions about clothes and colors by the women—older sisters and mothers who had grown up on a tropical island where the natural environment was a riot of primary colors, where showing your skin was one way to keep cool as well as to look sexy. Most important of all, on the island, women perhaps felt freer to dress and move more provocatively, since, in most cases, they were protected by the traditions, mores, and laws of a Spanish/Catholic system of morality and machismo whose main rule was: *You may look at my sister, but if you touch her I will kill you.* The extended family and church structure could provide a young woman with a circle of safety in her small pueblo on the island; if a man "wronged" a girl, everyone would close in to save her family honor.

This is what I have gleaned from my discussions as an adult with older Puerto Rican women. They have told me about dressing in their best party clothes on Saturday nights and going to the town's plaza to promenade with their girlfriends in front of the boys they liked. The males were thus given an opportunity to admire the women and to express their admiration in the form of *piropos:* erotically charged street poems they composed on the spot. I have been subjected to a few piropos while visiting the Island, and they can be outrageous, although custom dictates that they must never cross into obscenity. This ritual, as I understand it, also entails a show of studied indifference on the woman's part; if she is "decent," she must not acknowledge the man's impassioned words. So I do understand how things can be lost in translation. When a Puerto Rican girl dressed in her idea of what is attractive meets a man from the mainstream culture who has been trained to react to certain types of clothing as a sexual signal, a clash is likely to take place. The line I first heard based on this aspect of the myth happened when the boy who took me to my first formal dance leaned over to plant a sloppy overeager kiss painfully on my mouth, and when I didn't respond with sufficient passion said in a resentful tone: "I thought you Latin girls were supposed to mature early"—my first instance of being thought of as a fruit or vegetable—I was supposed to *ripen*, not just grow into womanhood like other girls.

It is surprising to some of my professional friends that some people, including those who should know better, still put others "in their place." Though rarer, these incidents are still commonplace in my life. It happened to me most recently during a stay at a very classy metropolitan hotel favored by young professional couples for their weddings. Late one evening after the theater, as I walked toward my room with my new colleague (a woman with whom I was coordinating an arts program), a middle-aged man in a tuxedo, a young girl in satin and lace on his arm, stepped directly into our path. With his champagne glass extended toward me, he exclaimed, "Evita!"

Our way blocked, my companion and I listened as the man half-recited, half-bellowed "Don't Cry for Me, Argentina." When he finished, the young girl said: "How about a round of applause for my daddy?" We complied, hoping this would bring the silly spectacle to a close. I was becoming aware that our little group I was attracting the attention of the other guests. "Daddy" must have perceived this too, and he once more barred the way as we tried to walk past him. He began to shout-sing a ditty to the tune of "La Bamba"—except the lyrics were about a girl named María whose exploits all rhymed with her name and gonorrhea. The girl kept saying "Oh, Daddy" and looking at me with pleading eyes. She wanted me to laugh along with the others. My companion and I stood silently waiting for the man to end his offensive song. When he finished, I looked not at him but at his daughter. I advised her calmly never to ask her father what he had done in the army. Then I walked between them and to my room. My friend complimented me on my cool handling of the situation. I confessed to her that I really had wanted to push the jerk into the swimming pool. I knew that this same man—probably a corporate executive, well educated, even worldly by most standards—would not have been likely to regale a white woman with a dirty song in public. He would perhaps have checked his impulse by assuming that she could be somebody's wife or mother, or at least *somebody* who might take offense. But to him, I was just an Evita or a María: merely a character in his cartoon-populated universe.

Because of my education and my proficiency with the English language, I have acquired many mechanisms for dealing with the anger I experience. This was not true for my parents, nor is it true for the many Latin women working at menial jobs who must put up with stereotypes about our ethnic group such as: "They make good domestics." This is another facet of the myth of the Latin woman in the United States. Its origin is simple to deduce. Work as domestics, waitressing, and factory jobs are all that's available to women with little English and few skills. The myth of the Hispanic menial has been sustained by the same media phenomenon that made "Mammy" from *Gone with the Wind* America's idea of the black woman for generations; María, the housemaid or counter girl, is now indelibly etched into the national psyche. The big and the little screens have presented us with the picture of the funny Hispanic maid, mispronouncing words and cooking up a spicy storm in a shiny California kitchen.

This media-engendered image of the Latina in the United States has been documented by feminist Hispanic scholars, who claim that such portrayals are partially responsible for the denial of opportunities for upward mobility among Latinas in the professions. I have a Chicana friend working on a Ph.D. in philosophy at a major university. She says her doctor still shakes his head in puzzled amazement at all the "big words" she uses. Since I do not wear my diplomas around my neck for all to see, I too have on occasion been sent to that "kitchen," where some think I obviously belong.

One such incident that has stayed with me, though I recognize it as a minor offense, happened on the day of my first public poetry reading. It took place in Miami in a boat-restaurant where we were having lunch before the event. I was nervous and excited as I walked in with my notebook in my hand. An older woman motioned me to her table. Thinking (foolish me) that she wanted me to autograph a copy of my brand new slender volume of verse, I went over. She ordered a cup of coffee from me, assuming that I was the waitress. Easy enough to mistake my poems for menus, I suppose. I know that it wasn't an intentional act of cruelty, yet with all the good things that happened that day, I remember that scene most clearly, because it reminded me of what I had to overcome before anyone would take me seriously. In retrospect I understand that my anger gave my reading fire, that I have almost always taken doubts in my abilities as a challenge—and that the result is, most times, a feeling of satisfaction at having won a convert when I see the cold, appraising eyes warm to my words, the body language change, the smile that indicates that I have opened some avenue for communication. That day I read to that woman and her lowered eyes told me that she was embarrassed at her little faux pas, and when I willed her to look up at me, it was my victory, and she graciously allowed me to punish her with my full attention. We shook hands at the end of the reading, and I never saw her again. She has probably forgotten the whole thing, but maybe not.

Yet I am one of the lucky ones. My parents made it possible for me to acquire a stronger footing in the mainstream culture by giving me the chance at an education. And books and art have saved me from the harsher forms of ethnic and racial prejudice that many of my Hispanic *compañeras* have had to endure. I travel a lot around the United States, reading from my books of poetry and my novel, and the reception I most often receive is one of positive interest by people who want to know more about my culture. There are, however, thousands of Latinas without the privilege of an education or the entrée into society that I have. For them life is a struggle against the misconceptions perpetuated by the myth of the Latina as whore, domestic, or criminal. We cannot change this by legislating the way people look at us. The transformation, as I see it, has to occur at a much more individual level. My personal goal in my public life is to try to replace the old pervasive stereotypes and myths about Latinas with a much more interesting set of realities. Every time I give a reading, I hope the stories I tell, the dreams and fears I examine in my work, can achieve some universal truth which will get my audience past the particulars of my skin color, my accent, or my clothes.

I once wrote a poem in which I called us Latinas "God's brown daughters." This poem is really a prayer of sorts, offered upward, but also, through the human-to-human channel of art, outward. It is a prayer for communication, and for respect. In it, Latin women pray "in Spanish to an Anglo God / with a Jewish heritage," and they are "fervently hoping / that if not omnipotent, / at least He be bilingual."

## DISCUSSION QUESTIONS

1.  What stereotypes does Cofer find in her everyday life? How do they show the mingling of race and gender in the way stereotypes are expressed?

2.  What generational differences have emerged for Puerto Ricans living in the U.S. mainland?

# 9

# Mixed-Race Chic

RAINIER SPENCER

*Spencer documents the increasing number of people who identify themselves as of two or more races. Although some think this will blur racial distinctions, thus making racism less prevalent, Spencer argues that the social structural origins of racism still remain.*

Popular wisdom suggests that we are in the midst of a transformation in the way race is constructed in the United States. Indeed, so strong and so inevitable is this shift said to be that longstanding racial dynamics are purportedly being dismantled and reconstructed even as you read these words.

According to this view, individuals of mixed race, particularly first-generation multiracial people, are confounding the American racial template with their ambiguous phenotypes and purported ability to serve as living bridges between races. This perspective is reflected in television and magazine advertising and coverage and in books both academic and nonacademic. As long as a decade ago, the sociologist Kathleen Odell Korgen wrote in *From Black to Biracial: Transforming Racial Identity Among Americans* (Praeger, 1998) that "today mixed-race Americans challenge the very foundation of our racial structure."

From his well-received speech on race, in which he positioned himself as having a direct understanding of both black and white anger, to his reference to himself as a "mutt," Barack Obama and his historic election have significantly boosted this view. Many Americans hail his background as portending our post-racial future. We hear that self-styled multiracial young adults accept their mixed identity far more than did their pre-civil-rights-era predecessors; but precisely what they are actually assenting to and what it means may be little more than a fad.

People who see us accepting a new multiracial identity have long argued that it is destructive of race: that recognition and acceptance of multiracialism will bring about the demise of the American racial model. The American Multiracial Identity Movement thereby suggests that multiracial identity possesses an insurgent character, a militant stance against the idea of recognizing race in the United States.

Regardless of their contemporary popularity, such claims are without merit. Indeed, they are self-contradictory. If one holds that multiracial identity is a real

SOURCE: 2009. *The Chronicle Review* 55(May 29): B4.

and valid identity, then it can be sensible only as a biological racial identity. If words are to mean anything, and they should, it quite obviously cannot be that a multiracial identity is somehow not a biological racial identity. Rather, multiracial identity merely falls in place to join other, already existing racial categories.

If the issue were ethnicity, we would be debating the idea of multiethnic identity. If the issue were nationality, we would be debating multinational identity. If the issue were cultural affinity, we would be debating multicultural identity. But in the debate over multiracial identity, the nature of that identity is made clear by the very wording in which the debate is framed. Nor is there any escape through asserting that racial identities are socially designated, since it is the sexual (and thereby biological) union of parents of ostensibly different races that produces multiracial children.

There are a variety of names in the popular literature referring to those children: Generation Mix, Generation M (multiracial), Generation EA (Ethnically Ambiguous), Remix Generation, and other just as authentically hip terms. The very hipness should alert us to the pop-cultural bandwagon mixed-race identity currently rides. Those (very typically, but not necessarily, young) people who consider themselves to be the mixed offspring of parents of different biological racial groups are proclaimed the path to America's postracial future. Will, as we are assured, the (multi) racial ambiguity of Generation Mix usher in a new American racial order? Will it undermine centuries of racial hierarchy?

Those questions hardly ever receive even the minimal sort of critical analysis one might expect for so important a topic—especially because some academics assume too readily the role of scholarly cheerleaders, as opposed to the intellectual referees they should be. That group includes the sociologist G. Reginald Daniel, the psychotherapist Ursula M. Brown, the anthropologist Marion Kilson, and others. They in turn provide quotations and sound bites to popular news-media writers more interested in hip story lines than in objective journalism. Consider that no less a source than National Public Radio avers that "race, culture and identity are changing in America." We are witnessing nothing less than a self-inflicted and self-authorized societal hoodwinking of the first order.

That may seem harsh, undercutting the progressive role that has been assigned to multiracialism of late. But it would be just as unwise to underestimate the degree to which we Americans allow ourselves to be influenced by the manipulations of the popular media; just as it would be to underestimate the ferocious tenacity of those who are invested heavily in maintaining the current racial order, even if it means making concessions in the form of minor adjustments to the American racial model that appear to privilege this newfound multiracial ambiguity. The vital point that seems to be missed is that racial ambiguity, in and of itself, is no guarantee of political progressiveness, racial destabilization—or anything in particular.

As Catherine R. Squires, a professor of journalism, writes in *Dispatches From the Color Line: The Press and Multiracial America* (State University of New York Press, 2007), multiracialism is fundamentally ambiguous: "This ambiguity is about exoticism and intrigue, providing opportunities for consumers to fantasize and speculate about the Other with no expectations of critical consideration of power and racial categories." Squires makes an important point, for it is crucial to be able to separate racial ambiguity that might be utilized to work consciously against racial hierarchies from racial ambiguity that is simply a form of self-interested celebration that ends up reinforcing those racial hierarchies.

Dealing with mixed-race identity in that context requires us to consider "hypodescent," the term for the longstanding practice by which mixed-race individuals have long been relegated to the identity of the parent whose racial status is lowest. Although hypodescent developed in various ways throughout the Americas as a result of European colonization and slavery, it achieved its most extreme formulation in British North America, where (aside from notable exceptions in places like New Orleans and Charleston, S.C.) the particular evolution of slavery relegated all degrees of black-white mixture to the black category. We must recognize as well that biological race, hypodescent, mono-racial people, and multiracial people are all figments of the American imagination, as they have been for centuries.

Although hypodescent applies de facto nearly exclusively to black people, it is the basis for structuring the American racial paradigm, since it ensures that black-ness remains separate from whiteness, providing the fixed endpoints of our racial schema. The category at the top of the racial hierarchy, white, has no way to con-stitute itself absent its relation to blackness, a phenomenon that has been well es-tablished by the (related) fields we now call whiteness studies and critical-race theory and exemplified by the work of scholars like the historian David Roediger and the law professor Cheryl Harris. One might reasonably inquire as to whether something other than blackness could serve this function, but I do not, at this point in our history, believe so. Hispanic, Asian, or American Indian categories do not yet possess the level of revulsion that continues to mark blackness.

For several centuries, the primary racial dynamic in America has been the black-white one, with American Indians being too few in number and also too somatically close to whiteness to stand as the kind of other that Africans and African-Americans have represented. Recognition of that somatic nearness is reflected in the failed, late-19th-century attempt to assimilate American Indians fully into American (white) culture—a horrifyingly destructive mission that in-cluded removing children forcibly from their parents and sending them to board-ing schools where their language and religion were punished out of them. Certainly no such effort has ever been mounted to assimilate black people. And until the relatively recent wave of Asian and Hispanic immigration, over the past 40 or so years, those latter two categories were not only too geographically lim-ited to regions of the United States but also too small in number to produce the fear that blackness did and still does.

Moreover, despite a current rise in anti-immigrant feeling that is largely anti-Hispanic, there is also a countermovement, what the sociologist Eduardo Bonilla-Silva calls "the Latin-Americanization of racial stratification in the U.S.A." As in the case of some Asians, many Hispanics are taking advantage of multiracialism to transition to a state of "honorary whiteness," a phenomenon whose corollary is that, once again, African-Americans are seen as the group to avoid.

A key failure of people who oppose the American pattern of race relations—including those who see the multiracialism as a useful weapon against it—has been an inability to articulate correctly the most critical aspect of the pattern's nature. We like to think that it consists of four or five races, with whites at the top of the hierarchy, followed by Asians, Hispanics (if conceived of racially), American Indians, and then black people. (While commentators might debate the relative placement of the middle groups, the endpoints—white and black—are fixed.) That conception is accurate, but it doesn't go far enough: It doesn't problematize the fact that whiteness—while it has been forced by recent population trends into expanding its boundaries by accepting specific and limited numbers of people with Asian, Hispanic, or American Indian ancestry—cannot admit the public entry of blackness and still remain white. That is surely not to say that members of the other groups live unproblematic lives untouched by racism, but rather that there simply is no American racial paradigm without blackness serving as the antipode of whiteness. Unless something arrives to radically deconstruct the status of whiteness at the top of the hierarchy, there is no hope for real change.

What that means is that those who proclaim that multiracial identity will destroy race are living a lie. Some people whose race is ambiguous and who possess the "right look" may well be able to distance themselves from blackness, but such movement has no ultimate effect on the general American racial scheme; it merely adds an additional category. For how can multiracial identity deconstruct race when it needs the system of racial categorization to even announce itself?

We may shuffle the intermediate categories, we may add nonwhite categories (including a multiracial category), and we may even see whiteness expand a bit to include some previously excluded people. But as long as black remains at the bottom of the hierarchy—and as long as it remains impossible for a black woman to be seen as giving birth to a white child (while the reverse case continues its unproblematic acceptance), nothing has changed.

The solution to our national racial madness does not lie in altering the racial paradigm so that it is somehow more equal or so that it includes more groups; the solution lies in rejecting both the idea of biological race and the hypodescent that flows from that idea. What popular wisdom tells us is the supposed twilight of how Americans have thought about race is merely a minor tweaking of the same old racial hierarchy that has kept African-Americans at the bottom of our paradigm since its very inception. Multiracial ideology simply represents the latest means of facilitating and upholding that hierarchy—while claiming quite disingenuously to be doing the opposite.

## DISCUSSION QUESTIONS

1. What is meant by the term *hypodescent*, and why is it significant to the social construction of race?

2. Why does Spencer think that the category "mixed-race identity" does not necessarily destroy notions of race?

# 10

# My Freshman Year:

## What a Professor Learned by Becoming a Student

REBEKAH NATHAN

*Nathan is an anthropologist and college professor who did field research on a college campus by passing as a college student for one year. This section of her book explores the experience of students of color who often find themselves in a position of "aloneness" on predominantly White campuses.*

## DIVERSITY AT ANYU?

Student networks may be able to explain, at least in part, the failed diversity efforts at many universities, and certainly at AnyU. About 22 to 25 percent of AnyU students are considered "minority" by federal standards, and minority students appear approximately in these percentages in AnyU dorms and classes. What makes diversity a "success" in a state university, however, is not only that the university population reflects the diversity of the general population but also that students become more involved in the lives and issues of that diverse population. Part of that diversity ideal is the hope that all students will develop friends and have important conversations with those of backgrounds and ethnicities different from their own.

The National Survey of Student Engagement tries to capture this information by asking a student to self-report as to whether he or she has "had serious conversations with students of a different race or ethnicity than your own." In 2003, fifty percent of college seniors nationwide indicated that they "often" or "very often" had such conversations, while only 13 percent said they did not.

This jibed with the information I initially was getting from any interviews about social networks, where I was finding that many students named someone from a different ethnic group within their close circle of friends. The interview information, though, did not match my direct observations, and this led me to probe further by fiddling with my interview questions and format. I soon realized that if I started, as I had, by asking informants whether they had close friends

SOURCE: Rebekah Nathan. 2005. *My Freshman Year: What a Professor Learned by Becoming a Student.* New York: Penguin.

from other ethnic groups, the majority of students would say that they did. If I questioned them further, they would name that man from a class, or woman on the same intramural volleyball team, with whom they had close contact and describe how they met.

If, however, I started by asking informants to name their closest friends and then later asked them to identify the ethnicity of the named people, it turned out that most students, but white students predominantly, ended up becoming close friends with people of their own ethnicity. Since I thought that this "names first, ethnicity later" approach was more accurate, I changed the order of my questions and arrived at a very different picture. Five out of six white students I interviewed in this way about their networks had no members of another racial or ethnic group in their close social close circle; the networks of five of the six minority students contained one or more minorities (more on the details of this later).

One can see from the descriptions of how networks form why this might be true. Many students are building on contacts developed before they entered college, contacts that have strong demographic and social components. If many student networks begin with hometown contacts, what is the likelihood that they will cross class, ethnic, race, or even religious lines when the United States is demographically divided along precisely these lines? Although there was one instance in my data of a cross-racial network pair with its origin in high school, the probabilities in this country work strongly the other way because of de facto neighborhood and school segregation. All other examples I found of high school or hometown friends in an AnyU network involved a woman or man of the same ethnicity as the person interviewed.

Even many relationships developed early in college contain a built-in bias. Although classes and interest clubs may be ethnically well mixed, this is not where students make their earliest school contacts. Freshman dorms are generally well integrated, but not several of the early programs and events that help introduce and acclimate new students, including Previews weekends designated for particular ethnic groups, pre-college "outdoor adventure" trips that cost extra money, a summer program for first-generation college students, or the opening round of sorority and fraternity mixers. Some institutional structures like these may encourage the early formation of same-ethnicity relationships.

There is no doubt that active racism also plays a part in the lack of diversity on college campuses. Yet, race or ethnicity is typically ignored as a topic of conversation in mainstream college culture, treated as an invisible issue and with silence. As Levine and Cureton (1998) found in their nationwide survey, students were "more willing to tell intimate details of their sex lives than discuss race relations on campus." When the subject *is* raised, as in the occasional class, students of color report being continually expected to educate whites about minority issues or speak "as a representative of their race."

Despite the general invisibility of the subject of race in informal student culture, there was not a single minority student I interviewed who hadn't experienced racism. Few openly complained, but everyone had at least one story to tell of comments made in class, rude remarks on the street, or just hostile looks.

When I asked Pat, a Hispanic-Native American woman, whether she had ever considered rushing a sorority, she told me that she had in her freshman year, but "I could see that it wasn't really right for me, because I'd pass by all the sorority tables—you know how they call out to girls to come over and take a look—well, I saw they called out to other girls but not to me. They just kinda ignored me, not hostile or anything, but not interested either."

"It's just how it is," another female student explained. "There are some good people and some not so good people, and you deal with it."

## WHO EATS WITH WHOM: A STUDY OF STUDENT DINING

My very small sample of student networks and interviews was suggestive to me but not convincing that diversity in student relationships was in serious question. So I decided to conduct a larger observational study of students' informal social behavior. I chose eating as the focus, one of my favorite social activities, and asked the research question "Who eats with whom?" This seemed a fair and appropriate inquiry into diversity, to determine the range of people with whom one breaks bread.

It was my most extensive and longest-running "mini-study" of campus life. For five months I directly observed and recorded the dining behavior of fellow students during randomly selected periods of the day at optional dining areas on campus. Although some patrons carried out their food, returning to their dorms or outside benches to eat, many ate and drank singly or in groups at the various tables provided in one of five eating areas I surveyed. Sitting at a different table in the room, I would record who sat at each table by gender and, as much as outward appearances can signal, ethnicity.

It is always problematic to do research like this, because there is a wide range of appearances for all ethnicities, and many sticky issues. My interest however, *was* in appearances, and in seeing to what extent students chose to share food and conversation with people who looked like them (or, more accurately, seemed to belong in the same broad ethnic category that an observer would attribute to them). Although there are others kinds of diversity (e.g., age), I recorded only the data reflecting each person's gender and, to the extent possible, his or her category of ethnicity such as white non-Hispanic, Hispanic, Asian, African American, Native American, and so on. These were not easy calls. Sometimes I could tell only that someone was not a white non-Hispanic but couldn't identify the more specific group to which she or he belonged; at other times I could not tell whether a person was white and non-Hispanic or something else.

In gathering this information I had these questions in mind: To what extent did informal university activities (e.g., eating together) convey diversity? Did students eat in same or mixed ethnic and gender groups? Were there differences in the eating patterns of dominant (defined as white non-Hispanic) non-dominant (defined as people of color) ethnic groups? Did any ethnic group or category eat alone more often than others?

I analyzed the data with regard to these questions but took care to analyze by person rather than by table in order to try to see the data through the eyes of the particular diner. For instance, if there were a table consisting of four people—a white male, two white women, and one Hispanic woman—each would have a different reality at the table: the male is eating with a table of all women of mixed ethnicity; both white women are eating at a table of mixed gender and mixed ethnicity; and the Hispanic woman is eating at a mixed-gender table where everyone is of a different ethnicity from herself. I recorded the data, preserving the perspective of each diner, and then analyzed the data in ten different categories that allowed me to examine the relationship of each table diner to the rest. In this way, I tracked almost 1,500 examples of dining behavior.

What I found was interesting. It showed not only an overall lack of diversity, as national studies report, but also the existence of huge differences in the diversity experiences of dominant and non-dominant groups. Minorities (people of color) ate alone only slightly more often: one-quarter of minority women and more than one-third of minority men sitting in public spaces ate alone, a rate greater than that for white women and men by 3 percent and 5 percent, respectively. But of all those who ate with others, only 10 percent of white men and 14 percent of white women ate at a table where there was anyone of a different color from themselves. Only 2.6 percent and 3.5 percent of white men and women, respectively, ate at a table of two or more where they were the only white person. The statistics were strikingly different for people of color: 68 percent of women and 58 percent of men ate with "mixed groups." People of color were ten times more likely than whites to eat in a group in which they were the only person of a different race/ethnicity at the table.

The same patterns I saw in the dining spaces proved true in the composition of personal networks when I compared a group of twelve students on my hall, six whites and six students of color. Although the networks of Caucasian students included more whites, and those of people of color more minorities, the total networks of minority of students were primarily "mixed", comprising people of various ethnicities, including whites. One student of color was in an all-white network, while another had friends of only her own ethnicity. By contrast, five of six white students had networks that were solidly white; only one white student had a mixed network, and none was the only Caucasian.

Seen in this context, minority ethnic clubs, dorms, and student unions have a clearer meaning. Ethnic-based groups are often clouded by perceptions that they, like the Greek system, remove their members from the mainstream and surround them with people of the same background. What the data suggest to me is that people of color are already heavily involved in interethnic and interracial relationships on campus. In fact, most of their informal dining contacts as well as personal networks included people who were ethnically different from them. Under these circumstances, an ethnic-based club—which half of the minorities in my sample thought was important in their lives—is better understood as a needed respite from difference, a chance to rest comfortably with others who share similar experiences in the world.

It was white students, most markedly males, whose social lives suffered from a lack of diversity. When white men *did* eat with those of different ethnicities, the majority of tables were "cross-gender." In other words, white men socialized at meals to a greater degree with nonwhite women or with groups that included nonwhite women. There was extremely little contact between white and nonwhite men. Only 4 of 489 white males, fewer than 1 percent, ate with (only) males of a different ethnicity, but 31 ate in different or mixed ethnic groups in which women were present. Men of color, while much more diverse in their dining, followed a similar pattern, tending to have fewer cross-ethnic male-only eating partners (7 of 79) while favoring cross-ethnic tables where women were eating too (24 of 79). The same pattern was not true of women. For both white women and women of color, their cross-color contact was primarily with other women.

One of the more disturbing but confusing findings was how few people of color, proportionally, used the common eating spaces. Only 13 percent of the entire dining sample was nonwhite, while 22 percent of full-time students were nonwhite. This left more than 40 percent of the minority population unaccounted for. There are certainly many ways to interpret what was going on. Perhaps this eyeballing approach to minority status simply fails to recognize many who are legal minorities. Perhaps there are economic factors at work that bear on having a meal ticket or buying food during the day that disadvantage some minority students. Perhaps the difference is explained by the larger percentage of minority students who enroll in off-campus programs. But there is another possibility that I entertained, which was related to my finding that more minority students eat alone.

My evidence is only anecdotal because I didn't formally monitor what I thought I was seeing, but this is what I noticed. I would observe the food court area as people got their food and stood in line to pay, watching each person leave the register to see where he or she went to sit in order to mark it in my book. I often found that instead of going to a table, however, a person of color would go to condiment area, pack up a napkin and the food in a bag, and leave. It seemed to me that a greater proportion of minorities was leaving.

One day, as I was just finishing an observation session, an African American woman left the register and headed for a table. She would be the last diner to enter my monitored space in the set time period. I prepared to mark her table choice, but instead of sitting down, she readjusted her backpack, took her food, and left. Where is she going? I asked myself. To meet a friend in a different area? To eat outside? I felt a bit like a stalker as I followed her out of the dining area and out of the building. She passed the outdoor tables and kept walking until she entered one of the freshman dorms, went through the lobby, and up the stairs. My guess was that she had returned to eat in her dorm room.

I will never know for sure what lay behind that one observation, or what I perceived to be the larger proportion of students of color who did not stay to eat. But it left me with the uncomfortable feeling that I was witnessing the effect of a "white space"—which I had never noticed because I am white—where people of color could eat alone publicly, or eat with people different from themselves, or go home to their rooms. Perhaps, many times, the dorm room just seems the most comfortable option, and this may have explained some of the missing 40 percent of minority students in the dining areas.

The ideals of community and diversity are certainly in place at AnyU and remain important components of stated university policy. Yet neither is fully realized in university culture, as I believe I have illustrated in this chapter. What I also hope I've illustrated, though, is what anthropologists mean when they say that a culture cannot really be divided into its parts; one part of culture cannot be understood in isolation from its other parts, and all must be contextualized within the larger whole. Culture, we argue, is both integrated and holistic.

In just this way "community" and "diversity" are parts of university culture, but they are not intelligible on their own. As the descriptions of student life attest, diversity is one part of college culture that is intimately tied to community, another part. And both parts are ultimately conditioned by structures in the larger American society—including values of individualism and choice, materialism, and the realities of U.S. demographics—that may seem, at first, to have little bearing on whether college diversity increases because freshmen Joe and Juan truly become friends, or whether Jane strengthens community by deciding to attend Movie Night. But they do. Not understanding this leads to a reality about diversity and community in university culture that does not match its rhetoric, and a persistent confusion about why this is so.

## DISCUSSION QUESTIONS

1.  What did Nathan find when she observed eating patterns on the campus where she posed as a student? Would you see the same thing on your campus? Why or why not?

2.  How does Nathan's study shape your understanding of what it means to be "alone" on a college campus?

# Student Exercises

1. For one week, keep a written log of every time the subject of race comes up in the conversations you hear around you. Make note of what people said and the tone in which they said it. You should also note, if possible, the age and race of the person making the comments. At the end of the week, review your log and answer the following questions:

   a. What evidence of racial prejudice did you find?

   b. Is the prejudice you observed related to racial discrimination? If so, how? If not, why?

   c. What do your observations reveal to you about the everyday reality of racism?

2. The readings in section II identify *color-blind racism* as a new form of racism in which dominant group members (and some subordinate group members) think that race no longer matters and that to recognize race is to be racist. Some of the authors claim that when you delve under the surface of these beliefs, you will find that people still harbor stereotypical ideas about racial minorities. Design a series of interview questions, perhaps modeled on some of the research reported, and then interview a small sample of people. Do you find evidence of color-blind racism among those you interviewed? How do the race and ethnicity of those you interviewed influence your findings?

# Section III

# Representing Race and Ethnicity:
## The Media and Popular Culture
### Elizabeth Higginbotham
### and Margaret L. Andersen

How many hours per week do you spend observing or otherwise engaged with the media? If you are anything like the national average, the answer is fifty hours per week, or nine to ten hours per day, counting television, radio, magazines, music, and books, among other forms of media (U.S. Census Bureau 2009). Note that this measure does not include the time people spend with on-line social networks, such as Facebook, Twitter, and other online media—where time spent is escalating tremendously every year. In an era dominated by the media, it is indeed impossible to escape the media's influence—even though most people think that it does not really affect how they think. But affect people it does; why else would advertisers spend so much money to target media audiences? The media is now much of the basis of what we could call a common culture. People talk together about their favorite shows, model themselves after favorite celebrities, and sing the lyrics of popular songs. Indeed, one could say that most of popular culture *is* the media.

The media in this nation are also based on an economic model of capitalism. That is, media companies are typically privately owned by corporations whose purpose is not just the dissemination of information and entertainment, but also—and perhaps primarily—the pursuit of profit. Huge corporate monopolies dominate the media, basing their profit in large part on selling things to the public. The things sold include not just commercial products, but also (perhaps less directly) ideas, images, and ideals. This means that much of what we think, how we see people, and how we interpret what is happening in our society is driven by the media, in all its forms.

Representations of race and ethnicity—things seen as well as not seen—are a huge part of what you see in the media, and of its shaping of popular culture. People generally engage in popular culture to be entertained (such as by listening to music, going to the movies, seeing art exhibits, and so forth) or to be informed about

79

national and world events (such as by linking to an online news source, reading a newspaper, watching the evening news, or tuning in to talk radio). But as they do so, people are also exposed to powerful images of race and ethnicity. And because the level of racial segregation in our society is so high, for many these images may be the primary, and sometimes the only, basis for how they see others.

Ask yourself, as one example, who you see in films and how different groups are portrayed. If you look carefully, you are likely to find that racial and ethnic stereotypes pervade popular films. Films, for example, tend to celebrate and promote White people and simultaneously obscure White privilege by making it seem natural or "just normal" (Vera and Gordon 2003). The next time you see a movie or a television show, read a popular magazine, or watch the news, go beyond the surface. Rather than just taking what you see at face value, try to examine how both White people and people of color are portrayed. Think about how these portrayals shape people's understandings of race in society.

As you begin to think critically about the images of different groups, you may be surprised at how pervasive they are. Even walking through a grocery store, you can see Native American women depicted as beguiling and inviting on boxes of butter! Latinas are depicted on tequila mix with long, flowing hair and big, red lips. Even White men are subject to such stereotyping—appearing on paper towel packaging as brawny, muscular, and tough. Indeed, if you observe the store carefully, you will likely find that the only place where interracial images are found is in the diaper and cereal aisles—as if interracial relationships are only acceptable for babies and children.

Looking at the everyday world more critically can help you challenge the representations of people that are routinely found in the media and popular culture. The articles in this section, which explore different arenas of the media and popular culture, show how such stereotypes produce myths about others, but also show how people can be and are active agents in the creation of new cultural forms.

We begin with a more theoretical article that provides an analytical frame for thinking about how the media and popular culture are both sources of domination and also sites where subordinated groups can challenge and resist controlling images. Craig Watkins ("Black Youth and the Ironies of Capitalism") argues that, even with its power to control people's ideas, the influence of the media and popular culture is not unidirectional. That is, the media *do* shape dominant values, ideas, and representations, but *at the same time* people have the capacity to challenge and subvert these constructions. Watkins uses the example of Black youth and their creation of alternative cultural forms to explore how people use culture to resist conditions of oppression and subordination. Caught in structural conditions of high unemployment and poor life chances, poor, urban Black youth construct styles and music that produce new cultural forms. But, as Watkins shows, these innovative cultural forms that emerge from conditions of oppression are then, within a capitalist economy, subsequently transformed and marketed by others to broader audiences. Watkins is showing some of the contradictions inherent in the power of the media. Thus, even while the media depict young Black men as dangerous and a threat to society, young Black men resist such images by being active agents in the creation of their own culture.

Hip hop is a good example of how cultural creations have been transformed by a capitalist culture. Hip hop and rap music emerged initially from the critical stance of urban Black youth. In the early years of their development, hip hop and rap (and some current underground music) expressed strong criticism of dominant social institutions. But, as hip hop has become more commercialized and dominated by capitalist interests (and purchased mostly by White youth), the messages in this popular music have changed, becoming not only more sexist but also less critical of current social injustices.

Fatimah N. Muhammed ("How to NOT Be 21st Century Venus Hottentots") uses a similar framework in discussing the place of women in hip hop music and culture. As many have done, she describes how Black women in hip-hop videos are framed by racism, sexism, sexuality, and class, limiting the images of Black women that you see. But she also argues that Black women perform hip hop as a way of expressing resistance to the very images that dominate their appearance in hip hop culture.

Looking at other dimensions of popular culture, Rosie Molinary ("Maria de la Barbie") explores how images of beauty in both the dominant culture and within Latina culture shape the self-images of Latinas. She notes that dominant images of beauty—that are projected for *all* women—are, in fact, unattainable, no matter how hard women try to achieve them. But there is little validation for Latina identity within these dominant ideals of beauty. Even with new Latina stars like Jennifer Lopez or Eva Longoria, images of beauty in the media are narrow and "White." Molinary conducted a research study (the "Growing Up Latina Survey") based on interviews with a sample of young Latinas. She finds that Latinas struggle with the images they see, both trying to achieve a beauty ideal and trying to develop a self-image that contests these narrow ideals.

Another arena of popular culture that has become quite controversial is the use of Native American mascots as symbols for sports teams. Charles Springwood and C. Richard King ("'Playing Indian': Why Native American Mascots Must End") examine the use of Native American icons in popular culture. These icons, mostly in the form of caricatures, are everywhere—as part of the names of sports teams, on products in the grocery store, and in school textbooks. Such images usually distort understanding of the realities of Native American life and stereotype Indians as either noble warriors or cartoonish savages—and as almost entirely men. Note how gender is displayed in such images. Although some defenders claim that such portrayals are all in fun, you might ask: Would it seem so light-hearted if your own group were depicted in such stereotyped, repetitive, and joking ways?

Springwood and King's article shows the extent to which racial stereotypes are part of the everyday landscape of our society. In another example, Jennifer C. Mueller, Danielle Dirks, and Leslie Houts Picca ("Unmasking Racism: Halloween Costuming and Engagement of the Racial Order") demonstrate how simple customs, such as celebrating Halloween, embrace racial representations. Their analysis of interviews with a sample of college students explores how race becomes represented through Halloween costumes. Along with the other articles in this section, their research helps you see the many racial representations that you see around you. But the point is not just to notice and observe them, but also to think about how they might be transformed or challenged.

## REFERENCES

U.S. Census Bureau. 2009. *Statistical Abstract of the United States*. Washington, DC: U.S. Department of Commerce.

Vera, Hernán, and Andrew M. Gordon. 2003. *Screen Saviors: Hollywood Fictions of Whiteness*. Lanham, MD: Rowman and Littlefield.

## FACE THE FACTS: RACE AND MEDIA VIEWING

|  | Total U.S. | African American |
|---|---|---|
| Have two or more sets | 76% | 83% |
| Have three or more sets | 43 | 53 |
| Have a VCR | 87 | 82 |
| Have a DVD | 62 | 62 |
| Video game owner | 35 | 39 |
| Receive wired cable | 66 | 64 |
| Receive cable plus with pay | 41 | 48 |
| Have a home PC | 71 | 57 |
| Have a home PC with Internet | 61 | 44 |
| Total TV Viewing Hours per Week | 8 hrs. 15 mins. | 11 hrs. 4 mins. |

*Think about it*: This table shows significant differences in the viewing habits of African Americans and the general U.S. population. What are some of the differences that you see and how would you explain this? What effects might such differences in exposure to the media have for different groups?

# 11

## Black Youth and the Ironies of Capitalism

### S. CRAIG WATKINS

*Watkins shows how youth culture is often derived from the creativity of Black urban youth who, through their cultural inventiveness, often produce new cultural forms. But he also shows how the workings of capitalism typically appropriate that culture and turn it into something different. Still, his argument emphasizes the agency of Black youth, even when living in conditions of disadvantage.*

… According to sociologist David Brain, cultural production is the "collective production of skills and practices which enable social actors to make sense of their lives, articulate an identity, and resist with creative energy the apparent dictates of structural conditions they nonetheless reproduce."[1] The cultivation of skills that allow them to participate in a rapidly expanding and global communications media culture enables black youth to produce a broad range of cultural products. The most arresting features of black youth popular cultural productions represent distinct forms of agency, struggle, and social critique. But the vigorous commodification of African American cultural productions also develops complicated features....

The study of popular media culture generally oscillates between two opposing poles: containment or resistance. Whereas the former maintains that the ideas, values, and representations that shape popular media discourses are determined by the dominant classes, the latter argues, alternatively, that popular cultures have the capacity to subvert dominant ideologies and regimes of representation. Yet popular media culture is remarkably more complex than the containment/resistance binary opposition implies. Similar to the social world from which it is produced, popular media cultures are marked by instability and change. It is, in fact, one of the main locations where the struggle for ideological hegemony is waged.... Popular media culture is perhaps best understood as a perpetual theater of struggle in which the forces of containment and resistance remain in a constant state of negotiation, never completely negating each other's presence or vigor.

While the different spheres of commercial media culture—television, film, music, video, and the Internet—function as sources of pleasure and entertainment,

SOURCE: S. Craig Watkins. 1998. *Representing: Hip Hop Culture and the Production of Black Cinema.* Chicago: University of Chicago Press.

they also perform a pivotal role in patterning the cultural and ideological landscape. The popular media productions created by black youth represent a distinct sphere of cultural production. Any serious consideration of black cultural productions must examine the relationship between several interlocking factors: the specific culture industries within which these productions are organized; the changing landscape of communications media technologies; emergent mood shifts and sensibilities that lead to the creation of new collective identities; and finally, the unsettled social world within which black youth cultural practices take shape....

Commercial forms of popular culture are a rapidly growing field of study. Scholars and social historians are beginning to understand it as a plentiful and remarkably revealing reservoir of practices and formations that are inextricably linked to the changing contours of American life: urbanization/suburbanization, technological innovation, and shifting conceptions of racial, gender, class, and sexual identities. Commercial forms of popular media culture, for example, are central to how we (re)produce and experience socially constructed formations like race.

More precisely, my aim is to more fully explain the increasingly complex ways in which young African Americans have mobilized around a changing racial and popular media landscape. Moreover, it is a story about how the pulsing gestures, performances, and representations practiced by black youth are structured, in large part, by the profound ways in which they experience the changing contours of American life....

The buoyant surge in black youth popular cultural production raises important questions about the evolving disposition of cultural and representational politics in a media-saturated universe. Early critics of "mass culture" demonstrated concern that popular media culture was controlled by and for the dominant classes. But this view fails to consider how popular media culture functions as a site of intense ideological struggle. Quite simply, can the commercial media, long regarded by many critical theorists as the modern-day "opium of the masses," function as a location of counterideological struggle? Similar to other institutional milieus, commercial media also develop specific antagonisms. So as new subjects gain access to the most prominent sites of media and representation, the possibilities for new collective identities, social movements, and distinct modes of struggle are also established....

... Why have cultural innovation and production among black youth exploded or, as they might boast, "blown up"? Even more to the point, how has the social, political, and historical terrain on which black youth cultural productions do their work enabled them to intervene in the remaking of society in ways that are more visible, invigorating, and problematic?

## THE NEW WORLD ORDER: BLACK YOUTH AND THE RACIALIZATION OF CRISIS

... A cursory glance at the cultural landscape—music, video, film, television, advertising and sports—reveals that the expressive cultures created by African

Americans play a lively role in patterning the racial and gender identities of youth as well as the general popular culture scene…. Historian Robin D.G. Kelley reminds us that, "unlike more mature adults, young people are in the process of discovering the world as they negotiate it. They are creating new cultures, strategies of resistance, identities, sexualities, and in the process generating a wider range of problems for authorities whose job it is to keep then in check."[2]

Admittedly, it is difficult to pinpoint with precision when and why a distinctive mood shift or transition in youth cultural production originates. However, it is possible to identify those factors that work, more or less, to establish the circumstances from which youth popular culture formations emerge. Certainly, any discussion of late twentieth-century black youth cultural practices that does not consider the social context that situates their agency would be severely impaired.

… The ideological and political formations of the postindustrial United States are marked by profound social, economic, and cultural transition. Moreover, this period of transition has established the conditions for the construction of different crisis scenarios, both real and imagined. In the process, crisis-tinted discourses are mobilized to make sense of and effectively manage the flux and uncertainty that abound. Even in cases where crises may in fact be real, they are typically *made* intelligible and, as a result, are defined, shaped, interpreted, and explained. For instance, a complex assemblage of crisis discourses revolves around the postindustrial ghetto. The ghetto has become an intensely charged symbol, particularly as it patterns discourses about crime and personal safety, welfare, familial organization, and the disintegration of American society.

African American (and Latino) youths are prominently figured in the crisis scenarios that stage some of the more contentious social and political episodes of the late twentieth century. Some researchers contend that increases in violent crimes, teen pregnancy, female-headed households, and welfare dependency can be *partially* explained by the sheer growth in the number of young people, particularly black and Latino, residing in many cities across the United States. Moreover, the concentration of black and Latino youth in postindustrial cities corresponds with structural changes in the postindustrial economy, especially the movement of industry and meaningful employment opportunities away from the communities in which they are most likely to live.

… One of the persistent tensions in the postindustrial economy is the widespread erosion of meaningful employment opportunities for poor, inner-city youth. As the labor-force participation of black youth hovers around chronically low levels, both their real and perceived prospects for upward mobility become more grim. Indeed, a tenacious set of factors restricts the social and economic mobility of poor youth: inadequate schools lower levels of educational attainment, low self-esteem and personal confidence, discriminatory hiring practices, and racially-based tensions on the job site….

The "underclass" is customarily portrayed as one of the most distressing social problems facing the United States….

It only takes a quick glance at legislative and electoral politics, the news media, public opinion polls, or popular entertainment culture to recognize that the "underclass," and poor youth especially, has attained the dubious distinction of being a celebrity social problem. The absorption of the "underclass" label into mainstream vocabulary corresponds with the social and economic transformations that configure postindustrial life. And even though the label circulates as if it were ideologically neutral, representations of the "underclass" are sharply coded in both racial and gender terms....

Take, for example, the proliferation of news media discourses that play a leading role in framing public perceptions of postindustrial ghetto life. Perhaps even more than social scientists or politicians, the news media industry has played a crucial role in coloring the public discourses that render the "underclass" seemingly more intelligible. The television news industry is a distinct sphere of commercial media and discourse production. Unlike most of television entertainment, it is nonfictional—in other words, real. But the news media is a peculiar blend of fact and artifice. Thus, while news media journalists deal with real-world phenomena, they do so in a way that is always selective and interpretive....

The news media are also an important site of racial discourse. In fact, part of the evolving role of the news media industry has been to determine what is most newsworthy about race, construct images and definitions of race, and pattern the range of potential connotations the idea of race produces. For example, television news discourse typically constructs African Americans as conflict-generating and problematic. And though it would be faulty to conclude that the news media are the primary agent in the racial fissures that percolate throughout the late twentieth century, the way in which television news frames race certainly occupies a crucial position on the embattled terrain of racial conflict.

The news media serve several functions at once. A primary purpose is to provide their mass audience with information and descriptions of events that take place in the world. However, another less obvious function is the news media's role as a mechanism of social control. The news media, to be sure, can be viewed as a central component of the social control processes that define and produce meaning about what constitutes difference and deviance. In this particular role, ... the news media are a kind of "deviance-defining elite" that play a key role in constituting visions of order, stability, and change and in influencing the control practices that accord with these visions. News media organizations specialize in visualizing—and accordingly, defining—deviant behavior for their audience. In the process, the news media also reproduce commonsense notions of civility, social order, and community consensus. Moreover, the focus on deviance develops an entertainment angle that appeases the commercial interests of news media organizations. Cognizant of its role in commercial television entertainment and the competition for ratings, the television news industry relies heavily on dramatic, sensational, and titillating images in order to attract and hold a wide viewing audience....

A main set of organizing themes in the "underclass" discourse is the alleged social pathologies of ghetto youth. To be sure, the connection of black youth with illegal drugs, gangs, and violence performs a distinct role in shaping how many of the crisis scenarios of the period were understood. More crucially, inner-city youth arouse public anxiety and precipitate what Diana Gordon describes as "the return of the dangerous classes."[3] Members of the dangerous classes, she argues, are believed to pose a threat to the personal safety of law-abiding citizens and, if unchecked, to the social, economic, and moral order of the larger society. Accordingly, black and Latino youth are prominently figured in the widely shared notion that inner cities—and by association, their racially coded populations—constitute a fiscal and moral strain on national resources. Subsequently, some of the salient crisis scenarios coloring the postindustrial United States have been redefined. In the process, meanings about race, class, gender, and youth undergo substantial revisions....

It is within this social context that the cultural productions of black youth amass energy and ever-increasing ingenuity. The transformations of urban ghetto life situate different formations of racial discourse and enable them to take shape. One aim of black youth popular culture is to redefine the crisis scenarios that prominently figure young African Americans. The symbolic practices of black youth develop distinct styles, moods, and imaginative contours that engage a broad spectrum of cultural producers—journalists, politicians, scholars—about African American life. The explosive surge in popular cultural productions by black youth prompts a reconsideration of how unsettled times reinvigorate not only social control discourses but resistive discourses, too. This is not to suggest that social and economic dislocations are *the* determinate causes of black youth cultural productions. Instead, I am suggesting that the ways in which black youth experience a rapidly changing society and how they practice cultural politics to express these experiences correspond.

Paradoxically, the intensification of racial and economic polarization in the United States produces space for the emergence of cultural practices that derive much of their symbolic efficacy from locations of marginality. The popularization of black youth expressive cultures is an excellent case in point. Despite high rates of poverty, joblessness, and criminal arrests, black youth occupy a dynamic role in the shaping of the popular cultural landscape. Many of the major culture industries—sports, television, advertising, music, cinema—incorporate the innovative styles and expressive cultures of black youth in order to appeal to their respective markets and revitalize their own commercial viability. Ironically, social isolation and economic marginalization contribute to the energy and imaginative capacities that enable black youth to participate effectively in the ever-expanding universe of popular media culture. In the process, black youth have accumulated significant amounts of symbolic capital.

So despite the currency of conservative discourses, black youth have mobilized their own discourses, critiques, and representations of the crisis-colored scenarios in which they are prominently figured. More important, young African Americans are acutely aware of the social world in which they live and the vast structural inequalities that impose severe restrictions on their economic mobility.

All members of society exercise some measure of agency—that is, capacity to exert some degree of power over the social arrangements and institutions that situate their lives. Faced with the increasing trend toward structurally enforced idleness and state-sanctioned coercion, black youth have fought diligently to create spaces of leisure, pleasure, and opposition from the social structures and institutional arrangements that influence their life chances.

## NOTES

1. Brain, D. 1994. "Cultural Production as 'Society in the Making': Architecture as an Exemplar of the Social Construction of Cultural Artifacts." In *The Sociology of Culture: Emerging Theoretical Perspectives*, edited by Diana Crane. Cambridge, Mass.: Blackwell.
2. Kelley, Robin D. G. 1994. *Race Rebels: Culture, Politics, and the Black Working Class.* New York: Free Press.
3. Gordon, Diana R. 1994. *The Return of the Dangerous Classes: Drug Prohibition and Policy Politics.* New York: Norton.

## DISCUSSION QUESTIONS

1. What social conditions does Watkins identify as shaping the cultural forms produced by Black youth? How does this formulation challenge the idea that culture is mainly something imposed on youth groups?
2. What does Watkins mean by arguing that the news media are an "important site of racial discourse"? What evidence of this do you see in the news that you observe?

# 12

# How to NOT Be 21st Century Venus Hottentots

FATIMAH N. MUHAMMAD

*Sara Baartman was a 20 year-old woman from South Africa who in 1810 was taken from South African to London and displayed in public as a "freak." Becoming known as the Venus Hottentot, the racist and sexist images of her were widely popularized in Britain as "scientific evidence" of racial inferiority. It is an early example of how racist beliefs about Black women's sexuality have perpetuated racism. Here Fatimah N. Muhammed discusses how contemporary racism still targets Black women's sexuality, but she also shows the resistance of Black women to such exploitation.*

... Hip-hop culture is a multi-media genre. Whether packaged in sound recordings, television, movies, magazines, books or the Internet, hip-hop culture's generic formula is defined by irreverent and/or ghetto aesthetics and excessive stylistic conventions that innovate and freely incorporate other cultural material, using various technological inventions (such as digital sampling). These stylistic practices, identified as flow (sustained motion and energy), layering (of sounds, ideas and images), and rupture (breaks in movement and music), are rooted in Black cultural practices and remain as hip-hop's cultural templates. Upon examination of the social, cultural, and industrial contexts of hip-hop's evolution, one sees that urban de-industrialization and social policy changes in the 1970s and 1980s presented a rupture in the trajectory for Blacks' upward mobility promised by the Civil Rights Movement and other social movements of the 1950s and 1960s. This rupture is similar to mainstream American society's cultural rupture in the 1960s that produced a generation gap where younger Americans converted to rock 'n' roll and its counter-establishment points of view. In the same way that the Vietnam war divided generations, President Reagan's War on Drugs and Trickle Down economic policy engendered a youth-led cultural revolt in Black America. Prosecution of drug-related crimes skyrocketed imprisonment rates among younger Blacks and dwindling employment and educational options contributed to the development of a Black "underclass." Younger Blacks' counter-establishment response was hip-hop culture. The proliferation

SOURCE: Pough, Ewendolyn D. et al. (eds.) 2007. *Home Girls Make Some Noise: Hip Hop Feminism Anthology.* Mira Loma, CA: Parker Publishing.

of media industry outlets magnified and broadened hip-hop's cultural reach. Through it all young Black women saw themselves here, there, and everywhere: as participants in rap music's production, hip-hop fashion and dance; as silenced objects of Black males' narratives and fantasies; as well as contributors to hip-hop's dialogues on poverty, racism, sexism, family life and sexuality.

## WHEN AND WHERE BLACK WOMEN ENTER HIP-HOP (CULTURE AND MEDIA)

To look into how Black female spaces are constructed within hip-hop's culture and media, I engage the theatrical trope of roles and its attendant concepts of identity and standpoint. As used in sociological terms, role "represents a cluster of customary ways of thinking on an individual level... [a set of] cultural expectations."[1] Individuals in daily interactions and social situations perform numerous roles. As a mother, sister, co-worker, or church-goer, we act in ways congruent with cultural expectations of these roles, "us[ing] the vocabulary, [and other behaviors and strategies] that go with our role."[2] The ways that Black women perform roles in hip-hop culture reflect both their adoption and creation of Black female identities.

Discerning roles can be accomplished by "ordering facts" from ongoing activities in reality, or as ethnographic sociologist Erving Goffman termed, by using a "framework."[3] Different frameworks define a social situation from various perspectives. The perspectives related to this discussion come from role expectations circulating in mainstream society, those derived from music industry demands, and in hip-hop's cultural values....

My purpose is to analyze the performative aspects of a cultural phenomenon—hip-hop and the rap music industry—in order to highlight how young Black women construct Black female social spaces and identities. Using Goffman's concept of frame analysis, this section surveys Black female cultural roles in hip-hop, from celebrated rappers and song producers, nameless video "actresses/models" to audience members and consumers....

Black women experience roles and identities within hip-hop culture as they are framed by systems informed by racism, sexism and economic oppression. In her book *From Mammy to Miss America and Beyond: Cultural Images and the Shaping of US Social Policy*, K. Sue Jewell asserts that Black women occupy the lowest place in the US social hierarchy, which is informed by White supremacy.[4] She theorizes:

"The determination of positioning on the social hierarchy of discrimination is based on how closely individuals approximate the race and gender of the privileged class... Clearly, African American women have the least common physical attributes in terms of race and gender, compared to White men who belong to the privileged class."

Discourses about race, class, and abridged economic opportunity predominate in hip-hop culture. Taking up issues of gender and sexual politics, Black women construct Black female spaces and identities while standing at all fronts of social struggles....

This focus also underscores performance as an epistemologically productive process. According to Goffman's frame analysis theory, the frames—in this case, social, cultural and industrial—establish what roles are possible. Within these frames, roles are both virtuoso styles and functional identities.[5] A look at the roles they perform within hip-hop's culture and industries can reveal their concerns and how they address them. Seen as performances, these women's actions also contain projections about the possible identities of their female audiences: that is, their works and careers proffer Black female identity narratives.

However, the use of frame analysis in this essay contrasts with the sense in which Goffman used it. Where Goffman's role metaphor is based on Western narratives in which the outcomes are predestined or built into roles, my look at roles will be informed by Black culture—understood as a hybrid of Afrodiasporic cultural practices and Western experiences. And where Goffman's use seems to assume that roles in social interaction are performed to maintain social structures, this analysis also looks for the ways Black women performed their roles to change social structures.

While their roles may share certain themes, no single standpoint is expected: performance of Black female identity occurs within a field of styles.... Black female's roles in hip-hop can be understood in light of the political economies that inform its cultural industries. With different roles, Black female artists perform their identities in response to differing material pressures. Moreover, the differing composites of roles they play result in different positions in society.

## HIP-HOP'S YOUNG BLACK FEMALE ROLES & IDENTITIES BECOME TWENTY FIRST CENTURY VENUS HOTTENTOTS

Moving from silence to speech is for the oppressed, the colonized, the exploited, and those who stand and struggle side by side a gesture of defiance that heals, that makes new life and new growth possible. It is that act of speech of "talking back"... that is the expression of our movement from object to subject—the liberated voice.

Hip-hop culture is organized around this practice of commanding autonomy and visibility in the face of material conditions that oppress, erase, and silence described by bell hooks.... Based on Afrodiasporic practices, identity is established through stylistic virtuosity and uniqueness; hip-hop artists (for example, rappers, DJs, song producers, graffiti writers, and dancers) represent who they are and where they come from by "develop[ing] a style nobody can deal with."[6] ...

In rap music, rappers narrate their identities through stories that focus on urban life and the material conditions that they face. The storytelling practices include boasting about individual characteristics and more overtly politically tinged messages that link conditions of urban violence or poverty to their root causes, including racism and class oppression. Graffiti writers and break dancers stake out territories in public spaces in the face of constant policing and erasure of their presence. Black women face all these conditions as well as the exclusion, marginalization, misogyny, and patronizing codes of American femininity. These are framed within the sexism and male-dominated street culture perpetrated in hip-hop, its media industries, and in mainstream society. Moreover, social double standards accommodate males hanging out in the streets while limiting the domain of respectable women to the home. The ways Black women are incorporated into hip-hop's cultural production are shaped by these social and cultural attitudes as well as industry dynamics.

Even though commercial representations of hip-hop culture denied their presence or limited them to the sidelines and background as onlookers or as the focus of male sexual attention/attraction, Black women were an integral part of hip-hop's formation. This denial and erasure of contributions is similar to that faced by women in other forms of popular culture....

Black female rappers participate in rap music's discourses on race and class oppression. Moreover, because rap music is very much about speaking up to command acknowledgement, Black female rappers have always dealt head-on with the blocks to their productive and creative involvement. The first major female rappers came to the fore with their records gaining radio airplay and club use, primarily through outspoken "answer records," which challenged male rappers' stories about a female's rejection of their romantic pursuits....

The act of rapping in this sense was essentially about "talking back" to the male dominance that circulates through American society as well as in hip-hop culture. More specifically, these records dealt with "the problems and issues that face young inner-city women who routinely battle what the fictional Roxanne rhymes about—sexual harassment and leering by arrogant male suitors."[7]

Black female rappers have also expanded rap's discourses to articulate issues that male rappers usually don't or won't, including, domestic violence. For example, Queen Latifah's 1993 song "U.N.I.T.Y." narrates an abused woman's decision to leave an abusive relationship, by discussing the role model functions of motherhood and sharing her conclusions with listeners as advice. Black female rappers perform gender narratives and act as advisors in the same way as Black female blues singers did according to Angela Davis's analysis of how these singers spoke to a larger female community in working-class terms. And according to hip-hop's generic conventions, Queen Latifah's "U.N.I.T.Y." rhetorically samples a line from an empowering moment of another Black woman's narrative, *The Color Purple* (1985). In the film, the "ain't nothin' good gon' come to ya, until you do right by me" quote was said by Whoopi Goldberg's character Celie when she stood up to her abusive husband.

With their presence in rap music, Black female rappers also explicitly draw attention to women as a significant group in rap's audience and market. Many are consciously aware of their role model status and engage in articulating a Black female standpoint....

Informed by a concern for Black women and the status of the Black "community," Black female rappers like Queen Latifah and Lauryn Hill extended the history of Black women's articulation of racism, sexism and class oppression as converging social forces that must be equally addressed....

In the last twenty-five years, the music recording industry, in general, has shifted to using music videos as the first way for labels to market and promote their artists; labels rely on visual tools (including artists' looks) to get their artists noticed in a crowded television and magazine environment....

Female rappers have most substantially felt this pressure, where "sex appeal is now the currency by which women in the music business are valued and devalued." Those like Lil' Kim and Foxy Brown, who have emerged within this "sex sells" framework, have not only had sex appeal as part of their marketing but also as a part their work, building their song repertoires around blatant sexuality. Additionally, as general American pop culture became more sexualized in the late 1990s, many female rappers found entry points through the framework of hip-hop culture's hyper-sexuality and discourses of pimping and sexuality as power. Reversing the player/pimping scenarios, female rappers rap about using their sexuality to control men and direct their life choices. Their versions of being "hard" with street credibility promote themselves as sexual predators, using sex to negotiate financial, social, and male/female issues. Such sexually-charged roles have been alternately seen as empowering for women as well as just an advertising and marketing gimmick banking on the shock value of women being as sexually aggressive and explicit as men in their performances. It also presents them as objects for male fantasies. Thus, not every female standpoint is feminist....

Finally, the majority of Black women images in hip-hop culture fit into the video actress role. According to prominent music video director Little X, "until *Vogue* starts putting regular women on covers, and Hollywood starts using people off the street for films, we will still need beautiful women to really sell that song. The video girl isn't going anywhere."[8] Twenty-three-year-old video model/actress Liris Crosse, known as "The Butt," says "to make it as a video girl, you have to have a pretty face, the ability to seduce the camera, personality and presence."[9] Getting the job as a video model/actress depends on getting noticed, and "the more revealing the outfit, the better chance girls have of showing that they don't have many inhibitions," according to male video director Rashidi Harper.[10] The video girl doesn't rap or dance; her role is to be the perfect physical specimen for most rap videos' T & A aesthetic.... The video girls represent hip-hop's female beauty codes: if a 40-ounce beer bottle is an identifying marker for hip-hop's male, then a big butt represents the identifying marker for hip-hop's female. The layered cultural and media industrial forces have combined to reduce young Black female hip-hop identity to a body part.

The focus in hip-hop culture and media on the Black female posterior is eerily reminiscent of the Venus Hottentot circus act framing of a Black woman in the 1820s. In the name of science, a European doctor awestruck by Khosian (of South Africa) women's posteriors and genitalia paid one of the Black women there to travel to England, then France, with him as a specimen for display and study as the "Hottentot Venus."[11] (The Europeans had named her tribe of heritage, Hottentot, and dubbed her a Venus.) Images of her display/visit in England and France depict her poised on a pedestal: her buttocks exposed, and men and women staring intensely at her and commenting "Oh! God Damn what roast beef!" and "Ah! how comical is nature."[12] Even if today's video hotties appear to have been initiated by Black men and co-signed by Black women themselves—instead of White supremacy-informed Europeans like those responsible for the "Venus Hottentot" spectacle—the predominance of such representations again circulates the Black female body and sexuality in a de-humanizing manner with far-reaching consequences. Turning to the audience factors of the equation offers a cautionary look into how young Black women use these roles and identities from hip-hop culture.

## YOUNG BLACK WOMEN AS AUDIENCE TO HIP-HOP'S MEDIA

One of the lesser-explored questions in most feminist writings on hip-hop is: if/when hip-hop as a youth culture is persistently packaged as male-oriented (and bent on negatively speaking of and objectifying women), how do young Black women actually participate in it as an audience? In the same way that hip-hop culture is tempered by a set of factors, audience participation in hip-hop culture is also guided by similar media industry dynamics, mass-media representations and socio-cultural factors like gender, race, class and age. These factors work in tandem to determine what is available to be engaged by young Black women as a media audience.

While my generation started an identity movement at a socio-cultural break, the succeeding millennial generation's media environment inherited our mix, amplified by technological changes and greater consumption capacity. While we went from listening to music in limited environments to listening to it on boom boxes and Sony Walkmans, and making audio cassettes, millennials only know about being able to customize their music choices and consumption, including burning their own CDs. We went from only seeing Black music acts Saturdays on *Soul Train* to seeing Black musicians every day in music videos. Millennials have no media experience without hip-hop's influence. Even oldies music for them is rap.

Then what of hip-hop culture's impact on them as audience members? One way to discern impact is through asking about identity. Who are you? What are you about? Most answers include a life story or "my story is" response. Identity understood as a story, then allows us to see how media representations, cultural discourses and life stage issues provide the model scripts and potential roles integrated in one's identity.

I have conducted two focus group studies that apply this encounter of audience/media identity stories approach, exploring hip-hop culture's impact on young Black women's identity. With the purpose of tracing links between participants' media habits and their identity strategies, the study asked eighteen- to twenty-year-olds (millennials) about their identity processes and how they would represent themselves in a media form to see how this hip-hop influenced media environment combines with their life stage issues. How do they see Black women represented in the media? How do they identify themselves? I had each brainstorm about the form, content and issues that she would include in a media representation of herself. I analyzed each participant's media representation idea and questionnaire responses to find traces of hip-hop culture's content, themes/issues and discourses.

Each focus group study consisted of three one-hour, questionnaire-guided discussion sessions and one brief one-on-one interview over three days with self-selected participants who attended the site schools where the studies were conducted. While the two groups had similar social backgrounds (all grew up in Illinois and most in the Chicago area), the main distinction between the two focus groups was educational achievement. This factor somewhat affected their engagement of mass media and hip-hop culture.

The first focus group study with eight participants was conducted in Fall 2002 at an alternative high school where I worked. The alternative school served students who, for reasons ranging from teen parenting and criminal cases to school truancy and academic failures, were not successful in or were expelled from Chicago's public school system high schools. At the time of the study, seven of the participants were in their last year or semester of high school (one had finished the prior summer). All but one of participants were from households whose major income came from financial assistance social programs. Every member of this focus group had experiences with drug use, violence, early sexual activity, parenting (of their own children or younger siblings) and early self-sufficiency due to little parental supervision.

The second focus group study with six participants was conducted in Fall 2005 at a university in Illinois where I worked. All were in their first semester of college and expressed feeling overwhelmed by the changes that being away from home in college made in their personal and social lives. All but two of the students were from the Chicago area, and these two were from urban areas in central Illinois. From the discussions and comments, I surmised that paying for college was a major financial burden.

Overall, the "trying to grow-up too fast" characteristic applied to millennials is reflected both in their social and their media environments: they're exposed to too much too soon. Social circumstances ruled by public scandal and trauma frame their life stage issues. The Columbine High School shootings and other such events cloud their roles as students with violence, suspicion and surveillance. The prevalence of dating violence and high rates of HIV infection raise the stakes of courtship. Through mass-mediated hip-hop culture they've consumed graphic violence, aggressive sexuality and glamorization of drug and alcohol use. Because of their adultlike life circumstances, it's like they're wearing their mothers' or big sisters' clothes and acting the parts. Without other strong ways to establish identity (e.g., roles as accomplished student or daddy's little girl), the sexual "speak" used by Black female rappers to assert power in male-female relationships becomes a real behavior choice for young Black women.

Differences in the findings for both groups' self-identity strategies were primarily based in subjectivity, i.e., social position, community/class affiliations and world outlook. While all participants in both groups listed rap/hip-hop as the type of music they thought their peers listened to and named similarly formatted radio stations as their favorites, none of those in the college focus group and all of those in the high school focus group included "sexy" as a self-identity value. Owing to the relative academic achievements that got them into college, all of the college group participants used terms like "talented," "skillfull" and "successful" in their self-identity lists; only two of them mentioned anything related to appearance at all, and they used the word "beautiful."

In both focus groups' self-identity media project outlines, I found that Black female millennials do, in fact, incorporate hip-hop into their identity stories. Hip-hop's discourses, themes and styles were relevant to their life stage issues and defined Blackness for them. Some explicitly engaged hip-hop culture as a means to be heard and affirmed by their social and cultural peers when they might otherwise be alienated and ignored.

Among the high school focus group participants, hip-hop's modes of creativity and expression were most salient with those not invested in heterosexual identities. They used rap as an expressive venue for posing critical looks at Black life and mainstream society. Their interpretation of what are usually perceived as hip-hop male-centered themes (e.g., life is hard, have a hustle to survive or die discourses) in the foreground and viewed street life and poverty as the authenticating Black experience.

While the college focus group's self-identity media projects directly applied hip-hop's "tell your life story" mode of expression, most other uses of hip-hop's expressive modes were in their use of ghetto iconography: their project descriptions indicated that gang members would be seen hanging out and gunshots would be heard. Their takes on hip-hop's "life is hard" theme also included discourses about struggling materially but included succeeding against odds that claimed others. Only half of this group equated street life and poverty as authenticating Black experiences.

Combining this essay's critical analysis and the focus group study results indicates hip-hop culture contributes to highly aggressive and sexualized young Black female identities. Young women see violence as a ready tool to deal with their conflicts. In my years at an alternative high school on Chicago's Southside, nearly all the in-school fights had been between and/or started by females. With the college focus group, aggression took the form of "looking" between Black women: most of these participants had experiences caused by other Black women looking at them and/or them looking at other Black women leading to some kind of verbal and/or physical conflict.

"I've been freaking since I was twelve" as well as comments about their initiation of contraception discussions with parents also came up in the focus group study discussions. As mentioned earlier, each of the alternative high school participants listed one or more references to being sexy as part of her identity. In their identity descriptions and in other research on young Black women, there is a tension between wanting independence and wanting a man to take care of them (not unlike Gen X women): reflecting the notion promoted by rappers like Lil' Kim and other aspects of hip-hop culture that sex is a means to female power and independence. Disturbingly, their identities harbor Venus Hottentot potentials: wanting to be paid for being sexy and the object of sexual interest by any and all beholders. Moreover, studies suggest that rap music videos with an overabundance of half-naked women dancing and partying with a male rapper as the center of attention have a problematic consequence for some youth.

[Some young women] reported greater acceptance of teen dating violence than females who were not exposed to the videos…[possibly because] that exposure to the rap videos featuring women in sexually subordinate positions increased the accessibility of constructs associated with female inferiority.

On images of Black women in mass media and hip-hop, both focus groups were somewhat dismissive of the effect of video dancers and sexualized female rappers, commenting, "It's just always been like that." In the college focus group's discussion, one participant summed up a description of female rapper Trina's sexually driven lyrics and persona by saying, "that's just her 'doing Trina' to get paid." For the most part, neither focus group questioned hip-hop's representation of women. The closest they came to that was in their descriptions of Black women in the media. They recognized that most images were of "half-naked women."

All the participants in both studies claimed a vague "strong Black woman" identity as an empowering way to define their individual and social roles. That identity is called upon whenever they spoke of being self-determined, worthy of others' respect and able to handle stress and crises. In relation to social roles, study participants saw their roles in the Black community in terms of helping children and poor women/mothers. In their discussions, there was recognition of various problems that endanger families, including drug trafficking and gun violence in their neighborhoods.…

The work of hip-hop feminist critics is to question such cultural practices and social conditions, send up the warning, and call out the cultural, social, historical and material factors involved.

The current social conditions, with so many young people practically raising themselves, make mass media [including their engagement of hip-hop culture] more salient an influence for them. Feeling like they have to settle conflicts violently and/or participating along with thug boyfriends in criminal behavior has increased the imprisonment rates for young Black women. The sexual "speak" of women rappers like Missy Elliot and Lil' Kim has not fostered young Black women to protect themselves from HIV infection as "a shocking 75% of the new infections are attributed to Black women having unprotected sex with men." Hip-hop's do-it-all female identities of independence often combine with the inherited "mammy" stereotype, in which Black women are born to nurture and serve others. As more young Black women take on the do-it-all roles of single parenting as well as education/career pursuits, the stressful reality of the "strong Black woman" syndrome is taking a toll: "after the 1980s [suicide] rates more than doubled, going from 3.6 deaths per 100,000 people to 8.1." While suicide remains rare, stress and depression severely tax Black women's health and physical well-being. Current trends will either claim us or push us to a critical breaking point to start a new flow. As this essay's audience factors discussion highlights, hip-hop's Black female identities are both part of the problem and resolution to the pressures confronting young Black women.

## NOTES

1. Scott G. McNall, *The Sociological Experience*, 2nd Edition (Boston: Little, Brown and Company, 1971), 63.

2. McNall, 71.

3. Erving Goffman, *The Presentation of Self in Everyday Life* (Garden City, NY: Doubleday & Co, 1959), 240–241.

4. K. Sue Jewell, *From Mammy to Miss America and Beyond: Cultural Images and the Shaping of US Social Policy* (New York: Routledge, 1993), 6.

5. Erving Goffman, *Frame Analysis: An Essay on the Organization of Experience* (New York: Harper Colophon Books, 1974), 22–237.

6. Tricia Rose, *Black Noise: Rap Music and Culture in Contemporary America* (Hanover, NH: Wesleyan University Press), 38.

7. William Eric Perkins, "The Rap Attack: an introduction," *Droppin' Science: Critical Essays on Rap Music and Hip-Hop Culture*, ed. William Eric Perkins (Philadelphia, PA: Temple University Press, 1996), 18.

8. Minya Oh, "Casting Couch," *Honey* (November 2000, vol. 1 no. 8), 37.

9. Minya Oh, "Confessions of a Video Hottie," *Honey* (November 2000, vol. 1, n. 8), 37.

10. Oh, "Casting Couch," 37.

11. John R. Baker, "The 'Hottentot Venus'" <www.heretical.com/miscella/baker4.html>

12. From "Human Exhibition: Hottentot Venus Illustration" http://www.english.emory.edu/Bahri/Hott.html

## DISCUSSION QUESTIONS

1. What contradictions do there seem to be in Black women's relationship to hip-hop culture? Are Black women active subjects or merely objects within hip-hop culture—or both? In what ways?

2. How does Muhammad compare Black women's depiction in contemporary hip hop to the historical phenomenon of the Hottentot Venus?

# 13

# María de la Barbie

ROSIE MOLINARY

*Rosie Molinary conducted a national study of Latinas, called the "Growing Up Latina Survey." In this part of her research, she discusses the prevailing images of Latinas within which they much try to negotiate a strong identity against the current of the dominant culture.*

What does it mean to be beautiful in America? The answer to this question brings up all kinds of clichés. Reality shows, magazines, sitcoms, movies, and music videos propagate a certain image—tall, thin, and blond—that's unattainable for most women. But pop culture isn't the only place where beauty matters. Female news anchors, executives, and CEOs, even high-power board members, find that beauty plays a role in breaking through the glass ceiling. Beauty is so valued, it's become a commodity.

The question that comes up for me when I think about the power beauty has over our lives and experiences is this: What happens to girls whose self-image is shaped by other people's beauty standards, and whose own features aren't reflected back to them in the everyday images the media promotes? For today's young Latinas, there are some Latina role models, but most of those role models represent yet another unattainable beauty standard. Many Latinas live with a feeling of not being accepted because of how they look. For those of us who have dark skin, raven hair, and a short stature, there isn't much external validation. The average woman of any ethnicity is different from what's celebrated on television, in magazines, and in life. But often, in Latin culture, these differences are exacerbated by the fact that the families' input offers such a distinct point of view—one that's often at odds with the larger culture.

Thus, as Latinas we can be caught in between two standards of beauty—not feeling beautiful in either culture, or feeling beautiful in one but not the other. No matter where we stand, we're on the precipice of judgment, with one set of values that informs out lives shaped by American pop culture and another set shaped by our families' culture and traditions.

Oftentimes there's another perspective as well—our own: a perspective that takes into account the impact of the first two and how they push and pull at our self-image and feelings of who we are. This last perspective is the place where we can find peace in ourselves, which is what ultimately leads us to our real beauty.

SOURCE: Molinary, Rosie. 2007. *Hijas Americanas: Beauty, Body Image, and Growing up Latina.* Emeryville, CA: Seal Press.

I have memories from childhood that indicate just how much American pop culture influenced my worldview. In one memory, I'm sitting in front of the TV, lamenting the size of my nose compared to the girls' noses on-screen. I can remember my mother telling me that I could pinch my nostrils and pull my nose outward to change its shape. And I followed that advice. Pinch, pull, pinch, pull. But my nose never budged.

As early as third grade, my best friend, Jenny, and I used to lament our weight in comparison to other girls'. I remember walking to her house one day as we schemed about the perfect plan to rid ourselves of those extra pounds.

"We could just pinch our fat off and put it on somebody else without her knowing it," we laughed. As an adult with some extra pounds to spare, I look empathetically back at my third-grade self, wondering how it was possible that I thought I was fat. We were just eight, and wiry. When Jenny and I remember it now, we are both dumbfounded by the fact that popular culture was sending us the message that we needed to be smaller.

I also remember watching *The Dukes of Hazzard* on Friday nights, and how much we loved that show, and how Daisy Duke would strut her stuff in those oh-so-short shorts. There are plenty of times when I might have learned that being skinny was how you should look, and that showing off your body is how to get attention.

One summer when I was a preteen, one of my cousins from Puerto Rico came to visit for a couple weeks. We spent our days playing a game we called Model. We did our hair, tried on various outfits, and posed awkwardly next to plants or on the edge of a bed with one hand under our chin while snapping photos of each other with my Kodak 110 camera. I still have those pictures. We both had thick black hair that was unmanageable, despite our attempts to iron it (this was before the availability of flatirons). We were emulating the white girls we saw in *Seventeen* and *YM*, even though we were a far cry from them.

Later I became so absorbed in other things that beauty was not my primary—or even secondary—obsession. As a middle-schooler, I ran for seventh-grade president in an effort to keep my family from moving. Either it worked or they decided they didn't want to move, because I won and we stayed.

I was involved in student government and community service throughout middle and high school. During that time, I wasn't at all stylish. Each morning, I woke up fifteen minutes before my ride came. I arrived at school with my tumble of curls sopping down my back, a Pop-Tart in hand. I wore jeans and T-shirts. And the first major blow to my self-confidence didn't come until ninth grade, when my football-star boyfriend, a junior, dumped me for a cheerleader I had introduced him to at the homecoming dance (because I thought I was too short to dance with him). She was 5'7" to his 6'4", a much better match than my 5'1". The cheerleader was almost my antithesis: Tall and thin, she had porcelain-white skin with pretty black hair and light eyes. I pined and I pitied myself, but I managed to rebound.

By the next fall, Wade and I were serious, and it was a relationship that would last until the spring of eleventh grade. My junior year, I threw myself into school and started losing weight. Still, I wasn't thinking too much about my looks. I wore no makeup, dressed casually, hung out with boys that were friends rather than potential flames. By my senior year, I was dating someone from another high school, and even though I sometimes looked around at my peers and felt not feminine enough, it wasn't enough to change my behavior.

The real shock to my worldview came in college, where the diversity of my high school no longer shielded me from seeing how different I was. At a school where the tuition was high and where less than one-third of the students were on financial aid (I being one of them), I was not just an ethnic minority. I was a socioeconomic minority, and both of those differences felt compounded in my daily life. It was then and there that I began to wonder whether my beauty and body were inadequate. It was then and there that I started to feel isolated, alone, unrecognizable—where I realized that I would have to reconcile my gringa and my Latina.

## THE SCENE MAY BE BROWNER, BUT ...

When J.Lo hit the scene, it seemed that Hollywood suddenly up and took notice of Latinas in a different way. As a Fly Girl on *In Living Color*, her dark, curly hair and strong body reflected her roots. Jenny from the Block was sturdy, not starved. Then, when she played Selena, the slain Chicana singer who was the number-one Latina star in the United States and Mexico before she died, she was introduced to a much broader audience because of the number and diversity of Latinas who flocked to see her on the silver screen.

Soon Jennifer Lopez had caught our collective imagination. She exemplified the modern Latina, but it wouldn't be long before her urban Bronx edge was traded in for a more sophisticated image that turned her into a national—not just Latina—beauty icon. Still, the attention she was getting affirmed Latinas all over the country. Suddenly there was someone who looked a lot more like one of us than the standard blond beauties of the 1980s and most of the '90s. She became the poster girl for Latina beauty: olive skin, full butt, feminine curves. It wasn't long before other Latinas entered the scene: Eva Mendes, Michelle Rodriguez, Zoe Saldana, Eva Longoria.

In the Growing Up Latina Survey, respondents were asked to list in rank order up to five well-known women of any ethnicity whom they considered beautiful. Eleven of the fifteen top choices reflected women of color. Only two were blonds, which reflects the fact that Hollywood does have many more minority and women-of-color role models than it did fifteen or even ten years ago....

More and more, women of color, and Latinas specifically, are gracing the covers of magazines, acting as spokeswomen for beauty products, and starring in major motion pictures and television dramas. Beauty in the new millennium, it seems, is diversifying, and the women interviewed for this [project] have certainly noticed the change.

"Being a Latina has become more mainstream and acceptable, so we're popular right now," says thirty-two-year-old Mia, a Puerto Rican who lives in

New York City. "In the media there is more exposure for Latinas, because we're becoming a fad in American mainstream culture. You certainly see more Latinas than when I was growing up."

Thirty-four-year-old Claudia, who's of Ecuadorian descent and lives in New York, says she felt better as soon as J.Lo entered the public scene. "I grew up with a complex about my body, because I was pretty thick. I always considered myself fat based on what I saw as the perfect American girl. Watching TV and looking at magazines, you see these skinny girls. I used to wear baggy clothes because I felt fat. But once people like J.Lo came out, I started to feel better about my body."

Camille, a twenty-seven-year-old New Yorker who's of mixed Colombian and Dominican descent, agrees with this sentiment. "I just never saw many women like me on the TV when I was growing up. Now I am seeing more women who look like me, and I am making a connection. But when I was growing up, I wanted light eyes and a different body type."

Still, the diversification of the type of women being represented in pop culture and by the media has not completely alleviated the pressure that many Latinas feel about their bodies and looks.

"The media has had a huge, huge impact on what I see as 'pretty.' I am okay with myself, but when I am around other college girls, and they all look like what the media shows, I am uncomfortable," says eighteen-year-old Nora, who's *puertorriqueña,* grew up in Charlotte, North Carolina, and feels pressured by American pop culture to have big breasts and a small stomach. "It is all about the body. You don't even have to have a pretty face as long as you have the body."

You would think that the greater representation of Latinas on the small screen and the silver screen would reinvigorate the confidence of Latinas across the country. It's not that easy. Some protest how these women are presented and the implications those portrayals have on the women who are out in mainstream society on a daily basis.

Twenty-five-year-old Gabriela is of Colombian and Cuban descent and is from Houston, Texas. "On the one hand, they portray [Latinas] as hoochies, but on the other hand, they are the fiery and sexy characters who bring a *boom!* to the screen. You have your Salma Hayeks and J.Los, but they are thinner and different from other Latinas," she says.

I agree with Gabriela's observation. I have always marveled at how the mainstream media praise Latina actresses like Salma Hayek and J.Lo for embracing their curves—but compared to many, these are relatively petite women. Salma and J.Lo may have curves, but their bodies are far closer to the bodies of Hollywood's other "It" women than to the average girl on the street in middle America....

While J.Lo is often praised by the media for staying true to her roots, would argue that she's been altering herself little by little—her hair, her look, her attitude—since her days as a Fly Girl....

As Latinas, we're inevitably shaped by the images we see, but can choose whether and how we internalize these values. In my research, women's opinions ranged widely. Some felt that American culture has truly made progress based on the mere fact that there are so many more Latinas in the mainstream media than there were five years ago. Others felt that we are still marginalized, and the very notion of lifting up these few women is equivalent to treating them like tokens, or poster girls, for Latinas as a whole.

Me? I fall somewhere in between: I'm thrilled to see the diversity that exists in the world begin to take root in American media, but sometimes I'm mortified by things like the choice of title for the ABC *telenovela Ugly Betty*. I know it's meant to be sarcastic and ironic, since Betty is the most together, well-intentioned, and good-humored character on the show. It might even be a clarion call for us about how we see beauty. But there is still a poignancy about calling her ugly, naming an anxiety that's so acute. To me, it seems to speak to the fact that plenty of people don't flinch at calling women ugly, fat, and many other descriptors that are direct attacks on our physical appearances. What are we saying to young girls when we identify someone as ugly just because she has bad hair, braces, and glasses?

I know that not having Latina role models when I was growing up was difficult, but sometimes I wonder whether it's even harder for Latinas growing up today, because the Latina role models on display are really in a league of their own. For a while it seemed as though Latina role models were far underrepresented compared to the range of female African American role models who were becoming more visible: Black girls had Queen Latifah, Aisha Tyler, Halle Berry, Beyoncé, Kerry Washington, and Janet Jackson long before any real Latina presence could be seen in the media. Fortunately, women like America Ferrera are showing us a more representative view of Latina women. Her work in *Real Women Have Curves* and *The Sisterhood of the Traveling Pants* revealed the depth of the Latina experience.

Sara Ramirez's role as Dr. Callie Torres on *Grey's Anatomy* portrays a strong, smart, beautiful, and average-size Latina. And it works, except for the occasional script misstep, such as the scene where another doctor describes her as so sexy that she is "dirty sexy." Why does the Latina doctor have to be the one who's *dirty sexy?*

## WHAT ARE YOU LOOKING AT?

The media—fickle and constantly changing as it is—embraces beauty because of its emphasis on everything visual. American mainstream or pop culture ideals are often propagated by the way they're interpreted. As viewers and consumers, we are largely responsible for the very things that bother many of us....

"Nowadays, the media makes it seem like if you aren't thin or skinny—if you don't look a certain way—you aren't going to be completely accepted by society. Even when you are beautiful, they still try to find other things that are wrong with us. There are all these shows about wanting to look like other people. People automatically assume that you can look better," adds Marisa, thirty-three, who's of Puerto Rican and Colombian descent....

## A WIDER BERTH

The weight of the media and beauty perception in the lives of women is severe. Respondents to the Growing Up Latina Survey were asked whether they thought American society expects women to do whatever they can to be more attractive, and 94.4% of the women said yes. Also, 87% believe that attractive women are move valued in our society, and 70.3% believe that attractive women face fewer obstacles on a daily basis.

These perceptions significantly impact the lives of Latinas. They perpetuate feelings of otherness and create standards that seem unattainable; as a result, some women feel as if they are second rate because of their appearance....

## ON BEING "JUST RIGHT"

We can definitely turn to the wise women among us to guide us toward our goals. At just twenty-two, Jessica has already found a truth that others may never discover. "Confidence and having a good sense of self-worth are essential for feeling beautiful, because regardless of how you look or dress, there will always be someone who thinks you're not good enough. With confidence, you automatically demand that respect. It's not how hot I look that matters—it's how I display my confidence."

Brenda, a fifty-four-year-old grandmother of Mexican descent, has made it her goal to provide an example to her granddaughter. "Taking into mind my own negative experiences growing up Latina, I am already teaching my granddaughter to have a positive self-image. Latinas need to be proud of all aspects of being Latina. There are plenty of Latina role models and success stories in most areas of American society today—something that was not the case I was growing up. We need to encourage our *Latinitas* to dream big. We need to teach them how to network and become resourceful in their efforts to develop their talents and skills, and in pursuing their future careers. Every Latina should learn early to find role models and mentors in successful Latinas: older sisters, aunts, famous Latinas, teachers, counselors, family friends, et cetera. That way, they can hitch their own stars to the Latina pioneers who are paving the way. *Latinitas* should also be taught that each one of them has a uniqueness to contribute to their own and future generations."

## DISCUSSION QUESTIONS

1.  How does Molinary see Latina self-images as constructed via images of beauty in popular culture?

2.  What images of Latinas do you see in the popular culture outlets that you regularly observe (fashion magazines, films, music videos, etc.)? Are they based in liberatory views or racial stereotypes? What impact do you think these images have on young Latinas and other groups?

# 14

# "Playing Indian":
## Why Native American Mascots Must End

CHARLES FRUEHLING SPRINGWOOD AND C. RICHARD KING

*Springwood and King show how the popular use of Native American mascots leads to disrespect and stereotyping of Native American people. Such mascots typically identify Native Americans as fierce warriors, sometimes with comic–like features. Research has also shown that exposure to such mascots is damaging to the self-esteem of young Native Americans.*

American Indian icons have long been controversial, but 80 colleges still use them, according to the National Coalition on Racism in Sports and Media. Recently, the struggles over such mascots have intensified, as fans and foes across the country have become increasingly outspoken.

At the University of Illinois at Urbana-Champaign, for example, more than 800 faculty members ... signed a petition against retaining Chief Illiniwek as the university's mascot.[1] Students at Indiana University of Pennsylvania have criticized the athletics teams' name, the Indians. The University of North Dakota has experienced rising hostilities on campus against its Fighting Sioux. Meanwhile, other students, faculty members, and administrators have vehemently defended those mascots.

Why, nearly 30 years after Dartmouth College and Stanford University retired their American Indian mascots, do similar mascots persist at many other institutions? And why do they evoke such passionate allegiance and strident criticism?

American Indian mascots are important as symbols because they are intimately linked to deeply embedded values and worldviews. To supporters, they honor indigenous people, embody institutional tradition, foster shared identity, and intensify the pleasures of college athletics. To those who oppose them, however, the mascots give life to racial stereotypes as well as revivify historical patterns of appropriation and oppression. They often foster discomfort, pain, and even terror among many American Indian people.

The December 1999 cover of *The Orange and Blue Observer,* a conservative student newspaper at Urbana-Champaign, graphically depicts the multilayered

SOURCE: "Playing Indian: Why Native American Mascots Must End" by Charles F. Springwood and C. Richard King in *Chronicle of Higher Education,* Dec. 9, 2001. Reprinted by permission of the authors.

and value-laden images that American Indian mascots evoke. Beneath the publication's masthead, a white gunslinger gazes at the viewer knowingly while pointing a drawn pistol at an Indian dancer in full regalia. A caption in large letters spells out the meaning of the scene: "Manifest Destiny: Go! Fight! Win!" Although arguably extreme, the cover, when placed alongside what occurs at college athletic events—fans dressing in paint and feathers, halftime mascot dances, crowds cheering "the Sioux suck"—reminds us that race relations, power, and violence are inescapable aspects of mascots.

We began to study these mascots while we were graduate students in anthropology at the University of Illinois in the early 1990s. American Indian students and their allies were endeavoring to retire Chief Illiniwek back then, as well, and the campus was the scene of intense debates. Witnessing such events inspired us to move beyond the competing arguments and try to understand the social forces and historical conditions that give life to American Indian mascots—as well as to the passionate support of, and opposition to, them. We wanted to understand the origins of mascots: how and why they have changed over time; how arguments about mascots fit into a broader racial context; and what they might tell us about the changing shape of society.

Over the past decade, we have developed case studies on the role that mascots have played at the halftime ceremonies of the University of Illinois, Marquette University, Florida State University, and various other higher-education institutions. Recently, we published an anthology, *Team Spirits: The Native American Mascots Controversy,* in which both American Indian and European American academics explored "Indian-ness," "whiteness," and American Indian activism. They also suggested strategies for change—in a variety of contexts that included Syracuse University and Central Michigan University, the Los Angeles public schools, and the Washington Redskins. Our scholarship and that of others have confirmed our belief that mascots matter, and that higher-education institutions must retire these hurtful symbols.

The tradition of using the signs and symbols of American Indian tribes to identify an athletic team is part of a much broader European American habit of "playing Indian," a metaphor that Philip Joseph Deloria explores in his book of that title (Yale University Press, 1998). In his historical analysis, Deloria enumerates how white people have appropriated American Indian cultures and symbols in order to continually refashion North American identities. Mimicking the indigenous, colonized "other" through imaginary play—as well as in literature, in television, and throughout other media—has stereotyped American Indian people as bellicose, wild, brave, pristine, and even animalistic.

Educators in particular should realize that such images, by flattening conceptions of American Indians into mythological terms, obscure the complex histories and misrepresent the identities of indigenous people. Moreover, they literally erase from public memory the regnant terror that so clearly marked the encounter between indigenous Americans and the colonists from Europe.

That higher-education institutions continue to support such icons and ensure their presence at athletics games and other campus events—even in the face of protest by the very people who are ostensibly memorialized by them—

suggests not only an insensitivity to another race and culture, but also an urge for domination. Power in colonial and postcolonial regimes has often been manifested as the power to name, to appropriate, to represent, and to speak—and to use such powers over others. American Indian mascots are expressive practices of precisely those forms of power.

Consider, for example, the use of dance to feature American Indian mascots. Frequently, the mascot, adorned in feathers and paint, stages a highly caricatured "Indian dance" in the middle of the field or court during halftime. At Urbana-Champaign, Chief Illiniwek sports an Oglala war bonnet to inspire the team: at Florida State University, Chief Osceola rides across the football field, feathered spear held aloft.

Throughout U.S. history, dance has been a controversial form of expression. Puritans considered it sinful; when performed by indigenous people, the federal government feared it as a transgressive, wild, and potentially dangerous form of expression. As a result, for much of the latter half of the 19th century, government agents, with the support of conservative clergy, attempted to outlaw native dance and ritual. In 1883, for example, the Department of the Interior established rules for Courts of Indian Offenses. Henry Teller, the secretary of the department, anticipated the purpose of such tribunals in a letter that he wrote to the Bureau of Indian Affairs stating that they would end the "heathenish practices" that hindered the assimilation of American Indian people. As recently as the 1920s, representatives of the federal government criticized American Indian dance, fearing the "immoral" meanings animated by such performances.

The majority of Indian mascots were invented in the first three decades of the 20th century, on the heels of such formal attempts to proscribe native dance and religion, and in the wake of the massive forced relocation that marked the 19th-century American Indian experience. European Americans so detested and feared native dance and culture that they criminalized those "pagan" practices. Yet at the same time they exhibited a passionate desire for certain Indian practices and characteristics—evidenced in part by the proliferation of American Indian mascots.

Although unintentional perhaps, the mascots' overtones of racial stereotype and political oppression have routinely transformed intercollegiate-athletic events into tinderboxes. Some 10 years ago at Urbana-Champaign, several Fighting Illini boosters responded to American Indian students who were protesting Chief Illiniwek by erecting a sign that read "Kill the Indians, Save the Chief." And, in the wake of the North Dakota controversy, faculty members who challenged the Fighting Sioux name have reported to us that supporters of the institution's symbol have repeatedly threatened those who oppose it.

Although many supporters of such mascots have argued that they promote respect and understanding of American Indian people, such symbols and the spectacles associated with them are often used in insensitive and demeaning ways that further shape how many people perceive and engage American Indians. Boosters of teams employing American Indians have enshrined largely romanticized stereotypes—noble warriors—to represent themselves. Meanwhile, those who support competitive teams routinely have invoked images of the frontier, Manifest Destiny, ignoble savages, and buffoonish natives to capture the spirit of impending athletics contests and their participants. In our studies, we find countless instances of such mockery on the covers of athletics programs, as motifs for homecoming floats, in fan cheers, and in press coverage.

For example, in 1999, *The Knoxville News-Sentinel* published a cartoon in a special section commemorating the appearance of the University of Tennessee at the Fiesta Bowl. At the center of the cartoon, a train driven by a team member in a coonskin cap plows into a buffoonish caricature of a generic Indian, representing the team's opponent, the Florida State Seminoles. As he flies through the air, the Seminole exclaims. "Paleface speak with forked tongue! This land is ours as long as grass grows and river flows. Oof!"

The Tennessee player retorts. "I got news, pal. This is a desert. And we're painting it orange!" Below them, parodying the genocide associated with the conquest of North America, Smokey, a canine mascot of the University of Tennessee, and a busty Tennessee fan speed down Interstate 10, dubbed "The New and Improved Trail of Tears." What effect can such a cartoon have on people whose ancestors were victims of the actual Trail of Tears?

The tradition of the Florida State Seminoles bears its share of responsibility for inviting that brand of ostensibly playful, yet clearly demeaning, discourse. For, at FSU, the image of the American Indian as warlike and violent is promoted without hesitation. Indeed, the Seminoles' football coach, Bobby Bowden, is known to scribble "Scalp 'em" underneath his autograph.

Such images and performances not only deter cross-cultural understanding and handicap social relations, they also harm individuals because they deform indigenous traditions, question identities, and subject both American Indians and European Americans to threatening experiences. For example, according to a *Tampa Tribune* article, a Florida resident and Kiowa tribe member, Joe Quetone, took his son to a Florida State football game during the mid-1990s. As students ran through the stands carrying tomahawks and sporting war paint, loincloths, and feathers, Quetone and his son overheard a man sitting nearby turn to a little boy and say, "Those are real Indians down there. You'd better be good, or they'll come up and scalp you!"

Environmental historian Richard White has suggested that "[White Americans] are pious toward Indian peoples, but we don't take them seriously; we don't credit them with the capacity to make changes. Whites readily grant certain non-whites a 'spiritual' or 'traditional' knowledge that is timeless. It is not something gained through work or labor; it is not contingent knowledge in a contingent world." The omnipresence of American Indian mascots serves only to advance the inability to accept American Indians as indeed contingent, complicated, diverse, and genuine Americans.

Ultimately, American Indian mascots cannot be separated from their origins in colonial conditions of exploitation. Because the problem with such mascots is one of context, they can never be anything more than a white man's Indian.

Based on our research and observations, we cannot imagine a middle ground for colleges with Indian mascots to take—one that respects indigenous people, upholds the ideals of higher education, or promotes cross-cultural understanding.

For instance, requiring students to take courses focusing on American Indian heritage, as some have suggested, reveals a troubling vision of the fit between curriculum, historic inequities, and social reform. Would we excuse colleges with active women's studies curricula if their policies and practices created a hostile environment for women?

Others have argued that colleges with American Indian mascots can actively manage them, promoting positive images and restricting negative uses. Many institutions have already exerted greater control over the symbols through design and licensing agreements. But they can't control the actions of boosters at their institutions or competitors at others. For example, the University of North Dakota would probably not prefer fans at North Dakota State University to make placards and T-shirts proclaiming that the "Sioux suck." Such events across the nation remind us that mascots are useful and meaningful because of their openness and flexibility—the way that they allow individuals without institutional consent or endorsement to make interpretations of self and society.

American Indian mascots directly contradict the ideals that most higher-education institutions seek—those of transcending racial and cultural boundaries and encouraging respectful relations among all people who live and work on their campuses. Colleges and universities bear a moral responsibility to relegate the unreal and unseemly parade of "team spirits" to history.

## NOTE

1. Note that since the writing of this article, the University of Illinois retired its mascot, Chief Illiniwek, in 2007.

## DISCUSSION QUESTIONS

1. What arguments do different groups make in favor of eliminating the use of Native American mascots? What arguments are there to keep them?

2. What would happen if similar caricatures of other groups were used as school mascots?

# 15

# Unmasking Racism:

## Costuming and Engagement of the Racial Other

JENNIFER C. MUELLER, DANIELLE DIRKS, AND LESLIE HOUTS PICCA

*Racism is a feature of everyday culture and in this study, Mueller and her col-
leagues analyze how racist images pervade even ordinary celebrations, in this case,
Halloween costuming. Their research reveals how racism is perpetuated, often by
the seemingly ordinary behaviors in everyday life.*

In 2003, Louisiana State District Judge Timothy Ellender arrived at a Halloween
party costumed in blackface makeup, afro wig, and prison jumpsuit,
complete with wrist and ankle shackles. When confronted with his actions, he
noted wearing the costume was "a harmless joke" (Simpson, 2003). In 2002,
Massachusetts-based Fright Catalog marketed and sold the Halloween mask
"Vato Loco," a stereotyped caricature of a bandana-clad, tattooed Latino gang
thug, while retail giants Wal-Mart, Party City, and Spencer Gifts began sales
for "Kung Fool," a Halloween ensemble complete with Japanese kimono and a
buck-toothed, slant-eyed mask with headband bearing the Chinese character for
"loser" (Hua, 2002; Unmasking Hate, 2002). Additionally, there have been several Halloween party-related blackface incidents documented at universities
across the United States over the past several years. White college students have
donned blackface and reenacted images of police brutality, cotton picking, and
lynching at such parties, invoking degrading stereotypes and some of the darkest
themes in our nation's racial past and present.

Collectively, these incidents indicate that Halloween may provide a unique
opportunity to understand contemporary racial relations and racial thinking in
the U.S....

## HALLOWEEN AS A UNIQUELY CONSTRUCTIVE
## SPACE FOR ENGAGING RACIAL CONCEPTS

Holidays have been theoretically described as socializing agents, acting on members of the society to reinforce shared beliefs and reaffirm commitments to values

SOURCE: 2007. *Qualitative Sociology* 30: 315–335.

(Durkheim, [1912] 1995; Etzioni, 2000). While many holidays, such as Christmas and Easter, are thought to directly enforce shared commitments, holidays such as Halloween arguably serve as tension-management holidays. Such holidays fulfill the socialization process *indirectly*, by managing the tensions that result from close conformity to societal mores and their concomitant behavioral prescriptions (Etzioni, 2000).

Similarly, "constructed" events such as Halloween, New Year's Eve and Mardi Gras have also been examined as "rituals of rebellion"—culturally permitted and ritually framed spaces where the free expression of countercultural feelings are tolerated, and protected to some degree by the agents of the official culture (Gluckman, 1963; Yinger, 1977). As described by Gluckman, these ritually bounded periods allow for the reversal of social roles, wherein subjugated groups temporarily assume positions of power. This temporary inversion is thought to act as a safety valve for sentiments of injustice. As such, rather than permanently alter hierarchies, rituals of rebellion tend to paradoxically strengthen and reinforce the social structure, norms, and roles they seek to deride....

As a holiday that allows individuals to "let off a little steam" from the routine of everyday life, Halloween's potential for social disruption is similar to that for rituals of rebellion and the Carnival (Rogers, 2002, p. 163). Indeed, Rogers argues that Halloween was eventually promoted to national status in the U.S. in part because it fostered a context for social inversion during the mid- to late-nineteenth century era, when other holidays became more institutionalized and focused on the values of family, home, and respectability. As Skal (2002) contends, the tradition grew that for one night each year, individuals could enjoy "a degree of license and liberty unimaginable—or simply unattainable—the rest of the year" (p. 17). Both historically and contemporarily, this context of free license often creates the impression among revelers that all potential for insult is suspended. When considered collectively, alongside the holiday's ritual costuming and social setting, Halloween's historical and contemporary license set the stage for the easy engagement of racial concepts.

## HALLOWEEN RITUAL COSTUMING: ROLE-TAKING, ROLE-MAKING

Donning costumes has become a ritual component of the Halloween tradition in North America (Nelson, 2000; Santino, 1994). Over half of all U.S. consumers celebrate Halloween in some way, with sixty percent reporting that they will costume for the holiday (The Macerich Company, 2005). Several studies suggest that most college students participate in some form of costuming—whether donning store-bought or homemade costumes (McDowell, 1985; Miller, Jasper, & Hill, 1991, 1993). Indeed, "dressing up" according to one's "fantasy" is very much a part of the release afforded by the holiday and consumers spend millions of dollars each year on Halloween costumes (Belk, 1994; Rogers, 2002).

Halloween allows masqueraders to step out of their everyday roles, opening up a wide range of personas for adoption, if only temporarily. Indeed, even when costumes do not disguise their actual identity, playing different roles remains a major part of the appeal of Halloween among college students (Miller et al., 1991). Significantly, adopting new roles through costume is not merely about *playing* different roles, but may also involve *constructing* and defining those roles. As McDowell (1985) suggests, costuming is about creating inhabitable representations of the "Other"—that is, "metaphors that can be carried about on the mobile human frame" (p. 1). If one adopts this definition, it becomes clear how powerful the experience of costuming across racial or ethnic lines can be in creating, resurrecting and communicating generic and negative ideas about a "racial other"—those persons of color, particularly African Americans, defined in negative contrast to white normativity.

## SIGNIFICANCE OF THE SOCIAL CONTEXT

… Significantly, the goal of Halloween humor and play is often achieved at the expense of a target, for example, an individual or group that is mocked. While a costume may represent an ultimately aggressive judgment about its target, the joking nature of this practice makes acceptable the sharing of information, which in its unadulterated form might be considered unacceptable (Freud, 1960). Because both masquerader and his or her audience identify the humor as the principal feature of the costume, they are able to circumvent any judicious assessment of the negative images of the racial other being shared. It is for precisely these reasons that humor is such an effective tool in communicating racist thoughts, particularly in the contemporary post-Civil Rights era where open, frontstage expression of such ideas is considered socially taboo (Bonilla-Silva, 2003; Dundes, 1987; Feagin, 2006; Picca & Feagin, forthcoming). Collectively, individuals' behavior in the social setting is reinforced, encouraging both the continued reproduction of racially prejudiced ideas, as well as an uncritical appraisal of them….

## RACIAL RELATIONS AND HALLOWEEN

Examining both Halloween and racial relations in the U.S. requires a historical lens that considers the unquestionable relationship between the two. Children's antique Halloween costumes remain some of the most popularly collected items from the Jim Crow era (Pilgrim, 2001). For example, Sears' 1912 "Negro make-up outfit" allowed children to "play at being a 'Negro'" and was described as "the funniest and most laughable outfit ever sold" (Wilkinson, 1974, p. 105). To be sure, today's popular caricatures extend a long history of blackface minstrelsy and racist iconography, reconstructing deep-seated ideas of white superiority against the clear contrast of black inferiority (Feagin, 2000, 2006). This history is replete with numerous empirical examples to support Stone's

(1962) general assertion that acting out the role of the other allows individuals to develop and enhance conceptions regarding their own attitudes and roles as differentiated from the adopted role. Although occasionally less explicit, today it would appear that cross-racial costuming often serves the same purpose in accenting the goodness of whiteness through the relational devaluation of the racial other as did the minstrel shows of old. Caricatured imitations of people of color, as in costuming, are written off as harmless joking, but the method of parody seems to be nothing less than an updated version of the same old show....

## PARTICIPANT OBSERVATION JOURNALS

The current study uses data collected from two samples of college students, who contributed a total of 663 individually-written participant observation journals on racial events.... This original sample was collected between the spring of 2002 to the summer of 2003, with the majority of journals being written during the fall semester of 2003. As a result, Halloween costumes and experiences emerged as a recurring theme among the racial events recorded.

Stimulated by Halloween as a potential racial event, a supplementary sample was gathered during the fall 2004 semester among a demographically similar group of undergraduate students. This theoretical sample—a non-representative sample collected with the primary goal of capturing themes and developing an analytic framework (Glaser & Strauss, 1967)—was collected from students at a single, large Southeastern state university. Students in this sample were asked to specifically address Halloween as an event of sociological interest with regard to race/ethnicity, gender, sexuality, social class, and age. Like the first sample, these students received both oral and written instructions on how to do participant observation (while maintaining anonymity of those they observed) while writing their reactions and perceptions in their journals the two weeks before and after Halloween. Specifically, students were instructed to record the "what," "where," "when," and "who" of their observations—while indicating the age, gender, and race/ethnicity of the people around them. Importantly, all students were assured that "even 'no data' is data" in sociology to encourage writing about even the most mundane, everyday events around this time. For our analysis, we included as data both these observations and students' own reflections....

## PARTICIPANTS AND LOCATIONS

The original sample included the journals of 626 white students, 68% of whom were white women and 32% white men.... The majority of students in this sample were between the ages of 18–25, although there were many students in their late 20s and 30s, and a small minority of students were in their 40s and 50s. Despite aggressive efforts to collect journals from a geographically diverse sample of U.S. college students, the majority of these 626 journals came from five colleges and universities in the Southeast (63% of the sample). Additionally, 19% of the students were at schools in the Midwest; 14% were in the West; and 4% were in the Northeast. Students came from both small and large schools; private and public schools; and in both rural and urban settings in 12 states.

The supplementary "Halloween-only" theoretical sampling yielded 37 additional journals collected and approved for use by participating students. Seventy percent of this sample was women, 30% men. As in the first sample, the majority of students were between the ages of 18 and 25, with only one student in her late 20s. By race, nearly half of the participants self-described as white American/ Caucasian, nearly 19% Hispanic/Latino/a, close to 11% African American/black American, nearly 13% Asian American/Pacific Islander, and just over 8% multiracial. Numerically, 19 students in our supplementary sample were people of color, and thus only 19 of our 663 total participants (3%). Unlike the original geographically inclusive sample, this smaller, theoretical sample included students from just one of the Southeast universities included in the original sample. Given the collection methods described, most of the students in our sample attended schools in the Southeast....

## HALLOWEEN LICENSE: SETTING THE STAGE FOR ENGAGEMENT WITH RACIAL CONCEPTS

One of our primary arguments is that Halloween provides a uniquely constructive space for engaging race, in part because of the holiday's intuitive license, such that revelers assume the right to do, say, and be whatever they want. Indeed, college students in our sample consistently described Halloween as a holiday affording them freedom and a license to "take a break from" or even "defy" social norms. As one student observed of her friend's enjoyment of holidays like Halloween, "He calls them 'breaks from reality where he can just go wild'" (white female). In addition, for many college students, the freeing experience of Halloween costuming is intimately tied to breaking from their everyday roles as one student shared: "Halloween is a way for people to see themselves as something different and uninhibited, if only for a day. Instead of being tied by how they expect others to interact with them" (Hispanic/Latino male). Such comments suggest that being "tied" to a certain identity in everyday life creates limits and inhibitions that one feels compelled to abide by, and for which Halloween provides an appropriate release....

## STEREOTYPING AS THE PREDOMINANT GUIDE IN CROSS-RACIAL COSTUMING

While students discussed and employed cross-racial costuming in a variety of ways, our analysis of journals reveals the near universal guide of racial stereotype in directing their efforts. Student commentary suggests that capturing race, both "physically" and "behaviorally," is the core criterion for determining cross-racial costuming success and as a result, most portrayals play to stereotypical ideas about the racial other. Our analysis of this phenomenon within the journals led to an emergent typology, such that the cross-racial costuming discussed, described and engaged in by our participants tended to fall within three categories: celebrity portrayals, "role" portrayals, and generic/essentialist portrayals. In some respects, these "discrete" types capture overlapping concepts. Most critically, all three types rely on stereotype to guide their portrayals. As such, it is useful to conceptualize these categories as something of a continuum in this regard.

## CELEBRITY PORTRAYALS

In some cases, cross-racial/ethnic dressing occurred as a function of students masquerading as celebrities, television/movie personalities, and otherwise notorious individuals. For instance, one Asian American woman recorded seeing a black man dressed as the white rapper, Eminem. Another white woman found two white male friends "covered in black paint from head to toe" in preparation for their costuming as Venus and Serena Williams, describing the scene as "the funniest thing [she] had seen in a long time." Yet another student wrote of dressing with two friends as "Charlie's Angels." She, a white woman, dressed as Asian American actress Lucy Liu's character; Stacy, her African American roommate, dressed as white actress Cameron Diaz's character; and, another friend, Tina, who is white, dressed as white actress Drew Barrymore's character.

While in these instances dressing across racial lines would appear to be required solely as a function of the chosen personality, the attention devoted to properly capturing the celebrity's race was in most cases intentional and elaborate. Notably, this respondent recorded that while Tina's costume would be "easy" because she was white (like her character), Stacy would have to borrow a long-sleeved upper body leotard "to hide her skin color." Additionally, she detailed the need for further makeup, "We are going to put makeup on her [Stacy's] hands and face to try and make her look Caucasian. For my costume the girls are going to do my makeup, particularly my eyes to try and make me look Oriental." From this detailed narrative, matching one's skin color appears to be more important than wearing a similar outfit or portraying an Angel's demeanor.

Interestingly, even among students attempting to portray famous people or personalities, students suggest that the "success" of the costume is principally determined by how well the race of the individual is captured. Students frequently evaluated cross-racial costuming on this basis of how convincing masqueraders were in portraying their "new" race, to the exclusion of other evaluations regarding believability. For example, one student wrote that an Asian man who had dressed as one of the Blues Brothers happily reported to her that "at the party he went to several people thought he was white and did not recognize him" (white female). Following her own Halloween celebrating, this same student recorded the following:

> An Oriental male...was dressed as President Bush. One of the really good costumes was this black male, also in his twenties, who dressed as Osama bin Laden. He really looked like he was Middle Eastern. (white female)

Addition of the singular remark regarding what she considered a "really good" costume stands in contrast to the lack of such validation for the Asian male's impersonation of President Bush....

## "ROLE" PORTRAYALS

One student responded to a friend's use of blackface paint, saying, "[His] outfit would be perfect if he went out and stole something before we left" (white male). Costumes such as this are indicative of racial "role" portrayals, and highlight attempts to embody race through the use of demeaning stereotypical notions about people of color. Unlike celebrity portrayals however, "role" portrayals have no person-specific or "real" reference, leaving much room for white imagineering of racial others.

Mass marketing of items such as "Vato Loco," "Kung Fool," and numerous pimp, thug and American Indian-themed costumes suggest the prevalence of racial caricatures in the larger culture. Rather than purchasing ready-made costumes, however, most journals documented students employing their own creativity in fashioning stereotypical cross-racial/ethnic identities, a finding that echoes McDowell (1985). Particularly plentiful were descriptions of "gangstas," "thugs," pimps, and Mafiosos. While some might contend that such representations are not fixed to one particular race or ethnicity, in reality they are typically connected to stereotypical racial caricatures, a finding supported by the students' journals.

As such, when whites costume in "ghetto" dress (with low-slung baggy pants and thick gold chains) or as pimps (complete with gold teeth, afro-like wigs, and velvet suits) they are arguably attempting to parody stereotypical images of blacks, even if they do not make use of blackface. Many students were clear about this in their responses: "one of my white friends, Eric, wanted to be a ghetto pimp. He defined ghetto as acting or being black" (black female). Another student's journal echoes this theme in more detail:

> The theme (of the fraternity party) was 'thug holiday'...my friends and I were wondering what we were supposed to wear.... I was the first to admit that the image that popped into my mind when I thought about a 'thug' was a modern-day rapper wearing baggy jeans, big gold chains around his neck, and a football jersey.... Missy was laughing when I was describing what I thought we should all wear. She said, 'so basically we should dress like black rappers.' We were all laughing at the thought of

us, three preppy white girls, dressing as what we had just described. (white female)

According to her later journaling, this young woman, her friends, and by her account "everybody (at the party), without any exceptions" costumed as they had discussed. Although she does not report the use of blackface, the party's theme invoked images that in her own words "thoroughly involved race," and in particular made reference to blackness....

The student journals make clear that the frames for stereotypical racial images, like those conjured by the above students, are readily accessible in the social minds of individuals. One respondent wrote of a friend, Ken's, last minute decision to attend a costume party. His girlfriend "took Ken into his room and closed the door. Twenty minutes later, Ken the dead mafia godfather walked out" (Hispanic/Latina female). These two individuals were able to spontaneously create a costume based on an ethnic stereotype, suggesting that such ideas are readily present and available for use in the minds of such would-be costume creators. Students seek out and fulfill the generic requirements of these imagined images of racial and ethnic groups with relative ease, and with a disturbing level of unthinking....

## GENERIC/ESSENTIALIST PORTRAYALS

While stereotypical cross-racial costuming most often drew on caricatured images of the racial other, a number of students described costumes that represented completely generic representations, such that simply portraying "race," usually blackness, was considered costume. Such portrayals represent the most extreme employment of stereotypes guiding cross-racial costuming....

Consider two non-celebrity examples: one young woman recalled a discussion over costumes prior to Halloween, "We were all getting dressed up and one person said that they wanted to paint themselves black and wear a diaper and be a black baby" (white female). Another young woman recalled her and her boyfriend's interest in simply costuming "as a black couple" (Native American/white female). While it is not known whether these individuals actually decided to cross-racially costume, what is significant is the non-descript nature of the costumes suggested. We might imagine that individuals actually choosing to costume as the nonspecific "black person" would actively engage in some type of stereotypical behavior in assuming the role of their costume. In any case, it would appear that such generic ideas represent whites' most fundamental attempts to strip all unique identity from people of color, to reveal race as the only relevant marker of those they claim to represent in costume. It is also significant that all generic representations in our sample referred to blackness. Arguably, generic and essentialist portrayals such as these tap into the most debased of representations, invoking the historical and archetypal consideration of the racial other in the white mind—that of the inferior black (Feagin, 2000, 2006).

Collectively considered, each cross-racial costume type helps us understand how such costumes serve as vehicles for transmitting racial judgments about people of color, particularly in light of the fact that stereotype guides each to a greater or lesser degree. From the relatively "innocuous" celebrity portrayal, to the "role" portrayal, to the fundamentally degrading generic/essentialist portrayal, cross-racial costuming represents the effort to create inhabitable representations (McDowell, 1985) of the racial other and to indeed, engage costume as a metaphor for those depictions....

## CONCLUSION

... Our data refute the idea that whites do not engage in public acts of racism in contemporary society. Indeed, some whites are engaging in nothing less than blackface performances, an inarguable throwback to the ubiquitous minstrelsy of the nineteenth century. While many other whites never step over the "threshold" into blackface, their stereotypical and essentialist portrayals must be charged with a striking similarity. Truly, are not "Ghetto Thugs," "Project Chicks" and "Niggas" just not-so-distant cousins of "Step-n-Fetchit" and "Mammy"? The characters may be different, but the consequences, if not the motives, are the same.

While some, like Skal (2002), may reduce the holiday engagement of racial concepts to a matter of simple Halloween "fun," this practice must be viewed within a greater framework. Seemingly playful and innocuous cultural practices, such as cross-racial costuming, should be considered within the sociohistorical and ideological context of the society, as a reflection of dominant group values and doctrines (Wilkinson, 1974). We must put aside the "fun" of costumes, which can distract from the subtle and not-so-subtle messages conveyed about people of color, and recognize that costumes provide a format for engaging commentary on personal and social values (McDowell, 1985). Indeed, to render people into character pieces, they must already exist as characters in one's mind, and there are many social forces that drive our constructions of race and people of different racial groups toward such ends....

This research suggests that at a minimum we must take up Feagin's (2006) call, and locate ways to encourage a deeper critical assessment of historical and contemporary racial oppression, acknowledging both the material and ideological consequences of this structure. Interestingly, the unconventional qualitative methodology of journal-collection not only reveals this need by demonstrating the transformed persistence of racism in the post-Civil Rights era, but also serves as a unique pedagogical and consciousness-raising tool as students reflect on experiences where they might have normally remained non-reflexive.

# REFERENCES

Bonilla-Silva, E. (2003). *Racism without racists: Color-blind racism and the persistence of racial inequality in the United States.* Lanham, MD: Rowman and Littlefield.

Dundes, A. (1987). *Cracking jokes: Studies of sick humor cycles and stereotypes.* Berkeley, CA: Ten Speed Press.

Durkheim, E. ([1912] 1995). *The elementary forms of religious life.* Translated by K. E. Fields. Reprint, New York: Free Press.

Etzioni, A. (2000). Toward a theory of public ritual. *Sociological Theory, 18,* 44–59.

Feagin, J. (2000). *Racist America: Roots, current realities and future reparations.* New York: Routledge.

Feagin, J. R. (2006). *Systemic racism: A theory of oppression.* New York: Routledge.

Gluckman, M. (1963). *Order and rebellion in tribal Africa: Collected essays with an autobiographical introduction.* London: Cohen and West.

Goffman, E. (1959). *The presentation of self in everyday life.* New York: Anchor Books.

Hua, V. (2002). Bucktoothed Halloween mask bites the dust. *The San Francisco Chronicle,* October 17. Retrieved June 1, 2005 (http://sfgate.com/cgi-bin/article.cgi?file=/chronicle/archive/2002/10/17/BU168521.DTL).

The Macerich Company. (2005, September). *Shopping in America Halloween 2005: Shopper survey analysis.* Santa Monica, CA: August Partners, Inc.

McDowell, J. (1985). Halloween costuming among young adults in Bloomington, Indiana: A local exotic. *Indiana Folklore and Oral History, 14,* 1–18.

Miller, K. A., Jasper, C. R., & Hill, D. R. (1991). Costume and the perception of identity and role. *Perceptual and Motor Skills, 72,* 807–813.

Nelson, A. (2000). The pink dragon is female: Halloween costumes and gender markers. *Psychology of Women Quarterly, 24,* 137–144.

Picca, L. H., & Feagin, J. R. (forthcoming, expected April 2007). *Two-faced racism: Whites in the backstage and frontstage.* New York: Routledge.

Pilgrim, D. (2001). New racist forms: Jim Crow in the 21st century. Retrieved August 28, 2004 (http://www.ferris.edu/jimcrow/newforms/).

Rogers, N. (2002). *Halloween: From pagan ritual to party night.* New York: Oxford University Press.

Santino, P. (1994). *Halloween and other festivals of death and life.* Knoxville, TN: University of Tennessee Press.

Simpson, D. (2003). "White La. judge draws fire for costume." *The Associated Press,* November 10, Dateline: New Orleans, Domestic News. Retrieved June 5, 2005. Available: LEXIS-NEXIS Academic Universe, News Wires.

Skal, D. J. (2002). *Death makes a holiday: A cultural history of Halloween.* New York: Bloomsbury.

Stone, G. P. (1962). Appearance and the self. In A. M. Rose (Ed.), *Human behavior and social processes: An interactionist approach* (pp. 86–118). Boston, MA: Houghton Mifflin Company.

Unmasking hate at Halloween. (2002). *Tolerance.org,* October 24. Retrieved June 10, 2005 (http://www.tolerance.org/news/article_print.jsp?id=629).

Wilkinson, D. (1974). Racial socialization through children's toys: A sociohistorical examination. *Journal of Black Studies, 5,* 96–109.

Yinger, J. M. (1977). Presidential address: Countercultures and social change. *American Sociological Review, 42,* 833–853.

## DISCUSSION QUESTIONS

1. In what particular ways do the authors find that Halloween costumes reproduce racial stereotypes? What examples have you seen of this?

2. Are there other common rituals, like Halloween, where racial or other stereotypes frame how people participate in this event? What are they and how do they influence everyday understandings of different social groups?

3. Think about the last time you attended a Halloween party (or, if it is near Halloween, visit a Halloween costume store). Do you see evidence of racism in the costumes? How would you interpret this in light of the article by Mueller, Dirks, and Picca?

# Student Exercise

1. Identify a particular form of media that interests you—film, television, magazines, or books, for example—and design a research plan that will examine some aspect of the images you find of a racial-ethnic group. Narrow your topic so it won't be overly general. For example, if you choose films, pick only those nominated for the Academy Award for Best Film in a given year, or if you choose television, look only at prime-time situation comedies. Alternatively, you could examine images of women of color in top fashion magazines or watch Saturday morning children's cartoons to see how people of color are portrayed. Once you have narrowed your topic, design a systematic way to catalog your observations, such as counting the number of times people of color are represented in the medium you select, listing the type of characters portrayed by Asian men, or comparing the portrayal of White men and men of color in women's fashion magazines. What do your observations tell you about the representation of race in the form you chose? If you were to design your project to study such images as seen now and in the past, what might you expect to find? What impact do you think the images you found have on the beliefs of different racial-ethnic groups?

# Section IV

# Who Are You? Race and Identity
### Elizabeth Higginbotham
### and Margaret L. Andersen

This book shows how race and ethnicity are part of social structure. As such, you might think they are "out there"—but they are also in us and in our relationships with other people and groups. As Peter Berger (1963) once pointed out, people live in society but society also lives in people. Similarly, in the United States people live within a system of race and ethnic relations, but race and ethnic relations also live within us. How society has organized race and ethnicity is reflected in our identities and in our relationships with others. And, as society becomes more racially and ethnically diverse, so do people's identities become more diverse, and the possibilities for multiracial identities and relationships increase.

**Identity** means the self-definition of a person or group, but it is not free-floating. Identity is anchored in a social context: We define ourselves in relationship to the social structures that surround us. Moreover, identities are multidimensional and thus include many of the social spaces we occupy. At any given time, some identities may be more salient than others—age as you grow older, sexuality if you are questioning your sexual orientation, race as you confront the realities of a racially stratified society, and so forth.

Racial identity is learned early in life, although those in the dominant group may take it for granted. People of color likely learn explicit lessons about racial identity early on, as parents prepare them for living in a society in which their racial status will make them vulnerable to harm. As Beverly Tatum (1997) shows, forming a positive racial identity—for both dominant and subordinate groups—means having to grapple with the realities of race. For people of color, this can mean surrounding themselves with others of their group, even though they may then be blamed for "self-segregating" by Whites who do not understand or appreciate the support this affiliation can provide.

The formation of racial identity is especially complex when multiple races are involved. The children of biracial couples may define themselves as being of two or more races. Thus, a child born to a White parent and a Black parent may identify as Black, but appear White to others, and then identify as biracial or multiracial. As our society becomes more diverse, multiracial identities are becoming increasingly common. Racial and ethnic identities can also be complex because we have so many immigrants from nations in which race and ethnicity may be constructed differently than in the United States. Such complex identities hold out the possibility that the rigid thinking about race that has prevailed for so long might break down.

The articles in this section each explore different dimensions of racial identity. We open with the views of psychologist Beverly Tatum ("Why Are the Black Kids Sitting Together?," interviewed by John O'Neil), whose work is well known for explaining the formation of racial identity. Tatum explains how racial identity emerges within the context of racial inequality. She challenges us to think about what it means when people of color, especially young people, choose to be among people like themselves. Although people of color get accused of "self-segregating" (even while White people who do the same thing are rarely so accused), Tatum shows the affirmation that such behavior provides.

Priscilla Chan's essay ("Drawing the Boundaries") illustrates how people of color have to negotiate the boundaries that racist societies produce. As an Asian American student, she tracks the development of a positive identity, achieved within a context where her racial–ethnic identity crossed with her class identity, especially as she navigated her way through elite, predominantly White institutions.

Michael Omi and Taeku Lee ("Barack Like Me: Our First Asian American President") take the issue of racial identity in perhaps a surprising direction, suggesting that President Barack Obama is the first Asian American president. "What!" you might say, "I thought he was African American." Omi and Lee's discussion of President Obama's multiracial, multi-ethnic identity shows how complicated the matter of racial identity can be. Like an increasing number of Americans, Obama's identity is anchored in multiple racial–ethnic backgrounds, illustrating the increasingly mosaic–like character of the U.S. population.

It is important to remember, though, that racial identity is not just acquired by people of color. White people are generally not considered to have a racial identity, because they are not socially "marked" as are people of color. Because they are the dominant group, their identity has been considered transparent, taken for granted, and not marked as are the identities of racial and ethnic minorities. White people actually do have a racial identity, but it is often not salient until they encounter experiences wherein that identity is brought to light. Thus, White college students may confront their own racial experiences for the first time when they interact with students of color on campus. Tim Wise ("White Like Me: Reflections on Race from a Privileged Son") shows that being

"White" involves certain common experiences, even when the particular experiences of any given White person will vary because of other social factors, such as class, age, gender, and so forth. Recognizing "Whiteness" as race is not meant to essentialize race—that is, make it a fixed, unchanging, somehow "natural" phenomenon. Quite the contrary, it shows how racial identity is fluid and changing in society, but emerges in the context of racial inequality.

## REFERENCES

Berger, Peter L. 1963. *Invitation to Sociology: A Humanist Perspective*. New York: Doubleday-Anchor.

Tatum, Beverly Daniel. 1997. *Why Are All the Black Kids Sitting Together in the Cafeteria? And Other Conversations about Race*. New York: Basic.

## FACE FACTS: PERCENT CHANGE IN U.S. POPULATION 2007–2008

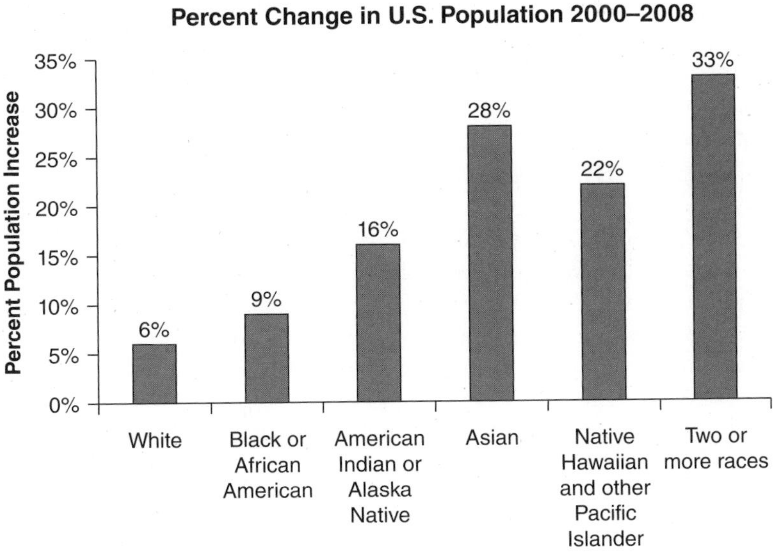

### Percent Change in U.S. Population 2000–2008

**FIGURE 1**   Changes in Racial Identity

SOURCE: U.S. Census Bureau. 2010. *Statistical Abstract of the United States*. Washington, DC: U.S. Department of Commerce. www.census.gov

***Think about it:*** This graph details the change in how people define their racial identity in the U.S. census, comparing from 2000–2008. Although a relatively small number report themselves as of two or more races (about 4 percent of the total U.S. population), change is clearly occurring. What changes do you think affect and accompany the trends you see here?

# 16

# Why Are the Black Kids Sitting Together?

## A Conversation with Beverly Daniel Tatum

BEVERLY TATUM AND JOHN O'NEIL

*Beverly Tatum, well-known educational leader, tackles the often-asked question here of why Black students (and, by implication, other students of color) tend to stick together when they are in predominantly White settings, such as a high school or college campus. Rather than blaming the students, as many do, for supposed "self-segregating," she interprets this behavior as a matter of social support and racial identity formation. Her work has been very influential in antiracism education.*

Y*ou call your recent book,* Why Are All the Black Kids Sitting Together in the Cafeteria? *What's the significance of the title?*

The question is one I'm asked over and over again when I do a workshop on racial identity at a racially mixed school. Educators notice that kids often group themselves with others of the same race, for reasons I'll explain later. But people have other concerns as well. How do we talk to young children about race? How do we address these issues with our colleagues? How do you even engage in conversations about such hot topics as affirmative action without alienating one another?

*Why do we have such problems discussing racial issues? Is it because we don't really understand one another's experiences?*

I think that's part of it. It's interesting to watch people's reactions when they are really forced to experience being in the minority. One of the exercises that I ask white students and educators to engage in is to create a situation in which they will be in the minority, for a short period of time. A common choice, for example, is to attend a black church on a Sunday morning. Another is to go to a place where you know there's going to be a large Spanish-speaking population.

Usually, whites are very nervous and anxious about doing this. Some are even unwilling to do it alone, so they find a partner to go with, which is fine. But it's just interesting to me how fearful people are about this kind of

SOURCE: *Educational Leadership* 55 (December 1997): 12–15.

experience. When they come back, they often report how welcomed they felt, what a positive experience it was. But I do point out to them how worried they were about their own discomfort. And I hope that they develop a greater sense of understanding of how a person of color might feel in an environment that is predominantly white.

*Some people suggest that race relations among kids are much improved, compared to our generation or our parents'. What do you see?*

Young children do interact across racial lines fairly comfortably at the elementary school grade level. If you visit racially mixed schools at the elementary level, you will see kids interacting in the lunch room and on the playground. To the extent that neighborhoods are segregated, their interracial friendships might be limited. But you see much more cross-racial interaction at the elementary level than you do at the junior high or high school level.

*Why?*

I think the answer has to do with the child's transition into adolescence. Adolescents are searching for identity; they're asking questions like: Who am I in the world? How does the world see me? How do I see others? What will I be in the future? and Who will love me? All those questions of identity are percolating during that time period. And particularly for adolescents of color, these questions cannot be answered without also asking: Who am I ethnically? Who am I racially? and What meaning does this have for how people view me and interact with me?

*Can you give an example of how these issues might emerge in students of color?*

Sure. Imagine a 7-year-old black boy who everybody thinks is cute. So he's used to the world responding to him in a certain way: Look at that cute kid!

Now imagine that same kid at 15. He's six feet tall, and people don't think he's cute anymore. They think he's dangerous or a potential criminal; maybe people are now crossing the street to avoid him. So the way he sees himself perceived by others is very different at 15 than at seven. And that 15-year-old has to start figuring out what this means. Why are people crossing the street when I walk down it? Why am I being followed around by the security guards at the mall? And as that young person is trying to make sense of his experience, in all likelihood he is going to seek out and try to connect with other people who are having similar experiences.

As a result, even the young person who has grown up in a multiracial community and had a racially mixed group of friends tends to start to pull away from his non-black friends, his white friends in particular. This happens, in part, because their white friends are not having the same experiences; they're not having the same encounters with racism. And, unfortunately, many white youth don't have an understanding of how racism operates in our society, so they're not able to respond in ways that would be helpful.

The example I used was of a black boy, but a similar process unfolds among black girls or children of another race or ethnic heritage.

*What about white students? What are they experiencing during this time of self-identification?*

They can be confused and hurt by some of the changes. For example, it's not uncommon for a white student in my college class on the psychology of racism to say: "You know, I had a really close black or Latino friend in elementary school, and when we went to junior high she didn't want to hang around with me anymore." The student reporting the story usually is quite confused about that; it's often a very hurtful experience.

The observation I make is that, again, many white students are oblivious to the power of racism and the way that it's operating in society. And so when their friends are starting to have encounters with racism, they don't necessarily know how to respond. An example from my book is when a teacher makes a racist remark to a young black female. Afterward, her white friend comes up to her and says, "You seem upset, what's the problem?" So she explains what upset her to her friend, and the friend says, quite innocently, "Gee, Mr. Smith is such a nice guy, I can't imagine he would intentionally hurt you. He's not a racist, you know."

*So white students might discount it because they can't identify with it?*

Exactly. They can't identify, and also in many situations people who try to comfort often end up invalidating the person's feelings, by saying things like, "Oh, come on, it wasn't that bad." What happens is that you withdraw from the conversation. The feeling is "Well, you don't get it, so I'll find somebody who does." It would have been a very different response, however, if this young white student understood stereotypes and the reality of racism in society and told her friend, "You know, that was a really offensive thing he said."

*Should teachers or principals be concerned when students self segregate? Should they actively seek to integrate the groups?*

During "downtime" like lunch or recess, students should be able to relax with their friends, regardless of whether or not those friendship groups are of the same race or ethnic groups. However, it is important to create opportunities for young people to have positive interactions across group lines in school. So structuring racially mixed work groups—for example, by using cooperative learning strategies in the classroom—can be a very positive thing to do.

Similarly, intentionally working to recruit diverse members of the student body to participate in extracurricular events is worthwhile. We need to take advantage of every opportunity we have to bring young people together where they can work cooperatively as equals toward a common goal. Sports teams are a good example of the kind of mutually cooperative environment where young people often develop strong connections across racial lines, and we should look to create more such opportunities in schools. Unfortunately, school policies like tracking (which tends to sort kids along racial lines) impede rather than facilitate such opportunities.

*You've talked with students of color who attend integrated schools but find themselves isolated in honors or advanced classes. What are those students experiencing in terms of their identity?*

Even in racially mixed schools, it is very common for young people of color, particularly black and Latino students, to find that the upper-level courses have very few students of color in them. And, of course, honors chemistry is only offered during a certain time period in the day, which means you might also be taking English and other courses with the same kids. And so those students in honors chemistry or advanced algebra may find that their black or Latino peers accuse them of "trying to be white" because they're hanging around with all white kids. So to the extent that you're frequently in the company of white students, and your black classmates who are in the lower tracks see you as somehow separating yourself from them—it's a hard place to be.

*What can educators do to support the healthy development of kids as they work through these issues?*

I think students of color really need to see themselves reflected in positive ways in the curriculum. And that probably sounds very obvious, but the fact is that too often they don't see themselves reflected in the curriculum.

When and how they see themselves reflected in the curriculum is so important, though. To use African Americans as an example: Most schools teach about slavery, and for many black students that's a point of real discomfort. Their experience of that is that the teacher's talking about slavery and all the white kids in the class are looking at us, to see what our reaction is. I'm certainly not suggesting we shouldn't teach about slavery, but I think it's important to teach it in an empowering way. Teachers need to focus on resistance to victimization. Students of color need to see themselves represented not just as victims but as agents of their own empowerment. And there are lots of ways to do that. You can talk about Sojourner Truth, you can talk about Harriet Tubman, Frederick Douglass, and so on.

At the same time, I think white children need to be helped to understand how racism operates. Inevitably, when you talk about racism in a predominantly white society, you generate feelings of discomfort and often guilt among white people because they might feel that you're saying that white people are bad. What do we do? In these discussions, we need to include examples of white people as agents of change. Teach students about the abolitionists. Teach students about Virginia Foster Durr, who was so active during the Civil Rights movement. All children need to learn about those white folks who worked against oppression. Unfortunately, many white students don't have that information.

*Some people have suggested that the school curriculum be heavily focused on cultural heritage; that black students need an Afrocentric curriculum, and so on. What's your perspective?*

It's important to have as diverse a curriculum as possible because all students need to be able to view things through multiple perspectives.

A high school teacher told me recently that the young white men in her English classes were reluctant to read about somebody's experience other than their own. For example, she had the class read *House on Mango Street* (Cisneros 1994), a book about a young Chicana adolescent coming of age in Chicago. These young men were complaining: "What does this have to do with me? I can't identify with this experience." But, at the same time, they never wonder why the Latino students in the class have to read Ernest Hemingway. We need to help them develop that understanding. All of us need to develop a sense of multiple perspectives, regardless of the composition of our classrooms.

*The teaching ranks are predominantly white, even though the student population is becoming increasingly diverse. What does this mean for efforts to increase racial understanding?*

It makes it harder, but it's not impossible. We should be working very hard to increase the diversity of the teaching pool, and many teacher education programs are trying to do that. Still, we need to recognize that it's going to be a long time before the teaching population reflects the classroom population. So it's really important for white teachers to recognize that it is possible for them to become culturally sensitive and to be proactive in an antiracist way. Many white educators have grown up in predominantly white communities, attended mostly white schools, and may have had limited experiences with people of color, and that is a potential barrier. But what that means is that people need to expand their experiences.

*Dreamkeepers: Successful Teachers of African-American Children,* by Gloria Ladson–Billings (1994) is a great resource. I often encourage educators I work with to read that because she profiles several teachers, some of them white. Those are teachers who probably didn't grow up in neighborhoods or communities where they had a lot of interaction with people of color, but, one way or another, they have really been able to establish great teaching relationships with kids of color. So it certainly can be done.

*You train teachers to work on issues involving race in their schools. What kinds of things do they learn?*

For a number of years, I've taught a professional development course called *Effective Anti-Racist Classroom Practices for All Students.* It's basically designed to help teachers recognize what racism is, how it operates in schools, and what the impact of that is personally and professionally. So the focus is not just the impact of racism on the racial identity development of the students but also on the teachers. I've found that teachers who have not reflected on their own racial identity development find it very difficult to understand why young people are reflecting on theirs. So it's important to engage in self-reflection even as we're trying to better understand our students.

The course also looks at stereotypes, omissions, distortions, how those have been communicated in our culture and in our classroom, and then, what that means in terms of how we think about ourselves either as people of color or as whites. And, finally, what we can do about it. I talk in my book about racism as a sort of "smog." People who aren't aware of it can unwittingly perpetuate a cycle of oppression. If you breathe that smog too long, you internalize these messages. We can't really interrupt that process until it becomes visible to us. That's the first step—making the process visible. And once it is visible, we can start to strategize about how we're going to interrupt it.

*Many teachers have been caught short by a racist incident or comment in their classroom. It often happens suddenly, and the teacher may be at a loss for how to respond. How have you handled it?*

Well, I've been teaching a course on the psychology of racism since 1980, so I feel like I've probably heard it all.

It is a difficult situation, because you want the classroom to be a safe place, where students can say anything, knowing that only by opening up will they get feedback about their comments and learn another perspective. At the same time, you want the classroom to be a safe place for someone who may be victimized by a comment.

One time, a white student in my class made a very offensive remark about Puerto Ricans being responsible for crimes. Well, one of the things I've learned is that it really helps to validate somebody's comment initially, even if it is outrageous. So I said something like: "You know, I'm sure there are many people who feel that way, and if you've been victimized by a crime, that's a very difficult experience. At the same time, I think it's important to say that not all Puerto Ricans are car thieves." From there, you can move into how making such statements can reinforce stereotypes.

*It must be hard to make it a teachable moment for everyone in the class.*

Absolutely. One time I was observing somebody else's teaching when there was a similar kind of incident—a student made comments that the teacher thought were inappropriate, but she didn't know how to respond. So she didn't respond to them. After the class, we talked, and she said she felt terrible—she knew she should have done something, but she didn't know what to do. And we talked about what the choices might have been.

Even though she felt badly about how she handled it, those moments can be revisited. So in this particular case, the teacher opened her next class by saying: "You know, in our last class something happened that really bothered me, and I didn't say anything. I didn't say anything because I wasn't sure what to say, but in my silence I colluded with what was being said. So I would really like to talk about it now." And she brought the class back to the incident, and it was not an easy conversation. But I think it really deepened the students' understanding—both of the teacher herself and of how racism operates, because it showed how even well-intentioned people may unwittingly contribute to perpetuating the problem.

*Although integrated schools have been a goal for decades, current statistics show a growth in schools that are nearly all-black or all-Hispanic. What do you see as the likely impact of the trend toward even more racially identifiable schools?*

It's a very difficult issue from a number of perspectives. The continuing pattern of white flight is one of the main reasons that schools resegregate. A lot of money is put into a busing plan, and then white families leave the school. So now many people are asking questions about whether it's a good idea to spend all that money transporting kids instead of just using it to improve the neighborhood school, regardless of who attends it.

Many parents of color experience this as a double-edged sword. They're offended by the notion that children of color can only learn when they are in classes with white kids. They know there is nothing magical about sitting next to a white kid in class. On the other hand, the reality of school funding is that schools with more white students receive more financial support.

And so the question that I hear people asking now is: Can separate ever be equal? That's one I don't have the answer to!

… Real progress is being made in starting conversations at the local community level. Many grass-roots organizations are encouraging this kind of dialogue.

For example, an organization in Connecticut has a program of Study Circles.[1] They actually have a guidebook for facilitating conversations about race. Using the guides, people come together and begin to discuss the questions together to improve their understanding. Also, many houses of worship encourage cross-group dialogue, whether it means interfaith dialogue or cross-racial dialogue.

It's sometimes frustrating for people who have been doing this work for years, because it may seem like there's talk, talk, talk and it doesn't go anywhere. However, I do think that when you engage in open and honest dialogue, you start to recognize the other person's point of view, and that helps you see where your action might be needed most. So if people engage in dialogue with the understanding that dialogue is supposed to lead somewhere, it can be a very useful thing to do.

We can't afford to forget the institutional nature of racism. And so it's not just about personal prejudices, though obviously we want to examine those. We can't just aspire to be prejudice-free. We need to examine how racism persists in our institutions so we don't perpetuate it.

## NOTE

1. For more information, contact Study Circles Resource Center, P. O. Box 203, 697 Pomfret St., Pomfret. CT 06258: tel: 203-928-2616, fax: 20392N 3713.

## REFERENCES

Cisneros, S. (1994). *House on Mango Street*. New York: Random House.

Ladson-Billings, G. (1994). *Dreamkeepers: Successful Teachers of African-American Children*. San Francisco, Calif.: Jossey-Bass.

## DISCUSSION QUESTIONS

1. How does Tatum's analysis link the everyday behavior of minority students to the social structures in which they live?
2. What lessons are there in Tatum's discussion for reducing racial prejudice?
3. What are the implications of Tatum's argument for educating teachers about race?

# 17

# Drawing the Boundaries

PRISCILLA CHAN

*Chan's narrative about growing up Chinese American shows how feelings of exclusion develop from racial and ethnic inequalities. But she also shows the strength that evolves in her self-image as her life unfolds.*

## ...FRESHMAN YEAR AND THE CRISIS OF IDENTITY

I grew up in Chinatown, New York, in a tiny apartment roughly the same size as my freshman-year room. Everywhere I looked as a child, I was reminded of the monumental stereotypes of and pressure from my neighborhood and culture: poor, passive, non-English-speaking, reclusive, foreign. A significant portion of my adolescence revolved around the discomforting and awkward acceptance that I was Asian, from Chinatown, and that I would always be seen as such.

So I came to college with more baggage—and I do not mean just my suitcases—than a normal eighteen-year-old should hopefully have. Seeing my entryway-mates unpack their state-of-the-art computers, I quietly unpacked my own model, a rickety PC still running Windows 3.1. It was the first and only computer my family had ever purchased.

A friend, forgetting his tact, scoffed at my hardware. I came back with a line about how I appreciated its antique charm, but I was deeply embarrassed. What I really wanted to explain was how I was lucky to have a computer at all. But somehow, surrounded by hardware that put mine to shame, I just could not find the words.

Yet how did I even get there? From the Asian ghetto to the Ivy League—it is the road less taken. The journey began when I was given the opportunity to transfer into a more prestigious (and more white) middle school. This was the first time I had regularly left Chinatown, and saying that I had culture shock would be an understatement. I could not believe that some people who lived only minutes away from me by train could live such remarkably different lives.

Some kids took private charter buses to school; my family considered my free subway pass a godsend. Some kids blew ten dollars every day on lunch

SOURCE: Arar Han and John Hsu, eds. 2004. *Asian American X: An Intersection of 21st Century Asian American Voices.* Ann Arbor, MI: University of Michigan Press.

outside school grounds; most days, I made sure I had my tickets for school lunch. These were very hard times for me. My Chinatown had seemed so comfortable just a few weeks ago, yet traveling a little outside its boundaries, I quickly understood that being Asian in Chinatown meant climbing over more hurdles to even get to the same starting point as my new friends. No one likes to be told that their lives are not entirely in their control; imagine what it's like to realize this when you're starting middle school.

In this world, my issues with race blended with my issues with economics. I could not differentiate the discomfort that came from being Asian from that which came from being poor. The few other Asian students in the school did not go back to their own versions of Chinatown, so relating to them had its own difficulties. Awkwardness came therefore from the realization that not only was I in a subcategory within the school by being Asian but, by being from Chinatown, I was in a subset of a subcategory.

I often found myself remaining quiet when everyone spoke about their new CDs, or shows they watched on cable, or the latest movies, because I wasn't lucky enough to see or hear them. A naive belief that money would make me fit in characterized many of my middle and high school experiences. I hoped that upon entering college, this belief would take a backseat. I was so sure that here, in this wonderful world of higher education, intellectualism would always surpass materialism.

Unfortunately, I found myself only a few months after high school in another new world, far removed from Chinatown, at another moment of silence. I had become a person who allowed others to talk down to her—an Asian American who feared the outside, a less well-off individual who was ashamed of her means. Though disappointed, I did not come down too harshly on myself, thankfully. Freshman. So much to learn …

## SOPHOMORE YEAR AND PREPARING
## FOR THE ENEMY...

As I eased past the discomfort of freshman year on my journey to Asian America, I realized that helplessness was a pit stop where I did not want to rest long. Slowly, I discovered that the control that I had started losing in middle school was not lost forever. I felt confident that once I grasped why I had become *powerless* in the first place, I would be halfway to becoming *powerful*.

The path to regaining control was not easy; to understand the "enemy" was a skill. To hone my skills, like a boxer preparing for a championship bout, I mapped out some workout tips. I pictured the enemy as the all-encompassing "Ignorance"—another one of those words that would have incredible influence on my life.

**REGIMEN TIP ONE:** Ignorance comes in many forms, so be ready for any opponent. It never fails to irritate me when Asian "elements" are thrown upon an object for the sole reason of fashion. In a magazine, I once saw an outfit adorned with images of Mao Zedong. I immediately wondered if it came with matching Hitler and Castro accessories. Surprisingly, I didn't see them anywhere. In another magazine, a stylist described an exquisitely dressed Michelle Yeoh like so: "Forget Taiwan. How about Tai-wow!" First, Michelle Yeoh is from Malaysia. Second, imagine the parallel: A Mexican artist walks down the red carpet to similar commentary: "Forget Puerto Rico. How about Pretty Rico!" The comment's silliness was glaring.

Once, a friend of mine asked me to translate her Chinese-character tattoo. The cynic in me wanted to make up my own translation, but the human being in me conjured up my knowledge of characters. It wasn't too hard to spout off a rough translation, since there are only a few characters that have made it into the "Foreigners Guide to Things Asian." This handful will appear on various T-shirts, bracelets, slippers, and other trendy items (apparently including tattoos) and will say "love," "friendship," "success," or "strength" … Exoticism has hit an all-time high.

The objectification of a culture's traditions as marketable products allows someone to selectively choose what they see when viewing Asian America, a choice that is unacceptable to me. It offends me that modern-day "riches of the Orient" are embraced as a suitable substitute for a real exchange of culture, that racism is masked by dollars in the eyes of the Ignorant. It is as if the wearing of good-luck bracelets with Chinese characters means that the very arms wearing them are outstretched in a warm welcome. The hypocrisy makes me cringe.

**REGIMEN TIP TWO:** Know your weaknesses, and use them to your advantage, Recently, I learned the startling fact that Asian American youth have the highest rate of suicide in the country. I used to believe that as an Asian American, I knew better than anyone the issues affecting us as a group. Yet I soon realized that I knew little of the issues affecting Indian Americans, Filipino Americans, etc., and that I had a broad but shallow perspective on Asian America.

Knowing the disadvantages in the lumpy term "Asian American," however, does little justice to the benefits it also offers. I await eagerly the day when we will have the luxury of making distinctions between the various ethnicities covered under the term, but currently, our basic needs, hopes, struggles are too similar. But if I choose to fall under the "Asian American" category, I am expected to be a representative. But then we go full circle: How much can I claim to represent? It is a hard thing to know your weaknesses while at the same time using them to your advantage.

I fear that I appear to be hypocritical. While it is nearly impossible to know all of "Asian America," I refer to it as if somewhere a comprehensive list of components exists—as if it *were* indeed possible. "We." "Us." These words sprinkled throughout the essay refer to the "first-person plural," to something more than "me," but they are only words of hope. I could easily write "I" or "me," but there is an idealized community to which I envision belonging … *my* version of "Asian America," *my* concept of "we" and "us." My weakness and advantage lie in this hope. They have to.

**REGIMEN TIP THREE:** Stay on guard; the match isn't over until the opponent is dead. In other words, be wary of small concessions. During my sophomore year, a report entitled "Reaching the Top" concluded that since

Asian Americans outperform other minority groups, they have less need for the numerous "affirmative action"-type programs offered under the College Board, like private scholarships, government aid, and job assistance. For many, the report signified the "success" of Asian Americans—the long-awaited stamp of approval that they had reached the academic levels of whites. For others, the report was rather a stamp of death, the epitome of embracing the tragic "model minority" stereotype. How better to lose than to accept your opponent's false congratulations? How better to let down your guard than to blindly believe that "we have made it"? What about our headway in other areas? Recently Asian American candidates have been turning up on local ballots in record numbers, rallying "their own" to the polls. Yet, with so many names to choose from, even optimists fear that minority votes will cancel each other out and split voting strength. Similarly, the recent campaign finance scandal involving Asian donors also signified future Asian American participation in politics. Even on the rare occasion that politicians actively court our votes, they usually hold their promises until the checks clear. Our community is indeed making progress on the American political front, but we are far from victory.

Undoubtedly, it is a major effort to actively incorporate all these tips into our lives. But there is the reward of knowing that you will hopefully be equipped to make an impact with your words and actions, and that is no small achievement. No pain, no gain.

## JUNIOR YEAR AND AVOIDING
## THE SUPERFICIAL ...

Ask anyone to name the first thing that comes to mind when they think of Chinatown, and they will probably think of something along the lines of food, restaurants, etc. There's no shame in that. The shame lies in *only* seeing those things.

Chinatown is an expert at facades. What other community could lure hundreds of visitors every week into its boundaries without ever suggesting that people actually live in those tiny rooms above the restaurants? That factories (sweatshops) actually operate in those nondescript buildings?

The enemy, therefore, cannot just be Ignorance (not knowing). It must also be Facades (that which makes seeking the truth much harder).

I am reminded of an article in which a white student questioned why many negatively viewed her decision to consider herself ethnically Chinese. I admit to being part of this wary contingent. On the one hand, I was entirely a fan of allowing all people to choose their own "labels." (I would hardly be one to fixedly advocate being lumped into a category on the sole basis of your looks.) On the other hand, I was cynical about the fact that she had the luxury of *choosing* for herself what had *been chosen* for me. (I do not teach English, but I still know the difference between the active and passive tenses.) At the time, what I wanted to say to her was this: You cannot fully know what it means to be Chinese until you have walked one day in yellow skin.

But I was wrong. Perhaps the greatest progress I have made in my views on race is in considering this: If it is only possible to know what it means to be me by walking in my colored skin, then how deep is that reality? The most unique thing about what I imagine to be the Asian American community is its diversity. For every rich Asian American, there is a poor Asian American (if not more). For every Republican, there is a Democrat. For every college graduate, there is a high school dropout. For every yellow-skinned individual, there is a multitude of other physical appearances.

However, I am not just saying that whites can justifiably call themselves Asian American. I am only acknowledging that I have no more of a claim to the definition of Asian American than anyone else, even whites. There are some non-Asians who can speak Asian languages or who have studied Asian cultures better and more than I ever have. In a world filled with "half Asians," "quarter Asians," "eighth Asians," where is the line drawn for membership in this community I envision? Are whites any more outside the boundaries of Asian America than I am?

I dare not even wager an attempt at defining this community for anyone other than myself.

I was born Chinese, but I have been Asian American for most of my life. The first statement I did not choose; the second I eventually and proudly did. Yet, by speaking it aloud, I chose to enter a community (however imaginary and illusive) and to embrace its inevitable diversity.

## SENIOR YEAR AND LOOKING AHEAD ...

My thoughts return again to the instance when I felt momentary restitution in my student's encounter with reverse racism during my student-teaching. I thought I had encouraged my classes to dig deeper into their studies, yet I discovered that we all sorely needed more practice in looking beyond the superficial. For most of my life, "race" has simply denoted a group with the same geographic history and similar physical characteristics. Yet race cannot just scrape the surface. Whatever it means, it must be accompanied by a way of thinking, a sense of responsibility, an awareness of community.

I devoted most of my senior year to earning my teaching certification. One of the most important lessons I learned in my education classes involved a simple fraction. Ask any child to draw you a visual representation of "one-half," and more likely than not, she will draw you a circle with half of it shaded in. This image of a semicircle has been ingrained in so many of us as the archetypal symbol for "one-half." But I could easily draw any other polygon with half of the area shaded in, a black-and-white checkerboard, or even a yin-yang, and all of these would accurately represent "one-half." The lack of choice I felt I had during my identity crisis is much like the often singular concept of "one-half." There seemed to be only one option, and that was the choice that was made for me. I was an Asian American because I looked like one. Although I was thrust into Asian America, in my development as an adult, I definitely *chose* to remain there.

We only stand to gain by opening up the definitions of "race," "Asian American," "one-half." Every day as a teacher, I have the opportunity to help kids define these terms for themselves. Ironically, as I assist them, I assist myself. I have come long way in my views on race, but even as a recent college graduate, how much do I really know? It is humbling to realize that at twenty-three years old, I know little more than I did at thirteen.

Yet such knowledge has never really been my goal. I do not seek definitions. After all, knowing the meaning of the term "Asian American," if that is even possible, would only be the first step in making that term *meaningful* to me or any other inquisitive young person of color.

I now mentor a girl who recently emigrated from China. When we talk, we do not discuss our race, our culture, or our neighborhood. Instead, she wants to know about clothes, magazines, movies, boys—"regular teenage stuff." I often wonder about her feelings about racial identity; does she even question it, and if so, does she have the same questions I did when I was growing up? She is an Asian person in America, so some would say she is Asian American. But she is a non-English-speaking immigrant, so others would say she is not. Yet the distinction is often comical in the face of more pressing concerns.

To my mentee, knowing the definition of "Asian American" is a useless tool. This definition will not help her learn English; it will not get her parents better jobs; and it will no help her graduate from high school. She is Chinese, that she knows. One day, perhaps, she will choose to be Asian American. That is not a choice that I can or will make for her by defining an "Asian America" in which she is a citizen before she has chosen to be.

Definitions are often as advantageous as they are limiting. Sometimes they are desperately sought; other times, they are barely given a second thought. They will push, pull, and stop you all at once. In the end, do not just define "Asian American." *Define yourself* and you will have succeeded.

## DISCUSSION QUESTIONS

1.  Why does Chan say she experienced culture shock when she arrived on campus? How is this transition influenced by both her identity as Asian American and her social class background? Have you ever had similar experiences and, if so, on what basis?

2.  What did Chan gain from what she calls her "Regimen Steps"? How did they help her construct a positive identity?

# 18

# Barack Like Me:

## Our First Asian American President

### MICHAEL OMI AND TAEKU LEE

*Asian Americans have a long history in the United States, but are often viewed as aliens or foreign. Omi and Lee consider President Obama's racial and social heritage and his link to Asians and Asian Americans. They suggest that his election creates new possibilities for fostering the greater incorporation of Asian Americans in national political affairs.*

In an oft-quoted 1998 column for *The New Yorker*, Toni Morrison described Bill Clinton as "our first Black President." White skin notwithstanding, Morrison observed that, "Clinton displays almost every trait of Blackness: single-parent household, born poor, working-class, saxophone-playing, McDonald's-and-junk-food-loving boy from Arkansas."[1] Ten years after she wrote that Clinton was "Blacker than any actual Black person who could ever be elected in our children's lifetime," Barack Obama became the 44th President of the United States. The fantasy of another Clinton, funkmaster George Clinton who urged folks in 1993 to "Paint the White House Black,"[2] has finally been fulfilled.

While there has been near-universal agreement and acclamation about the election of the first-ever *African American* President of the United States, there has also been a remarkable diversity of additional identity claims in the 2008 election, each taking their own Morrisonian turn. Rabbi Arnold Wolf, whose synagogue lies across the street from Obama's Chicago home, talks about Obama's "Jewish side": "Obama is from nowhere and everywhere—just like the Jews. He's Black, he's White, he's American, he's Asian, he's African—and so are we."[3] Riffing off of this observation, we like to claim Barack Obama as our first Asian American President.

Morrison's original claim regarding Clinton was that distinctive cultural markers made him "Black." Similarly, Obama has cultural markers that make him "Asian." His life experiences have been fundamentally shaped in close association with Asians and Asian Americans. He was born at the Kapi'olani Medical Center for Women & Children in Honolulu, Hawai'i—the only state that has historically had a majority Asian American and Pacific Islander population.

SOURCE: Kirwin Institute: 2010. *Obama Reflections: From Election Day to Presidency: Social Justice Leaders Speak Out*. Columbus, OH: The Ohio State University. www.kirwininstitute.org

From ages six to ten, he attended local schools in Jakarta, Indonesia. He later returned to Honolulu and attended Punahou School from the fifth grade through his graduation from high school. The cultural influences are clearly evident. On inauguration day at Blair House, Amron-Paul Yuwono greeted Obama in Indonesian, and Obama demonstrated his command of the language by responding "in perfect Indonesian."[4] While Obama's daughters may prefer macaroni and cheese, Garrison Keillor says that Obama can cook Indian and Chinese food.[5]

In addition to the cultural influences are the experiential ones grounded in family and kin. *New York Times* reporter Jodi Kantor observes that, "The family that produced Barack and Michelle Obama is Black and White and Asian, Christian, Muslim, and Jewish. They speak English; Indonesian; French; Cantonese; German; Hebrew; African languages including Swahili, Luo and Igbo; and even a few phrases of Gullah, the Creole dialect of the South Carolina Low country."[6]

Journalist Bill Wong says that Asian Americans ought to rejoice in Obama's presidency since, "In many ways, he is like us" and that "he is certainly closer to us in terms of living experience, and background than any other U.S. President has been."[7] Of note is the fact that both his father and stepfather were immigrants and both men of color. Maya Soetoro-Ng, Obama's half-sister, is half-Indonesian and her husband, Konrad Ng, is Chinese Canadian. During a fundraiser with Asian American campaign contributors, Obama himself contended, "I consider myself to be an honorary [Asian Pacific Islander] member, and I think I've got some pretty good credentials."[8]

Representing Barack Obama as Asian American based on a pastiche of cultural practices, life experiences, and familial ties reproduces only one aspect of Morrison's racial marking of Bill Clinton as African American. While Morrison has drawn the greatest attention for calling out Clinton's raggedy upbringing and love for jazz and lowbrow gastronomy, the heart of her argument is that—as a result of the constant persecution and prosecution from House Republicans for Whitewater, Monica Lewinsky, and the like—Clinton understood what it felt like to be an African American male in the United States. That is, Clinton is Black for Morrison not just because he might have the experiential bona fides to pass himself off as an honorary; he is also Black for Morrison as an *ascribed* identity vis-a-vis the harsh and unjust treatment by others.

In this respect, we are especially struck by how distinctly *Asian American* Barack Obama's racial ascription appeared to be in the 2008 election. By distinctly Asian American we refer here to political scientist Claire Jean Kim's concept of "racial triangulation" to capture how Asian Americans, both historically and in the contemporary period, are located in a broader field of racial positions.[9]

Kim argues that Asian Americans are "triangulated" with respect to Whites and Blacks. On the one hand, Whites perceive Asian Americans as a racially subordinate group but one that is valorized as a "model minority" by Whites in contrast to Blacks. On the other hand, while racially positioned "above" Blacks by the White dominant group, Asian Americans are simultaneously ostracized in the civic and political arena and viewed as immutably foreign and unassimilable in contrast to both Whites *and* Blacks. Indeed, Asian Americans are collectively regarded as the symbolic "alien" and remain "perpetual foreigners" despite the very long and continued presence of Asians in America.

We find that Barack Obama was similarly "racially triangulated" as a candidate in the 2008 election. On the one hand, he was valorized in public discourse as the "exceptional" Black man whose saga from humble birth to the Office of the President speaks to the supposedly unparalleled social mobility afforded individuals in the United States. Obama himself did little to dampen this enthusiasm, noting repeatedly on the campaign trail, "In no other country is my story even possible."

On the other hand, a recurring theme in the McCain-Palin campaign was to incite suspicion of Obama as an "outsider"—the greenhorn "Manchurian candidate" and possible Islamic revolutionary whose actual loyalties and sympathies were subject to question. In this context, the McCain-Palin campaign theme of "America First" and advertisements that asked, "Who is Barack Obama?" raised suspicions, in a none-too-subtle fashion, of Barack Obama's primary national allegiances and political loyalties. Such characterizations of Obama sound a lot like the conflicted, ambiguous, and unsettled racial position of Asian Americans.

These features of President Obama's life and his 2008 candidacy allow us to claim—if cheekily—him as our first Asian American President. But did Asian American voters themselves find a compelling frame of commonality? Here it was not initially clear during the presidential campaign whether Obama would be actively soliciting the Asian American vote, and it was correspondingly unclear whether Asian Americans would vote for him.

Early on, he was criticized by the 80/20 Initiative, an Asian American political action committee, for not initially responding to a survey of Asian American equal opportunity issues that the committee had circulated to the major presidential contenders of both parties. In February 2008, a *Time* magazine article titled, "Does Obama Have an Asian Problem", stated that Asian Americans were the one racial/ethnic group who has voted consistently and overwhelmingly for Obama's rival, Hillary Clinton, and asked, *"Could some Asian Americans not be voting for Obama simply because he's black?"*[10] The issue of Asian anti-Black racism surfaced on a segment of CNN's *Anderson Cooper 360* that implied that Asian American immigrants were fearful of "change" and correspondingly suspicious of a Black candidate.

In fact, as the researchers of the 2008 National Asian American Survey (NAAS) found, racial considerations did play a role in shaping Asian Americans' vote choices between Hillary Clinton and Barack Obama.[11] This effect, however was modest and smaller than the role that other factors like age, gender, and ethnic group differences played in differentiating Clinton from Obama supporters.

It is also interesting that Obama's race as "Asian American" does not appear to have been a notable asset with Asian American voters. The data from the 2008 NAAS suggest that Asian Americans voted for very much the same reasons as the rest of the country. When the 2008 NAAS asked respondents whether Obama's childhood in Indonesia or his Indonesian half-sister increased their favorable impression of his candidacy, there was very little effect to speak of. When asked about the importance of the economy, pulling U.S. troops out of Iraq, and achieving universal access to health care coverage for all Americans, however, there was an overwhelming degree of support among Asian Americans and a clear perception that there was a difference between the Democrats and the Republicans on these defining issues of the 2008 election. Thus, while *we* may be persuaded to think of Obama as the first Asian American President, average Asian American voters themselves do not appear to share our gusto. Not yet, anyway. But here we close by noting reasons why Asian Americans might eventually be drawn to think of Obama as "their" President as a result of his symbolic and substantive actions in the White House. In particular, what Asian American voters could not have anticipated in the heat of the campaign was the extraordinary degree to which the Obama administration has relied on Asian Americans in his transition into the presidency and the unprecedented number of Asian Americans that Obama has nominated and appointed to key positions in his administration. In the weeks immediately following the election, Obama's former law school classmate, Chris Lu, was named executive director of his transition team, with Peter Rouse (who is of Japanese-American ancestry) added as his transition team co-chair and Sonal Shah as a member of the transition team.

Then, in his initial appointments, Obama has named three Cabinet-rank Asian Americans—Physicist Steven Chu as secretary of energy, Army General Eric Shinseki as secretary of Veterans Affairs, and former Washington Governor Gary Locke as secretary of commerce. Beyond the Cabinet ranks, there also has been a multitude of other high profile appointments such as Tammy Duckworth as secretary of Public and Intergovernmental Affairs within Veterans Affairs, Yale Law School dean Harold Koh as head legal advisor in the State Department, and Kal Penn (of *Harold and Kumar* fame) as associate director, White House Office of Public Liaison.

The heavy Asian American presence in the Obama administration, quite frankly, presents a high-stakes course for the future of Asian American politics. A misstep into political scandal or a downturn into economic or geopolitical scapegoating and fearmongering, and the visibility of Asian Americans in key positions could become a liability for President Obama and, ultimately, further ossify prevailing Yellow Perils and perpetual foreigners in our midst. Absent such missteps or downturns, however, Asian Americans will likely take great pride— a pride of ownership—in the successes of the Obama administration. If this more sanguine scenario should come to pass, future generations may well credit 2008 as a watershed election in the political maturation and Democratic consolidation of Asian Americans.

## NOTES

1.   Toni Morrison, "Talk of the Town: Comment," *The New Yorker*, October 5, 1998.

2.   George Clinton, "Paint the White House Black" from the Paisley Park/Warner Bros. album Hey Man, Smell My Finger released in 1993.

3.   Quoted in Tom Hundley, "Barack Obama: The First Jewish President?" *Chicago Tribune*, December 12, 2008.

4.  Quoted in Leah Garchik column "Inauguration Day Special." *San Francisco Chronicle*, January 21, 2009, A-11.

5.  Garrison Keillor, "Sitting On Top of the World." *International Herald Tribune*, November 13, 2008. iht.com/articles/2008/11/13/opinion/edkeillor.php.

6.  Jodi Kantor, "A Portrait of Change: In First Family, A Nation's Many Faces." *The New York Times*, January 21, 2009.

7.  Bill Wong, "Barack Obama: Almost like Us." November 6, 2008. ocadc.org/boardblog/William-Wong-Barack-Obama-Almost-Like-Us-.html.

8.  Jonathan Weisman, "Obama, at Fundraiser, Pronounces Himself an 'Honorary AAPI.'" The Trail Blog, posted on July 29, 2008, voices. washingtonpost.com/the-trail/2008/07/29/obama_at_fundraiser_pronounces.html (accessed November 11, 2008).

9.  Claire Jean Kim, "The Racial Triangulation of Asian Americans." *Politics and Society* 27 (1999), 105–138.

10. Lisa Takeuchi Cullen, "Does Obama Have an Asian Problem." *Time*, February 18, 2008.

11. Karthick Ramakrishnan, Janelle Wong, Taeku Lee, and Jane Junn. 2009. "Race-Based Consideration and the Obama Vote: Evidence from the 2008 National Asian American Survey." *Du Bois Review* 6 (Spring).

## DISCUSSION QUESTIONS

1.  On what basis do Omi and Lee argue that Barack Obama is the first Asian American president? Would you agree? Why or why not?

2.  What does it mean to say that Asian Americans are "triangulated" with respect to Whites and Blacks?

# 19

# White Like Me:
## Reflections on Race from a Privileged Son

TIM WISE

*Tim Wise, like many White scholars, is questioning his racial identity and the privileges that accompany this status. He provides a way for dominant groups to think about how they perceive themselves and the system of racial inequality.*

It's a question no one likes to hear, seeing as how it typically portends an assumption on the part of the questioner that something is terribly wrong, something that defies logic and calls for an explanation.

It's the kind of query one might get from former classmates on the occasion of one's twenty-year high school reunion: "Dear God, what the hell happened to you?" Generally, people don't ask this question of those whom they consider to have dramatically improved themselves in some way, be it physical, emotional, or professional. Instead, it is more often asked of those considered to be seriously damaged, as if the only possible answer would be, "Well, I was dropped on my head as a baby," to which the questioner would then reply, "Aha, I see."

So whenever I'm asked this, I naturally recoil for a moment, assuming that the persons inquiring "what happened" likely want an answer only in order to avoid, at whatever cost, having it (whatever "it" may be) happen to them. In my case, however, I'm usually lucky. Most of the persons who ask me "what happened" seem to be asking less for reasons of passing judgment than for reasons of confusion. They appear truly perplexed about how I turned out the way I did, especially when it comes to my views on the matter of race.

As a white man, born and reared in a society that has always bestowed upon me advantages that it has just as deliberately withheld from people of color, I am not expected to think the way I do. I am not supposed to speak against and agitate in opposition to racism and institutionalized white supremacy. Indeed, for people of color, it is often shocking to see white people even thinking about race, let alone challenging racism. After all, we don't have to spend much time contemplating the subject if we'd rather not, and historically white folks have made something of a pastime out of ignoring racism, or at least refusing to call it out as a social problem to be remedied.

SOURCE: Tim Wise. 2008. *White Like Me.* Brooklyn, NY: Soft Skull Press.

But for me, and for the white folks whom I admire in history, ignoring race and racism has never been an option. Even when it would have been easier to turn away, there were too many forces, to say nothing of circumstances, pulling me back, compelling me to look at the matter, square in the face—in *my* face, truth be told.

Although white Americans often think we've had few first-hand experiences with race, because most of us are so isolated from people of color in our day-to-day lives, the reality is that this isolation *is* our experience with race. We are all experiencing race, because from the beginning of our lives we have been living in a racialized society, where the color of our skin means something socially, even while it remains largely a matter of biological and genetic irrelevance. Race may be a scientific fiction—and given the almost complete genetic overlap between persons of the various so-called races, it appears to be just that—but it is a social fact that none of us can escape no matter how much or how little we may speak of it. Just as there were no actual witches in Salem in 1692, and yet anti-witch persecution was frighteningly real, so too race can be a falsehood even as racism continues to destroy lives, to maim, to kill, and, on the flipside, to advantage those who are rarely its targets.

A few words about terminology: When I speak of "whites" of "white folks," I am referring to those persons, typically of European descent, who are able, by virtue of skin color or perhaps national origin and culture, to be perceived as "white," as members of the dominant group. I do not consider the white race to be a real thing, in biological terms, as modern science pretty well establishes that there are no truly distinct races, genetically speaking, within the human species. But the white race certainly has meaning in social terms. And it is in that social sense that I use the concept here.

As it turns out, this last point is more important than you might think. Almost immediately upon publication, [my book's] first edition came under fire from various white supremacists and neo-Nazis, who launched a fairly concerted effort to discredit it, and me as its author. They sought to do this by jamming the review boards at Amazon.com with harsh critiques, none of which discussed the content—in all likelihood none of them had actually read the book—but which amounted, instead, to ad hominem attacks against me as a Jew. As several explained, being Jewish disqualifies me from being white, or writing about my experience as a white person, since Jews are, to them, a distinct race of evildoers that seeks to eradicate Aryan stock from the face of the earth.

On the one hand (and ignoring for a second the Hitler-friendly histrionics) of course, it is absurd to think that uniquely "Jewish genes" render Jews separate from "real" whites, despite our recent European ancestry. And it's even more ridiculous to think that such genes from one-fourth of one's family, as with mine, on my paternal grandfather's line, can cancel out the three quarters Anglo-Celtic contribution made by the rest of my ancestors. But in truth, the argument is completely irrelevant, given how I am using the concept of whiteness here. Even if there were something biologically distinct about Jews, this would hardly alter the fact that most Jews, especially in the United States, are sufficiently light skinned and assimilated so as to be fully functional as whites in the eyes of authority. This wasn't always the case but it is inarguably such now. American Jews are, by and large, able to reap the benefits of whiteness and white racial privilege, vis-à-vis people of color, in spite of our Jewishness, whether viewed in racial or cultural terms. My "claiming to be white," as one detractor put it, was not an attempt on my part to join the cool kids. I wasn't trying to fool anyone.

Whiteness is more about how you're likely to be viewed and treated in a white supremacist society than it is about what you *are*, in any meaningful sense. This is why even some very light-skinned folks of color have been able to access white privilege over the years by passing as white or being misperceived as white, much to their benefit. Whiteness is, however much clichéd the saying may be, largely a social construct....

... As for the concept of privilege, here, too, clarification is in order. I am not claiming, nor do I believe, that all whites are wealthy and powerful. We live not only in a racialized society, but also in a class system, a patriarchal system, and one of straight supremacy/heterosexism, able-bodied supremacy, and Christian hegemony. These other forms of privilege, and the oppression experienced by those who can't manage to access them, mediate, but never fully eradicate, something like white privilege. So I realize that, socially rich whites are more powerful than poor ones, white men are more powerful than white women, able-bodied whites are more powerful than those with disabilities, and straight whites are more powerful than gay, lesbian, bisexual or transgendered whites.

But despite the fact that white privilege plays out differently for different folks, depending on these other identities, the fact remains that when all other factors are equal, whiteness matters and carries great advantage. So, for example, although whites are often poor, their poverty does not alter the fact that, relative to poor and working-class persons of color, they typically have a leg up. In fact, studies suggest that working-class whites are typically better off in terms of assets and net worth than even middle-class blacks with far higher incomes, due to past familial advantages. No one privilege system trumps all others every time, but no matter the ways in which individual whites may face obstacles on the basis of nonracial factors, our race continues to elevate us over similarly situated persons of color.

The notion of privilege is a relative concept as well as an absolute one, a point that is often misunderstood. This is why I can refer to myself as a "privileged son," despite coming from a family that was not even close to wealthy. In relative terms, compared to persons of color, whites receive certain head starts and advantages, none of which are canceled out because of factors like class, gender, or sexual orientation. Likewise, heterosexuals receive privileges relative to LGBT folks, none of which are canceled out by the poverty that many straight people experience. So too, rich folks have certain privileges on the basis of their wealth, none of which vanish like mist just because some of those wealthy persons are disabled.

While few of us are located only in privileged groups, and even fewer are located only in marginalized or oppressed groups—we are all privileged in some ways and targets in others—the fact remains that our status as occasional targets does not obviate the need for us to address the ways in which we receive unjust advantages at the expense of others....

The first [lesson] is that to be white is to be "born to belonging." This is a term I first heard used by my friend and longtime antiracist white ally Mab Segrest, though she was using it in a different context. To be white is to be born into an environment where one's legitimacy is far less likely to be questioned than would be the legitimacy of a person of color, be it in terms of where one lives, where one works, or where one goes to school. To be white is, even more, to be born into a system that has been set up for the benefit of people like you (like us), and as such provides a head start to those who can claim membership in this, the dominant club.

Second, to be white means only that one will typically inherit certain advantages from the past, but also that one will continue to reap the benefits of ongoing racial privilege, which is itself the flipside of discrimination against persons of color. These privileges have both material components, such as better job opportunities, better schooling, and better housing availability, as well as psychological components, not the least of which is simply having one less thing to constantly worry about during the course of a day. To be white is to be free of the daily burden of constantly having to disprove negative stereotypes. It is to have one less thing to sweat, and in a competitive society such as ours, one less thing on your mind is no small boost.

Third …, in the face of these privileges, whether derived from past injustice handed down or present injustice still actively practiced, to be white is typically to be in profound denial about the existence of these advantages and their consequences. I say denial here, rather than ignorance, because the term *ignorance* implies an involuntary lack of knowledge, a purity, an innocence of sorts, that lets white Americans off the hook, even if only linguistically. The fact is, whites' refusal to engage the issues of race and privilege is due largely to a *willed* ignorance, a voluntary evasion of reality, not unlike the alcoholic or drug addict who refuses to face their illness. How else but as the result of willed ignorance can we understand polls taken, not today, but in the early sixties, which demonstrated that even then—at a time of blatant racism and legally accepted discrimination against black people— the vast majority of whites believed that everyone had equal opportunity?

The only way that one can be completely ignorant of the racial truth in the United States, whether in the sixties or today, is to make the deliberate choice to think about something else, to turn away, to close one's ears, shut one's eyes, and bury one's head in the proverbial sand.

Oh sure, there are young people, perhaps of high school age or even younger, who might truly be ignorant in the strictest sense of the word when it comes to issues of racism and privilege. But if so, this is only because their teachers, preachers, parents, and the mass media to which they are daily subjected have made the choice to lie to them, either directly or by omission. So even the ignorance of the young is willed, albeit by their elders, much to their own detriment.

Fourth, whites can choose to resist a system of racism and unjust privilege, but doing so is never easy. In fact, the fear of alienating friends and family, and the relative lack of role models from whom we can take direction, renders resistance rare, and even when practiced, often ineffective, however important it may be. Learning how to develop our resistance muscles is of vital importance. Thinking about how and when to resist, and what to do (and not do) is critical.

Fifth, even when committed to resistance, and even while in the midst of practicing it, we sometimes collaborate with racism and reinforce racial domination and subordination. In other words, we must always be on guard against our own screwups and willing to confront our failings honestly.

Sixth, whites pay enormous costs in order to have access to the privileges that come from a system of racism—costs that are intensely personal and collective, and which should inspire us to fight racism *for our own sake*.

And finally, in the struggle against injustice, against racism, there is the possibility of redemption.

Belonging, privilege, denial, resistance, collaboration, loss, and redemption: the themes that define and delineate various aspects of the white experience. The trick is getting from privilege, collaboration, denial, and loss to resistance and redemption, so that we may begin to belong to a society more just and sustainable than what we have now.

## DISCUSSION QUESTIONS

1.  What does it mean to say that Whites receive the benefits of White privilege, even when they differ on the basis of social class or other social facts?

2.  What costs does Wise identify as occurring for White people because of racism and unjust privilege?

# Student Exercises

1.  Think back to the first time you remember recognizing your own racial identity. What were the circumstances? What did you learn? Now ask the same question of someone whose race is different from your own. How do the two experiences compare and contrast? How do the answers illustrate how racial identity is formed in different contexts and with different meanings depending on the group's experience?

2.  Having read the interview with Beverly Tatum, observe patterns of seating in public spaces on your campus (such as in student centers, eating areas, and classrooms). What patterns do you see and how would you interpret them based on Tatum's arguments?

# PART II

# Continuity and Change: How We Got Here and What It Means

# Section V

# Who Belongs? Race, Rights and Citizenship

## Elizabeth Higginbotham and Margaret L. Andersen

As a social construction, race is more than beliefs and identities. Race is embedded in the larger social structures that form the organization of society. In the twenty-first century, we cannot escape the legacy of how race and ethnicity were critical factors in the early building of the United States, with different people being denied rights at various times. Rights are critical to belonging to a nation. As Supreme Court Chief Justice Earl Warren wrote in 1958, dissenting on a denationalization case, "Citizenship *is* man's basic right, for it is nothing less than the right to have rights" (Ngai 2003: 10).

Key to the realization of rights is the notion of citizenship, which enables people to participate fully in the society. The rules of citizenship—that is, who is granted this status and how they achieve it—shape the relationships among groups in a society. In the United States, rules of citizenship historically developed along lines of racial and ethnic hierarchies, giving some groups access to social and political power and others less. People who were not granted citizenship rights were pushed to the margins of the society, where they had little leverage in shaping the major social institutions that influenced their lives. Marginalized racial groups struggled for inclusion in the larger political body, a struggle that is ongoing in many respects. A legacy of exclusion has limited people's access to political, economic, and social resources and has shaped the majority group's thinking about racially different groups and their place in society.

Membership in different racial and ethnic groups was linked with rights and obligations in colonial America, but as the nation became independent from England, it developed its own economic, social, and political institutions. Race was central to thinking about the nature of the country and the development of its citizenry. "We the people" in the U.S. Constitution did not refer to an inclusive group; rather, the founding fathers were very deliberate in extending rights only to White men with property. As a group, American Indians—the indigenous populations—were seen as outside the developing society. The Constitution also identified "other persons," that is, enslaved people, most of them of African descent. This initial national vision of people of color as unworthy of citizenship meant that they had to battle for inclusion in the nation.

We can see the vision the founding fathers had for the future of this young nation in the procedures established for how new immigrants could become citizens. The 1790 Naturalization Act offered citizenship to "free white persons" who were of good moral character; the 1802 Act added a five-year residency requirement. The United States was viewed as a site of Anglo-Saxon settlement, and that view shaped its very construction, including the exclusion and differing treatment of racial and ethnic groups. Eastern and southern European immigrants, who came in significant numbers in the late nineteenth and early twentieth centuries, were viewed as racially inferior. While they were socially marginalized, these groups were White politically and were on a path to full participation in the nation (See Karen Brodkin in Section I). We can contrast the experiences of eastern and southern European immigrants with those of American Indians, African Americans, Asian immigrants, and other people of color who faced extended hardships because of their lack of rights. These excluded groups would have to change many laws to become citizens. For example, immigrants from Asia were ineligible for citizenship because they were not "free white persons." The denial of rights made racial minorities vulnerable to having their labor exploited and their land, if they had any, taken for the benefit of others.

In the early twentieth century, although most were literate, Japanese immigrants could not become citizens and vote. Consider the case of Takao Ozawa, originally from Japan. Ozawa arrived in California in 1894, graduated from high school, and attended the University of California, Berkeley. After that, he worked for an American company, living with his family in the U.S. territory of Honolulu. Because he was not legally classified as White, he could not become a U.S. citizen, though he very much wanted to and filed for citizenship in 1914. Ozawa argued that he was a "true American," a person of good character who neither drank, smoked, nor gambled. His family went to an American church; his children went to American schools; he spoke only English at home and raised his children as Americans. In 1922, the U.S. Supreme Court denied his eligibility for citizenship based on the claim that he was not "Caucasian" (Takaki 1989; *Ozawa v. United States*, 260 U.S. 178, 1922).

As a group, the Japanese found their employment options limited; many worked to serve their own community. Using their agricultural skills to develop farms outside of cities, many provided fruits and vegetables for the growing urban population. As their settlements grew, so did opposition from the native and immigrant White population. Californians first passed laws in 1913 limiting Japanese people to being able to lease land for only three years. Later the Alien Law Act of 1920 prohibited them from leasing or owning land, so Japanese people either changed occupations or had White people purchase land for them (Takaki 1989). These **racial projects** demonstrate the severe costs of lacking citizenship rights.

Citizenship means that individuals are participants in the development of a nation's political and social institutions. They can use this power to shape institutions for their own benefit, while groups without power have to work within social institutions they did not design. In terms of the economy, those with power have historically appropriated the labor of others for their own benefit. As the nature of the economy shifts, a few more people might gain access to political power—such as the extension of the vote to White men without property in the mid-nineteenth century. However, the practice of denying racial-ethnic groups the rights of citizenship continues into this new century.

You can think of citizenship in two ways. First, there is an actual system of rights (the right to vote, the right to sit on a jury, the right to own property, and so forth). Second, citizenship is also symbolic—meaning the right of belonging to a nation or a community (Glenn 2002). For example, at the ceremony that opened the Lincoln Memorial in Washington, DC in 1922, Black Americans had to witness the event from behind a rope. The pattern of segregation, like separate schools for people of color and the denial of access to libraries and museums, communicated that they were not full citizens in the nation.

In this section, we look at citizenship, patterns of exclusion, and their implications for people's participation in the nation. These themes help us think about critical questions. How can we respect individual human rights? How can we build a society of equals? How can we protect all of our citizens? We will likely grapple with these questions for decades, but looking at our past helps us think about how we want to address these questions today.

Evelyn Nakano Glenn ("Citizenship and Inequality") explores how people of color battled the limitations of the Constitution and opened up options for others. She finds that, even when laws are changed, racial practices continue to shape social policy. Overall, her selection provides us with insights into the contested nature of citizenship and its relationship to labor exploitation.

C. Matthew Snipp ("The First Americans: American Indians") explains how early national leaders considered American Indians to be candidates for extinction. Thus, Native Americans' path to incorporation into the nation involved removal from their original homelands and, later, forced assimilation. Their inability to speak for themselves as citizens of the United States made them vulnerable to laws that were supposed to help, but really harmed American Indian people. Only in the twentieth century could tribes gain some measure of control and real sovereignty to improve their status, but their rights are still contested in many states.

Now, not all the immigrants who enter our borders are treated fairly. Susan M. Akram and Kevin R. Johnson ("Race, Civil Rights, and Immigration Law after September 11, 2001: The Targeting of Arabs and Muslims") look at the response to the horrific events of 9/11, which have changed the lives of Arabs and Muslims living in the United States. Arabs and Muslims have found themselves racialized and deprived of basic rights. The negative acts by a few have been generalized to the larger group, resulting in racial and ethnic profiling. Akram and Johnson help us think about the global context that affects immigrants' quest for citizenship.

Many immigrants become citizens in the United States and still maintain ties to their own nations of origin; we call such immigrants *transnationals*. This pattern has become common in the twenty-first century as new communication technologies enable newcomers to remain closely connected to their distant families. Peggy Levitt ("Salsa and Ketchup: Transnational Migrants Straddle Two Worlds") offers a study of immigrants from the Dominican Republic and Gujarat (a state on the west coast of India) that demonstrates how these new citizens create community, even while maintaining strong ties to their homelands. These families negotiate their place in communities when their ethnic identity makes them part of two worlds—the community in which they live and their community of origin.

## REFERENCES

Glenn, Evelyn Nakano. (2002). *Unequal Freedom: How Race and Gender Shaped American Citizenship and Labor*. Cambridge, MA: Harvard University Press.

Ngai, Mae M. (2003). *Impossible Subjects: Illegal Aliens and the Making of Modern America*. Princeton, NJ: Princeton University Press.

Takaki, Ronald. (1989). *Strangers from a Different Shore: A History of Asian Americans*. New York: Penguin.

## FACE FACTS: THE AGING POPULATION: FOREIGN AND NATIVE-BORN

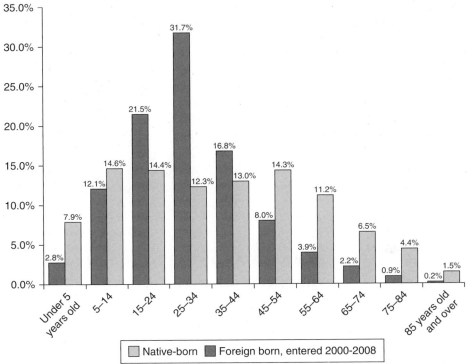

### Age Distribution of U.S. Foreign-born and Native-born Populations

Legend: Native-born   Foreign born, entered 2000-2008

| Age group | Native-born | Foreign born, entered 2000-2008 |
|---|---|---|
| Under 5 years old | 7.9% | 2.8% |
| 5–14 | 14.6% | 12.1% |
| 15–24 | 14.4% | 21.5% |
| 25–34 | 12.3% | 31.7% |
| 35–44 | 13.0% | 16.8% |
| 45–54 | 14.3% | 8.0% |
| 55–64 | 11.2% | 3.9% |
| 65–74 | 6.5% | 2.2% |
| 75–84 | 4.4% | 0.9% |
| 85 years old and over | 1.5% | 0.2% |

SOURCE: U.S. Census Bureau. 2009. *Statistical Abstract of the United States*. U.S. Department of Commerce. www.census.gov

***Think about it:*** These facts (based on U.S. census data) show you the age distribution of the foreign-born and native-born population of the United States. The median age for foreign born citizens entering between 2000 and 2008 is 29 years; for native-born citizens, 35.7 years. What implications might the differences in age distribution mean for such public policy needs for child care, schooling, and support for the elderly? Note that the foreign-born data do not include undocumented (that is, illegal) immigrants.

# 20

# Citizenship and Inequality

EVELYN NAKANO GLENN

*Glenn explores the concept of citizenship as it has developed in reference to racial and ethnic groups in American society. She points out that citizenship includes both formal rights and the informal sense of belonging—forms of citizenship that have historically been denied to different groups in different ways. Her article shows how central the construction of citizenship is to being seen as having full "personhood."*

## HISTORICAL DEBATES ABOUT CITIZENSHIP IN THE UNITED STATES

… The concept of citizenship is, of course, historically and culturally specific. The modern, western notion of citizenship emerged out of the political and intellectual revolutions of the seventeenth and eighteenth centuries, which overthrew the old feudal orders. The earlier concept of society organized as a hierarchy of status, expressed by differential legal and customary rights, was replaced by the idea of a political order established through social contract. Social contract implied free and equal status among those who were party to it. Equality of citizenship did not, of course, rule out economic and other forms of inequality. Moreover, and importantly, equality among citizens rested on the inequality of others living within the boundaries of the community who were defined as non-citizens. The relationship between equality of citizens and inequality of non-citizens had both rhetorical and material dimensions. Rhetorically, the "citizen" was defined and, therefore, gained meaning through its contrast with the oppositional concept of the "non-citizen" as one who lacked the essential qualities needed to exercise citizenship. Materially, the autonomy and freedom of the citizen were made possible by labor (often involuntary) of non-autonomous wives, slaves, children, servants, and employees.…

A specifically sociological conception of citizenship as membership is offered by Turner (1993:2), who defines it as "a set of practices (judicial, political, economic, and cultural) which define a person as a 'competent' member of society...." Focusing on social practice takes us beyond a juridical or state conception of citizenship. It points to citizenship as a fluid and decentered complex that is continually transformed through political struggle.

Membership entails drawing distinctions and boundaries for who is included and who is not. Inclusion as a member, in turn, implies certain rights in, and reciprocal obligations toward, the community. Formal rights are not enough, however; they are only paper claims unless they can be enacted through actual practice. Three leading elements in the construction of citizenship, then, are membership, rights and duties, and conditions necessary for practice.

These three elements formed the major themes that have run through debates, contestation, and struggles over American citizenship since the beginning. First, membership: who is included or recognized as a full member of the imagined community (Anderson 1983), and on what basis? Second, what does membership mean in terms of content: that is, what reciprocal rights and duties do citizens have? Third, what are the conditions necessary for citizens to practice citizenship, to actually realize their rights and carry out their responsibilities as citizens?

## MEMBERSHIP

Regarding membership, there are two major strains of American thought regarding the boundaries of the community. One tradition is that of civic citizenship, a definition based on shared political institutions and values in which membership is open to all those who reside in a territory. The second is an ethno-cultural definition based on common heritage and culture, in which membership is limited to those who share in the heritage through blood descent (Smith 1989; Kettner 1978).

Because of its professed belief in equality and natural rights (epitomized by the Declaration of Independence), the United States would seem to fit the civic model. However, since its beginnings, the U.S. has followed both civic and ethno-cultural models. The popular self-image of the nation, expressed as early as the 1780s, was of the United States as a refuge of freedom for those fleeing tyranny. This concept, later elaborated in historical narratives and sociological accounts of America as a nation of immigrants, blatantly omitted Native Americans, who were already here, Mexicans, who were incorporated through territorial expansion, and Blacks, who were forcibly transported. This exclusionary self-image was reflected at the formal level in the founding document of the American polity, the U.S. Constitution. The authors of the Constitution, in proclaiming a government by and for "we the people," clearly did not intend to include everyone residing within the boundaries of the U.S. as part of "the people." The Constitution explicitly named three groups: Indians, who were identified as members of separate nations, and "others," namely slaves; and finally, "the people": only the latter were full members of the U.S. community (Ringer 1983).

Interestingly, the Constitution was silent as to who was a citizen and what rights and privileges they enjoyed. It left to each state the authority to determine qualifications for citizenship and citizens' rights, e.g., suffrage requirements, qualifications for sitting on juries, etc. Individuals were, first, citizens of the states in which they resided and only secondarily, through their citizenship in the state, citizens of the United States. The concept of national citizenship was, therefore, quite weak.

However, the Constitution did direct Congress to establish a uniform law with respect to naturalization. Accordingly, Congress passed a Naturalization Act in 1790, which shaped citizenship policy for the next 170 years. It limited the right to become naturalized citizens to "free white persons." The act was amended in 1870 to add Blacks, but the term "free white persons" was retained. As Ian Haney Lopez (1996) has documented, immigrants deemed to be non-white (Hawaiians, Syrians, Asian Indians), but not Black or African, were barred from naturalization. The largest such category was immigrants from China, Japan, and other parts of Asia, who were deemed by the courts to be "aliens ineligible for citizenship." This exclusion remained in force until 1953....

It was Black Americans, both before and after the Civil War, who were the most consistent advocates of universal citizenship. Hence, it is fitting that the Civil Rights Act of 1865 and the Fourteenth Amendment, ratified in 1867 to ensure the rights of freed people, greatly expanded citizenship for everyone. Section 1 of the Fourteenth Amendment stated that "All persons born or naturalized in the United States and subject to the jurisdiction thereof, are citizens of the United States and of the State wherein they reside. No State shall make or enforce any law which shall abridge the privileges or immunities of citizens of the United States; nor shall any State deprive any person of life, liberty, or property, without due process of law; nor deny to any person within its jurisdiction, the equal protection of the law."

In these brief sentences, the Fourteenth Amendment established three important principles for the first time: the principle of national citizenship, the concept of the federal state as the protector and guarantor of national citizenship rights, and the principle of birthright citizenship. These principles expanded citizenship for others besides Blacks. To cite one personal example, my grandfather, who came to this country in 1894, was ineligible to become a naturalized citizen because he was not white, but his daughter, my mother, automatically became a citizen as soon as she was born. Birthright citizenship was tremendously important for second and third generation Japanese Americans and other Asian Americans who otherwise would have remained perpetual aliens, as is now the case with immigrant minorities in some European countries. Foner (1998) was on the mark when he said that the Black American struggle to expand the boundaries of freedom to include themselves, succeeded in changing the boundaries of freedom for everyone.

The battle was not won, once and for all, in 1867. Instead, the nation continued to vacillate between the principle of the federal government having a duty to protect citizens' rights and states' rights. By the end of the Reconstruction period in the 1870s, the slide back toward states' rights accelerated as all branches of the federal government withdrew from protecting Black rights and allowed southern states to impose white supremacist regimes (Foner 1988). In the landmark 1896 *Plessy v. Ferguson* decision, the Supreme Court legitimized segregation based on the principle of "separate but equal." This and other court decisions gutted the concept of national citizenship and carved out vast areas of life, employment, housing, transportation, and public accommodations as essentially private activity that was not protected by the Constitution (Woodward 1974). It was not until the second civil rights revolution of the 1950s and 1960s that the federal courts and Congress returned to the principles of national citizenship and a strong federal obligation to protect civil rights....

## MEANINGS OF RIGHTS AND RESPONSIBILITIES

Just as with the question of membership, there has not been a single understanding of rights and duties. American ideas on rights and duties have been shaped by two different political languages (Smith 1989). One, termed liberalism by scholars, grew out of Locke and other enlightenment thinkers. In this strain of thought, embodied in the Declaration of Independence, citizens are individual rights bearers. Governments were established to secure individual rights so as to allow each person to pursue private as well as public happiness. The public good was not an ideal to be pursued by government, but was to be the outcome of individuals pursuing their own individual interest. The other language was that of *republicanism*, which saw the citizen as one who actively participated in public life. This line of thought reached back to ancient Greece and Rome where republicanism held that man reached his highest fulfillment by setting aside self-interest to pursue the common good. In contrast to liberalism, republicanism emphasizes practice and focuses on achieving institutions and practices that make collective self-government possible.

There has been continuing tension and alternation between these two strains of thought, particularly around the question of whether political participation is essential to, or peripheral to, citizenship. In the nineteenth century, when many groups were excluded from participation, the vote was a mark of standing; its lack was a stigma, a badge of inferiority, hence, the passion with which those denied the right fought for inclusion. Three great movements—for universal white manhood suffrage, for Black emancipation, and for women's suffrage—resulted in successive extensions of the vote to non-propertied white men, Black men, and women, between 1800 and 1920 (Roediger 1991; Foner 1990; DuBois 1978). According to Judith Shklar (1991), once the vote became broadly available, it ceased to be a mark of status. Judging by participation rates, voting is no longer an emblem of citizenship. One measure of the decline in the significance of the vote has been the precipitous drop in persons voting. In 1890, fully 80% of those who were eligible, voted. By 1924, four years after the extension of suffrage to women, participation had fallen to less than half of those registered, and it has been declining ever since....

In the U.S., participation is discouraged by not making registration easy. Requiring re-registration every time one moves is only one example of the obstacles placed upon a highly mobile population. While Americans today would object if the right to vote was taken away, the majority don't seem to feel it necessary to exercise the vote. The U.S. obviously has the capacity to make registration automatic or convenient, but has not made major efforts to do so. Needless to say, it is in the interest of big capital and its minions that most people don't participate politically.

## CONDITIONS FOR PRACTICE

This leads to the third main theme, the actual practice of citizenship and the question of what material and social conditions are necessary for people to actually exercise their rights and participate in the polity.

The short answer to this question for most of American history has been that a citizen must be *independent*—that is, able to act autonomously.... Independence has remained essential, but its meaning has undergone drastic transformation since the founding of the nation.... In the nineteenth century, as industrialization proceeded apace and wage work became common even for skilled artisans and white collar men, the meaning of independence changed to make it more consistent with the actual situation of most white men. It came to mean, not ownership of productive property, but nothing more than ownership of one's own labor and the capacity to sell it for remuneration (Fraser and Gordon 1994). This formulation rendered almost all white men "independent," while rendering slaves and women, who did not have complete freedom to sell their labor, "dependent." It was on the basis of the independence of white working men that the movement for white Universal Manhood Suffrage was mobilized (Roediger 1991). By the 1830s, all states had repealed property requirements for suffrage for white men, while simultaneously barring women and Blacks (Litwak 1961).

By the late 19th century, capitalist industrialization had widened the economic gap between the top owners of productive resources and the rest, making more apparent the contradiction between economic inequality and political democracy. Rising levels of poverty, despite the expansion in overall wealth, raised the question of whether low-income, unemployment, and/or lack of access to health care and other services, diminished citizenship rights for a large portion of the populace. Growing economic inequality also raised the issue of whether some non-market mechanism was needed to mitigate the harshness of inequities created by the market.

One response during the years between World War I and World War II was rising sentiment for the idea of what T. H. Marshall called social citizenship. In Marshall's words (Marshall 1964:78), social citizenship involves "the right to a modicum of economic security and to share in the full social heritage and to live the life of a civilized being, according to the standards prevailing in the society." Economic and social unrest after World War I spurred European states to institute programs to ensure some level of economic security and take collective responsibility for "dependents"—the aged, children, the disabled, and others unable to work.... The significance of the "redistributive" mechanisms of the Welfare State, according to Marshall, was that they enabled working-class people to exercise their civil and political rights.... Compared to Western Europe, the concept of social citizenship has been relatively weak in the U.S. Welfare state researchers have pointed out that, although 1930s New Deal programs such as Social Security and unemployment insurance greatly increased economic security, they continued a pattern of a two-tiered system of social citizenship from the 1890s (Nelson 1990; Skocpol 1992; Fraser and Gordon 1993; Fraser and Gordon 1994). The upper tier consisted of "entitlements" based on employment or military service, e.g., unemployment benefits, old-age insurance, and disability payments, which were relatively generous and did not require means-testing. The lower tier consisted of various forms of "welfare," such as Aid to Dependent Children (changed to Aid to Families with Dependent Children—AFDC—in 1962), which were relatively stingy and entailed means testing and surveillance by the state.

White men, as a class, have drawn disproportionately on first tier rights by virtue of their records of regular and well-paid employment. White women, more often, had to rely on welfare, which is considered charity, a response to dependence, rather than a just return for contributions. Latino and African American men were generally excluded from employment-based benefits because of their concentration in agriculture, day jobs, and other excluded occupations; Latina and African American women, in turn, have often been denied even second tier benefits (Oliver and Shapiro 1995; Mink 1994). Moreover, in contrast to the situation in most European countries, there has been little sense of collective responsibility for the care of dependents. Thus, raising children is not recognized as a contribution to the society and, therefore, as a citizenship responsibility that warrants entitlements such as parental allowances and retirement credit, which are common in Europe (Glenn 2000; Sainsbury 1996).

As with the other issues of citizenship, the 1980s and 1990s [saw] a neoconservative turn with a concerted effort to roll back even attenuated social citizenship rights. Government funding of social services has been vilified for draining money from hardworking citizens to support loafers and government bureaucrats who exert onerous control over people's lives. And, after years of attacks on Black and other single mothers "dependent" on welfare, Congress passed the Personal Responsibility and Work Opportunity Act in 1996, which dismantled AFDC and replaced it with grants to the states that limited eligibility for federal benefits for a lifetime maximum of five years. The aim was, as Representative Richard Armey put it, that the poor return to the natural safety nets —family, friends, churches, and charities. States instituted stringent work requirements and limited total lifetime benefits. By limiting total years that they can stay on welfare, proponents argued the new regulations would wean single mothers from unhealthy dependence on the state. In contrast to an overly indulgent government, the market is expected to exert a moral force, disciplining them and forcing them to be "independent" (Roberts 1997; Boris 1998)....

Thus, we must ask: if people have a responsibility to earn, then, don't they also have a corresponding right to earn?—i.e., to have a job and to earn enough to support themselves at a level that allows them, in Marshall's (1964:78) words, "to participate fully in the cultural life of the society and to live the life of a civilized being according to the standards prevailing in the society."

## REFERENCES

Anderson, Benedict. 1983. *Imagined Communities: Reflections on the Origins and Spread of Nationalism.* London: Verso.

Boris, Eileen. 1998. "When work is slavery." *Social Justice* 25, 1:18–44.

DuBois, Ellen Carol. 1978. *Feminism and Suffrage.* Ithaca: Cornell University Press.

Foner, Eric. 1988. *Reconstruction: America's Unfinished Revolution.* New York: Harper and Row.

————— 1990. "From slavery to citizenship: Blacks and the right to vote." In *Voting and the Spirit of American Democracy*, ed. Donald W. Rogers. Urbana and Chicago: University of Illinois Press.

————— 1998. *The Story of American Freedom.* New York: Norton.

Fraser, Nancy, and Linda Gordon. 1993. "Contract versus charity: Why is there no social citizenship in the United States?" *Socialist Review* 212, 3:45–68.

————— 1994. "A genealogy of dependence: Tracing a keyword of the U.S. welfare state." *Signs* 19, 2:309–336.

Glenn, Evelyn. 2000. "Creating a caring society." *Contemporary Sociology* 29, 1:84–94.

Haney Lopez, Ian F. 1996. *White By Law: The Legal Construction of Race.* New York: New York University Press.

Kettner, James. 1978. *The Development of American Citizenship, 1608–1870.* Chapel Hill: University of North Carolina Press.

Litwak, Leon F. 1961. *North of Slavery: The Negro in the Free States, 1790–1860.* Chicago: University of Chicago Press.

Marshall, T. H. 1964. *Class, Citizenship, and Social Development.* Garden City, NY: Doubleday and Company.

Mink, Gwendolyn. 1994. *The Wages of Motherhood: Inequality in the Welfare State, 1917–1942.* Ithaca: Cornell University Press.

Nelson, Barbara. 1990. "The origins of the two-channel welfare state: Workman's compensation and mother's aid." In *Women, the State and Welfare*, ed. Linda Gordon, 123–151. Madison: University of Wisconsin Press.

Oliver, Melvin L., and Thomas M. Shapiro. 1995. *Black Wealth/White Wealth: A New Perspective on Racial Inequality.* New York: Routledge.

Ringer, Benjamin B. 1983. *We the People and Others: Duality and America's Treatment of Its Racial Minorities.* New York: Tavistock.

Roberts, Dorothy. 1997. *Killing the Black Body: Race, Reproduction, and the Meaning of Liberty.* New York: Pantheon.

Roediger, David. 1991. *The Wages of Whiteness: Race and the Making of the American Working Class.* London: Verso.

Sainsbury, Diane. 1996. *Gender, Equality and Welfare States.* Cambridge, UK: Cambridge University Press.

Shklar, Judith. 1991. *American Citizenship: The Quest for Inclusion.* Cambridge, MA: Harvard University Press.

Skocpol, Theda. 1992. *Protecting Soldiers and Mothers: The Origins of Social Policy in the United States.* Cambridge, MA: Harvard University Press.

Smith, Rogers M. 1989. "'One United People': Second-class female citizenship and the American quest for community." *Yale Journal of Law and the Humanities* 1:229–293.

Turner, Brian S. 1993. "Contemporary Problems in the Theory of Citizenship." In *Citizenship and Social Theory.* Ed. Brian S. Turner. 1–19. London: Sage Publications.

Woodward, C. Vann, ed. 1974. *The Strange Career of Jim Crow*, 3rd rev. New York: Oxford University Press.

## DISCUSSION QUESTIONS

1. What elements does Glenn identify as constructing citizenship?

2. During what era did social citizenship rights expand and why?

3. Thinking about your own citizenship status, what do you see as the rights and obligations linked to your position in the nation?

# 21

# The First Americans:

## American Indians

C. MATTHEW SNIPP

*Snipp's analysis of different periods in the treatment of American Indians by the U.S. government shows how citizenship rights have been denied to Native Americans, even though the specific processes by which this has happened have changed over time.*

By the end of the nineteenth century, many observers predicted that American Indians were destined for extinction. Within a few generations, disease, warfare, famine, and outright genocide had reduced their numbers from millions to less than 250,000 in 1890. Once a self-governing, self-sufficient people, American Indians were forced to give up their homes and their land, and to subordinate themselves to an alien culture. The forced resettlement to reservation lands or the Indian Territory (now Oklahoma) frequently meant a life of destitution, hunger, and complete dependency on the federal government for material needs.

Today, American Indians are more numerous than they have been for several centuries. While still one of the most destitute groups in American society, tribes have more autonomy and are now more self-sufficient than at any time since the last century. In cities, modern pan-Indian organizations have been successful in making the presence of American Indians known to the larger community, and have mobilized to meet the needs of their people (Cornell 1988; Nagel 1986; Weibel-Orlando 1991). In many rural areas, American Indians and especially tribal governments have become increasingly more important and increasingly more visible by virtue of their growing political and economic power. The balance of this [reading] is devoted to explaining their unique place in American society.

### THE INCORPORATION OF AMERICAN INDIANS

The current political and economic status of American Indians is the result of the process by which they were incorporated into Euro-American society (Hall 1989). This amounts to a long history of efforts aimed at subordinating an

SOURCE: From Silvia Pedraza and Rubén G. Rumbaut, eds. 1996, Belmont, CA: Wadsworth. *Origins and Destinies: Immigration, Race, and Ethnicity in America*, 1st ed.

otherwise self-governing and self-sufficient people that eventually culminated in widespread economic dependency. The role of the U.S. government in this process can be seen in the five major historical periods of federal Indian relations: removal, assimilation, the Indian New Deal, termination and relocation, and self-determination.

## Removal

In the early nineteenth century, the population of the United States expanded rapidly at the same time that the federal government increased its political and military capabilities. The character of Indian-American relations changed after the War of 1812. The federal government increasingly pressured tribes settled east of the Appalachian Mountains to move west to the territory acquired in the Louisiana Purchase. Numerous treaties were negotiated by which the tribes relinquished most of their land and eventually were forced to move west.

Initially the federal government used bargaining and negotiation to accomplish removal, but many tribes resisted (Prucha 1984). However, the election of Andrew Jackson by a frontier constituency signaled the beginning of more forceful measures to accomplish removal. In 1830 Congress passed the Indian Removal Act, which mandated the eventual removal of the eastern tribes to points west of the Mississippi River, in an area which was to become the Indian Territory and is now the state of Oklahoma. Dozens of tribes were forcibly removed from the eastern half of the United States to the Indian Territory and newly created reservations in the west, a long process ridden with conflict and bloodshed.

As the nation expanded beyond the Mississippi River, tribes of the plains, southwest, and west coast were forcibly settled and quarantined on isolated reservations. This was accompanied by the so-called Indian Wars—a bloody chapter in the history of Indian-White relations (Prucha 1984; Utley 1963). This period in American history is especially remarkable because the U.S. government was responsible for what is unquestionably one of the largest forced migrations in history.

The actual process of removal spanned more than a half-century and affected nearly every tribe east of the Mississippi River. Removal often meant extreme hardships for American Indians, and in some cases this hardship reached legendary proportions. For example, the Cherokee removal has become known as the "Trail of Tears." In 1838, nearly 17,000 Cherokees were ordered to leave their homes and assemble in military stockades (Thornton 1987, p. 117). The march to the Indian Territory began in October and continued through the winter months. As many as 8,000 Cherokees died from cold weather and diseases such as influenza (Thornton 1987, p. 118).

According to William Hagan (1979), removal also caused the Creeks to suffer dearly as their society underwent a profound disintegration. The contractors who forcibly removed them from their homes refused to do anything for "the large number who had nothing but a cotton garment to protect them from the sleet storms and no shoes between them and the frozen ground of the last stages of their hegira. About half of the Creek nation did not survive the migration and the difficult early years in the West" (Hagan 1979, pp. 77–81). In the West, a band of Nez Perce men, women, and children, under the leadership of Chief Joseph, resisted resettlement in 1877. Heavily outnumbered, they were pursued by cavalry troops from the Wallowa valley in eastern Oregon and finally captured in Montana near the Canadian border. Although the Nez Perce were eventually captured and moved to the Indian Territory, and later to Idaho, their resistance to resettlement has been described by one historian as "one of the great military movements in history" (Prucha 1984, p. 541).

## Assimilation

Near the end of the nineteenth century, the goal of isolating American Indians on reservations and the Indian Territory was finally achieved. The Indian population also was near extinction. Their numbers had declined steadily throughout the nineteenth century, leading most observers to predict their disappearance (Hoxie 1984). Reformers urged the federal government to adopt measures that would humanely ease American Indians into extinction. The federal government responded by creating boarding schools and the allotment acts—both were intended to "civilize" and assimilate American Indians into American society by Christianizing them, educating them, introducing them to private property, and making them into farmers. American Indian boarding schools sought to accomplish this task by indoctrinating Indian children with the belief that tribal culture was an inferior relic of the past and that Euro-American culture was vastly superior and preferable. Indian children were forbidden to wear their native attire, to eat their native foods, to speak their native language, or to practice their traditional religion. Instead, they were issued Euro-American clothes, and expected to speak English and become Christians. Indian children who did not relinquish their culture were punished by school authorities. The curriculum of these schools taught vocational arts along with "civilization" courses.

The impact of allotment policies is still evident today. The 1887 General Allotment Act (the Dawes Severalty Act) and subsequent legislation mandated that tribal lands were to be allotted to individual American Indians, ... and the surplus lands left over from allotment were to be sold on the open market. Indians who received allotted tribal lands also received citizenship, farm implements, and encouragement from Indian agents to adopt farming as a livelihood (Hoxie 1984, Prucha 1984).

For a variety of reasons, Indian lands were not completely liquidated by allotment, many Indians did not receive allotments, and relatively few changed their lifestyles to become farmers. Nonetheless, the allotment era was a disaster because a significant number of allottees eventually lost their land. Through tax foreclosures, real estate fraud, and their own need for cash, many American Indians lost what for most of them was their last remaining asset (Hoxie 1984).

Allotment took a heavy toll on Indian lands. It caused about 90 million acres of Indian land to be lost, approximately two-thirds of the land that had belonged to tribes in 1887 (O'Brien 1990). This created another problem that continues to vex many reservations: "checkerboarding." Reservations that were subjected to allotment are typically a crazy quilt composed of tribal lands, privately owned "fee" land, and trust land belonging to individual Indian families. Checkerboarding presents reservation officials with enormous administrative problems when trying to develop land use management plans, zoning ordinances, or economic development projects that require the construction of physical infrastructure such as roads or bridges.

## The Indian New Deal

The Indian New Deal was short-lived but profoundly important. Implemented in the early 1930s along with the other New Deal programs of the Roosevelt administration, the Indian New Deal was important for at least three reasons. First, signaling the end of the disastrous allotment era as well as a new respect for American Indian tribal culture, the Indian New Deal repudiated allotment as a policy. Instead of continuing its futile efforts to detribalize American Indians, the federal government acknowledged that tribal culture was worthy of respect. Much of this change was due to John Collier, a long-time Indian rights advocate appointed by Franklin Roosevelt to serve as Commissioner of Indian Affairs (Prucha 1984).

Like other New Deal policies, the Indian New Deal also offered some relief from the Great Depression and brought essential infrastructure development to many reservations, such as projects to control soil erosion and to build hydro-electric dams, roads, and other public facilities. These projects created jobs in New Deal programs such as the Civilian Conservation Corps and the Works Progress Administration.

An especially important and enduring legacy of the Indian New Deal was the passage of the Indian Reorganization Act (IRA) of 1934. Until then, Indian self-government had been forbidden by law. This act allowed tribal governments, for the first time in decades, to reconstitute themselves for the purpose of overseeing their own affairs on the reservation. Critics charge that this law imposed an alien form of government, representative democracy, on traditional tribal authority. On some reservations, this has been an ongoing source of conflict (O'Brien 1990). Some reservations rejected the IRA for this reason, but now have tribal governments authorized under different legislation.

## Termination and Relocation

After World War II, the federal government moved to terminate its longstanding relationship with Indian tribes by settling the tribes' outstanding legal claims, by terminating the special status of reservations, and by helping reservation Indians relocate to urban areas (Fixico 1986). The Indian Claims Commission was a special tribunal created in 1946 to hasten the settlement of legal claims that tribes had brought against the federal government. In fact, the Indian Claims Commission became bogged down with prolonged cases, and in 1978 the commission was dissolved by Congress. At that time, there were 133 claims still unresolved out of an original 617 that were first heard by the commission three decades earlier (Fixico 1986, p. 186). The unresolved claims that were still pending were transferred to the Federal Court of Claims.

Congress also moved to terminate the federal government's relationship with Indian tribes. House Concurrent Resolution (HCR) 108, passed in 1953, called for steps that eventually would abolish all reservations and abolish all special programs serving American Indians. It also established a priority list of reservations slated for immediate termination. However, this bill and subsequent attempts to abolish reservations were vigorously opposed by Indian advocacy groups such as the National Congress of American Indians. Only two reservations were actually terminated, the Klamath in Oregon and the Menominee in Wisconsin. The Menominee reservation regained its trust status in 1975 and the Klamath reservation was restored in 1986.

The Bureau of Indian Affairs (BIA) also encouraged reservation Indians to relocate and seek work in urban job markets. This was prompted partly by the desperate economic prospects on most reservations, and partly because of the federal government's desire to "get out of the Indian business." The BIA's relocation programs aided reservation Indians in moving to designated cities, such as Los Angeles and Chicago, where they also assisted them in finding housing and employment. Between 1952 and 1972, the BIA relocated more than 100,000 American Indians (Sorkin 1978). However, many Indians returned to their reservations (Fixico 1986). For some American Indians, the return to the reservation was only temporary; for example, during periods when seasonal employment such as construction work was hard to find.

## Self-Determination

Many of the policies enacted during the termination and relocation era were steadfastly opposed by American Indian leaders and their supporters. As these programs became stalled, critics attacked them for being harmful, ineffective, or both. By the mid-1960s, these policies had very little serious support: Perhaps inspired by the gains of the Civil Rights movement, American Indian leaders and their supporters made "self-determination" the first priority on their political agendas. For these activists, self-determination meant that Indian people would have the autonomy to control their own affairs, free from the paternalism of the federal government.

The idea of self-determination was well received by members of Congress sympathetic to American Indians. It also was consistent with the "New Federalism" of the Nixon administration. Thus, the policies of termination and relocation were repudiated in a process that culminated in 1975 with the passage of the American Indian Self-Determination and Education Assistance Act, a profound shift in federal Indian policy. For the first time since this nation's founding, American Indians were authorized to oversee the affairs of their own communities, free of federal intervention. In practice, the Self-Determination

Act established measures that would allow tribal governments to assume a larger role in reservation administration of programs for welfare assistance, housing, job training, education, natural resource conservation, and the maintenance of reservation roads and bridges (Snipp and Summers 1991). Some reservations also have their own police forces and game wardens, and can issue licenses and levy taxes. The Onondaga tribe in upstate New York have taken their sovereignty one step further by issuing passports that are internationally recognized. Yet there is a great deal of variability in terms of how much autonomy tribes have over reservation affairs. Some tribes, especially those on large and well-organized reservations have nearly complete control over their reservations, while smaller reservations with limited resources often depend heavily on BIA services....

## CONCLUSION

Though small in number, American Indians have an enduring place in American society. Growing numbers of American Indians occupy reservation and other trust lands, and equally important has been the revitalization of tribal governments. Tribal governments now have a larger role in reservation affairs than ever in the past. Another significant development has been the urbanization of American Indians. Since 1950, the proportion of American Indians in cities has grown rapidly. These American Indians have in common with reservation Indians many of the same problems and disadvantages, but they also face other challenges unique to city life.

The challenges facing tribal governments are daunting. American Indians are among the poorest groups in the nation. Reservation Indians have substantial needs for improved housing, adequate health care, educational opportunities, and employment, as well as developing and maintaining reservation infrastructure. In the face of declining federal assistance, tribal governments are assuming an ever-larger burden. On a handful of reservations, tribal governments have assumed completely the tasks once performed by the BIA.

As tribes have taken greater responsibility for their communities, they also have struggled with the problems of raising revenues and providing economic opportunities for their people. Reservation land bases provide many reservations with resources for development. However, these resources are not always abundant, much less unlimited, and they have not always been well managed. It will be yet another challenge for tribes to explore ways of efficiently managing their existing resources. Legal challenges also face tribes seeking to exploit unconventional resources such as gambling revenues. Their success depends on many complicated legal and political contingencies.

Urban American Indians have few of the resources found on reservations, and they face other difficult problems. Preserving their culture and identity is an especially pressing concern. However, urban Indians have successfully adapted to city environments in ways that preserve valued customs and activities—powwows, for example, are an important event in all cities where there is a large Indian community. In addition, pan-Indianism has helped urban Indians set aside tribal differences and forge alliances for the betterment of urban Indian communities.

These alliances are essential, because unlike reservation Indians, urban American Indians do not have their own form of self-government. Tribal governments do not have jurisdiction over urban Indians. For this reason, urban Indians must depend on other strategies for ensuring that the needs of their community are met, especially for those new to city life. Coping with the transition to urban life poses a multitude of difficult challenges for many American Indians. Some succumb to these problems, especially the hardships of unemployment, economic deprivation, and related maladies such as substance abuse, crime, and violence. But most successfully overcome these difficulties, often with help from other members of the urban Indian community.

Perhaps the greatest strength of American Indians has been their ability to find creative ways for dealing with adversity, whether in cities or on reservations. In the past, this quality enabled them to survive centuries of oppression and persecution. Today this is reflected in the practice of cultural traditions that Indian people are proud to embrace. The resilience of American Indians is an abiding quality that will no doubt ensure that they will remain part of the ethnic mosaic of American society throughout the twenty-first century and beyond.

## REFERENCES

Cornell, Stephen. (1988). *The Return of the Native: American Indian Political Resurgence.* New York: Oxford University Press.

Fixico, Donald L. (1986). *Termination and Relocation: Federal Indian Policy, 1945–1960*: Albuquerque, NM: University of New Mexico Press.

Hagan, William T. (1979). *American Indians.* Chicago, IL: University of Chicago Press.

Hall, Thomas D. (1989). *Social Change in the Southwest, 1350–1880.* Lawrence, KS: University Press of Kansas.

Hoxie, Frederick E. (1984). *A Final Promise: The Campaign to Assimilate the Indians, 1880–1920.* Lincoln, NE: University of Nebraska Press.

Nagel, Joanne. (1986). "American Indian Repertoires of Contention." Paper presented at the annual meeting of the American Sociological Association, San Francisco, CA.

O'Brien, Sharon. (1990). *American Indian Tribal Governments.* Norman, OK: University of Oklahoma Press.

Prucha, Francis Paul. (1984). *The Great Father.* Lincoln, NE: University of Nebraska Press.

Snipp, C. Matthew and Gene F. Summers. (1991). "American Indian Development Policies," pp. 166–180 in *Rural Policies for the 1990s*, edited by Cornelia Flora and James A. Christenson. Boulder, CO: Westview Press.

Sorkin, Alan L. (1978). *The Urban American Indian.* Lexington, MA: Lexington Books.

Thornton, Russell. (1987). *American Indian Holocaust and Survival: A Population History since 1942*. Norman, OK: University of Oklahoma Press.

Utley, Robert M. (1963). *The Last Days of the Sioux Nation*. New Haven: Yale University Press.

Weibel-Orlando, Joan. (1991). *Indian Country, L.A.* Urbana, IL: University of Illinois Press.

## DISCUSSION QUESTIONS

1. How do the four different periods of state policy that Snipp identifies reveal different ways that the state has managed American Indian affairs?

2. Explain what Snipp means when he concludes that American Indians have been resilient even in the face of massive state-based social control?

# 22

# Race, Civil Rights, and Immigration Law After September 11, 2001:

## The Targeting of Arabs and Muslims

SUSAN M. AKRAM AND KEVIN R. JOHNSON

*The events of September 11, 2001 dramatically changed the lives of many citizens and non-citizens in the United States, but the federal response has meant dramatic changes for non-citizens who are Arabs and Muslims. Regardless of their nations of origin, many find that they are suspects and can be targets of hate crimes by Americans who see them as terrorists. These non-citizens are also subjected to many federal legal actions that not only communicate their perceived lack of loyalty, but put them at risk for incarceration.*

## INTRODUCTION

Although only time will tell, September 11, 2001, promises to be a watershed in the history of the United States. After the tragic events of that day, including the hijacking of four commercial airliners for use as weapons of mass destruction, America went to "war" on many fronts, including but not limited to military action in Afghanistan.

As needed and expected, heightened security measures and an intense criminal investigation followed. Almost immediately after the tragedy, Arabs and Muslims, as well as those "appearing" to be Arab or Muslim, were subject to crude forms of racial profiling.... Immediately after September 11, hate crimes against Arabs, Muslims, and others rose precipitously. In Arizona, a U.S. citizen claiming vengeance for his country killed a Sikh immigrant from India based on the mistaken belief that this turban–wearing, bearded man was "Arab."

Supporters and critics alike saw the federal government as "pushing the envelope" in restricting civil liberties in the name of national security.... The federal government has acted more swiftly and uniformly than the states ever could, with harsh consequences to the Arab and Muslim community in the United States.

SOURCE: *NYU Annual Survey of American Law.* 58 No. 3 (2002): 295–356.

That the reaction was federal in nature—and thus national in scope as well as uniform in design and impact, and faced precious few legal constraints—increased the severity of the impacts.

## ...The Immediate Impacts

The federal government responded with ferocity to the events of September 11. Hundreds of Arab and Muslim noncitizens were rounded up as "material witnesses" in the ongoing investigation of the terrorism or detained on relatively minor immigration violations. The dragnet provoked criticism as a poor law enforcement technique as well as a major intrusion on fundamental civil liberties. Congress swiftly passed the USA PATRIOT Act, which, among other things, allowed the government to detain suspected noncitizen "terrorists" for up to a week without charges, and bolstered federal law enforcement surveillance powers over citizens and noncitizens associated with "terrorism." President Bush issued a military order allowing alleged noncitizen terrorists, including those arrested in the United States, to be tried in military courts while guaranteed few rights....

To the extent that the U.S. responses to September 11 can be characterized as regulating immigration, existing caselaw affords considerable leeway to the political branches of the federal government. The Supreme Court has upheld immigration laws discriminating against noncitizens on the basis of race, national origin, and political affiliation that would patently violate the Constitution if the rights of citizens were at stake. The so-called "plenary power" doctrine creates a constitutional immunity from judicial scrutiny of substantive immigration judgments of Congress and the Executive Branch. The doctrine thus allows the federal government, through the immigration laws, to lash out at any group considered undesirable. Such authority increases exponentially when, as in the case of international terrorism, perceived foreign relations and national security matters are at issue. When immigration law and its enforcement rests primarily in the hands of the federal government, uniform, national civil rights deprivations may result.

The laws supporting much of the immigration and civil rights incursions of the "war on terrorism," including the plenary power doctrine, have been subjected to sustained scholarly criticism. In important ways, contemporary immigration law ignores a constitutional revolution that occurred in the area of civil rights over the latter half of the twentieth century. Nonetheless, much of this body of law—or, more accurately, the perceived immunity from any legal constraints—has guided the Bush administration's domestic responses to the legitimate concerns with terrorism.

**The Dragnet**  The events of September 11, 2001 understandably provoked an immediate federal governmental response. Heightened security measures were the first order of the day. Within a matter of weeks, the U.S. government arrested and detained in the neighborhood of 1000 people as part of the Justice Department's investigation into the September 11 attacks. The mass dragnet of men from many nations, with the largest numbers from Pakistan and Egypt, apparently failed to produce any direct links to the terrorists acts; about 100 were charged with minor crimes and another 500 were held in custody on immigration-related matters, such as having overstayed their temporary nonimmi-grant visas. Attorney General John Ashcroft admitted that minor immigration charges would be used to hold noncitizens while the criminal investigation continues.

Information remains sketchy about the persons detained by the U.S. govern-ment because the Attorney General has refused to release specific information about them, prompting criticism from U.S. Senator Russell Feingold. The Justice Department issued a rule barring disclosure of information about INS detainees held by state and local authorities, which survived a legal challenge. After September 11 the immigration courts began holding secret hearings in immigration cases involving Arab and Muslim noncitizens. In sum, the federal government's treatment of the detainees, and its treatment of Arab and Muslim noncitizens in immigration proceedings, was shrouded in secrecy....

The nature and conditions of the initial wave of mass arrests and detentions warrant consideration. Arab and Muslim detainees were held for weeks—in some instances, months—without charges filed against them and without being provided information about why federal authorities continued to detain them....

The dragnet did not end there. The Justice Department also sought to inter-view about 5000 men—almost all of them Arab or Muslim—between the ages of 18 and 33 who had arrived on nonimmigrant visas in the United States since January 1, 2000. There was *no* evidence that any of the 5000 had been involved in terrorist activities. Although technically "voluntary," the interviews with law enforcement authorities undoubtedly felt compulsory to many. Arab and Muslim fears of detention and deportation were reinforced by the November 2001 arrest of Mazen Al-Najjar, who had previously been held on secret evidence and re-leased after the government failed to provide evidence that he was engaged in terrorist activity. In March 2002, Attorney General Ashcroft asked U.S. attorneys to interview another 3000 or so Arab and Muslim noncitizens. Around the time of this announcement, the federal government conducted raids on Arab and Muslim offices and homes in search of terrorist connections.

The questioning of noncitizen Arabs and Muslims could be expected to alienate those interviewed, as well as the communities of which they are a part....

The questions directed at the noncitizens suggested that the Arabs and Mus-lims were prone to disloyalty. One line of questioning was as follows: "You should ask the individual if he noticed anybody who reacted in a surprising or inappropriate way to the news of September 11th attacks. You should ask him how he felt when he heard the news." This tracks questions reportedly asked by federal investigators soon after the bombing....

By almost all accounts, Muslims perpetrated the terrorism of September 11. A few Arab and Muslim noncitizens in the United States might have information about terrorist networks.... Nonetheless, the dragnet directed at all Arabs and Muslims is contrary to fundamental notions of equality and the individualized suspicion ordinarily required for a stop under the Fourth Amendment. It exem-plifies the excessive reliance on race in the criminal investigation, a frequent law enforcement problem, and shows how, once race (at least of nonwhites) enters the process, it can come to dominate an investigation. To target an entire minority group across the country for questioning is obviously over-inclusive. Over one million of persons of Arab ancestry in the United States, all of whom may feel threatened and under suspicion, cannot miss the message sent by the nature of the federal government's investigation.

In important ways, the September 11 dragnet carried out by the federal government resembles the Japanese internment during World War II although detention fortunately does not appear to be a current part of the U.S. government's strategy. National identity and loyalty are defined in part by "foreign" appearance, ambiguous as that may be. In some ways, the current treatment of Arabs and Muslims is more extralegal than the internment. No Executive Order authorizes the treatment of Arabs and Muslims; nor has there been a formal declaration of war. Moreover, nationality, which is more objective and easier to apply than religious and racial classifications, is not used as the exclusive basis for the measures. Rather, the scope of the investigation is broad and amorphous enough to potentially include all Arabs and Muslims, who may be natives of countries from around the world....

In the aftermath of September 11, the U.S. government arguably over-reacted and appeared to place little value on the liberty and equality interests of Arabs and Muslims. The response may be motivated in part by invidious hostility based on race and religion. With few legal constraints, the federal government adopted extreme action, with a largely symbolic impact in fighting terrorism, while having devastating impacts on Arabs and Muslims in the United States.

Moreover, the dragnet might prove to be a poor law enforcement technique. Racial profiling in criminal law enforcement has been criticized for alienating minority communities and making it more difficult to secure their much-needed cooperation in law enforcement. In a time when Arab and Muslim communities might be of assistance in investigating terrorism, they are being rounded up, humiliated, and discouraged from cooperating with law enforcement by fear of arrest, detention, and deportation.

Ultimately, such tactics suggest to noncitizens and citizens of Arab and Muslim ancestry in the United States that they are less than full members of U.S. society. The various efforts by the U.S. government, even while it claims not to discriminate against Arabs or Muslims, marginalize these communities. Consequently, the legal measures taken by the federal government reinforce deeply held negative stereotypes—foreign-ness and possibly disloyalty—about Arabs and Muslims.

## DISCUSSION QUESTIONS

1.  What do the recent experiences of Arabs and Muslims in the United States suggest about the social construction of race?

2.  What is the impact of being racially profiled on Arabs and Muslims who are living in the United States?

# 23

# Salsa and Ketchup:

## Transnational Migrants Straddle Two Worlds

PEGGY LEVITT

*This study of two immigrant communities in Boston shows how immigrants form communities that sustain their ethnic identity as they become citizens in the United States. Even though the backgrounds of immigrants may be diverse, the formation of community ties also fosters a continuing link to their homes of origin, creating transnational identities that give them a dual identity as well as dual citizenship.*

The suburb, with its expensive homes with neatly trimmed lawns and sport-utility-vehicles, seems like any other well-to-do American community. But the mailboxes reveal a difference: almost all are labeled "Patel" or "Bhagat." Over the past two decades, these families moved from the small towns and villages of Gujarat State on the west coast of India, first to rental apartments in northeastern Massachusetts and then to their own homes in subdivisions outside Boston. Casual observers watching these suburban dwellers work, attend school, and build religious congregations might conclude that yet another wave of immigrants is successfully pursuing the American dream. A closer look, however, reveals that they are pursuing Gujarati dreams as well. They send money back to India to open businesses or improve family homes and farms. They work closely with religious leaders to establish Hindu communities in the United States, and also to strengthen religious life in their homeland. Indian politicians at the state and national level court these emigrants' contributions to India's political and economic development.

The Gujarati experience illustrates a growing trend among immigrants to the United States and Europe. In the 21st century, many people will belong to two societies at the same time. Researchers call those who maintain strong, regular ties to their homelands and who organize aspects of their lives across national borders "transnational migrants." They assimilate into the country that receives them, while sustaining strong ties to their homeland. Assimilation and transnational relations are not mutually exclusive: they happen simultaneously and influence each other. More and more, people earn their living, raise their family, participate in religious communities, and express their political views across national borders.

SOURCE: From *Contexts* 3(2):20–26. Copyright © 2004 by the American Sociological Association. All rights reserved. Reprinted by permission.

Social scientists have long been interested in how newcomers become American. Most used to argue that to move up the ladder, immigrants would have to abandon their unique customs, language and values. Even when it became acceptable to retain some ethnic customs, most researchers still assumed that connections to homelands would eventually wither. To be Italian American or Irish American would ultimately have much more to do with the immigrant experience in America than with what was happening back in Italy or Ireland. Social scientists increasingly recognize that the host-country experiences of some migrants remain strongly influenced by continuing ties to their country of origin and its fate.

These transnational lives raise fundamental issues about 21st century society. What are the rights and responsibilities of people who belong to two nations? Both home- and host-country governments must decide whether and how they will represent and protect migrants and what they can demand from them in return. They may have to revise their understandings of "class" or "race" because these terms mean such different things in each country. For example, expectations about how women should balance work and family vary considerably in Latin America and in the United States. Both home- and host-country social programs may have to be reformulated, taking into account new challenges and new opportunities that arise when migrants keep one foot in each of two worlds.

## TWO CASES: DOMINICANS AND GUJARATIS IN BOSTON

My research among the Dominican Republic and Gujarati immigrants who have moved to Massachusetts over the past three decades illustrates the changes that result in their origin and host communities. Migration to Boston from the Dominican village of Miraflores began in the late 1960s. By the early 1990s, nearly two-thirds of the 550 households in Miraflores had relatives in the Boston area, most around the neighborhood of Jamaica Plain, a few minutes from downtown. Migration has transformed Miraflores into a transnational village. Community members, wherever they are, maintain such strong ties to each other that the life of this community occurs almost simultaneously in two places. When someone is ill, cheating on their spouse, or finally granted a visa, the news spreads as fast on the streets of Jamaica Plain, Boston as it does in Miraflores, Dominican Republic.

Residents of Miraflores began to migrate because it became too hard to make a living at farming. As more and more people left the fields of the Dominican Republic for the factories of Boston, Miraflores suffered economically. But as more and more families began to receive money from relatives in the United States (often called "remittances"), their standard of living improved. Most households can now afford the food, clothing, and medicine for which previous generations struggled. Their homes are filled with the TVs, VCRs, and other appliances their migrant relatives bring them. Many have been able to renovate their houses, install indoor plumbing, even afford air conditioning. With money donated in Boston and labor donated in Miraflores, the community built an aqueduct and baseball stadium, and renovated the local school and health clinic. In short, most families live better since migration began, but they depend on money earned in the United States to do so.

Many of the Mirafloreños in Boston live near and work with one another, often at factories and office-cleaning companies where Spanish is the predominant language. They live in a small neighborhood, nestled within the broader Dominican and Latino communities. They participate in the PTA and in the neighborhood organizations of Boston, but feel a greater commitment toward community development in Miraflores. They are starting to pay attention to elections in the United States, but it is still Dominican politics that inspires their greatest passion. When they take stock of their life's accomplishments, it is the Dominican yardstick that matters most.

The transnational character of Mirafloreños' lives is reinforced by connections between the Dominican Republic and the United States. The Catholic Church in Boston and the Church on the island cooperate because each feels responsible for migrant care. All three principal Dominican political parties campaign in the United States because migrants make large contributions and also influence how relatives back home vote. No one can run for president in the Dominican Republic, most Mirafloreños agree, if he or she does not campaign in New York. Conversely, mayoral and gubernatorial candidates in the northeastern United States now make obligatory pilgrimages to Santo Domingo. Since remittances are one of the most important sources of foreign currency, the Dominican government instituted policies to encourage migrants' long-term participation without residence. For example, under the administration of President Leonel Fernández (1996-2000), the government set aside a certain number of apartments for Dominican emigrants in every new construction project it supported. When they come back to visit, those of Dominican origin, regardless of their passport, go through the customs line for Dominican nationals at the airport and are not required to pay a tourist entry fee.

## RELIGIOUS TIES

The people from Miraflores illustrate one way migrants balance transnational ties and assimilation, with most of their effort focused on their homeland. The Udah Bhagats, a sub-caste from Gujarat State, make a different set of choices. They are more fully integrated into certain parts of American life, and their homeland ties tend to be religious and cultural rather than political. Like Gujaratis in general, the Udah Bhagats have a long history of transnational migration. Some left their homes over a century ago to work as traders throughout East Africa. Many of those who were forced out of Africa in the 1960s by local nationalist movements moved on to the United Kingdom and the United States instead of moving back to India. Nearly 600 families now live in the greater Boston region.

The Udah Bhagats are more socially and economically diverse than the Mirafloreños. Some migrants came from small villages where it is still possible to make a good living by farming. Other families, who had moved to Gujarati towns a generation ago, owned or were employed by small businesses there. Still others, from the city of Baroda, worked in engineering and finance before migrating. About half of the Udah Bhagats now in Massachusetts work in factories or ware-houses, while the other half work as engineers, computer programmers or at the small grocery stores they have purchased. Udah Bhagats in Boston also send re-mittances home, but for special occasions or when a particular need arises, and the recipients do not depend on them. Some still own a share in the family farm or have invested in Gujarati businesses, like one man who is a partner in a computer school. Electronics, clothing, and appliances from the United States line the shelves of homes in India, but the residents have not adopted western lifestyles as much as the Miraflorenos. The Gujarati state government has launched several initiatives to stimulate investment by "Non-Resident Gujaratis," but these are not central to state economic development policy.

In the United States, both professional and blue-collar Gujaratis work along-side native-born Americans; it is their family and religious life that is still tied to India. Some Bhagat families have purchased houses next door to each other. In an American version of the Gujarati extended family household, women still spend long hours preparing food and sending it across the street to friends and relatives. Families gather in one home to do puja, or prayers, in the evenings. Other families lives in mixed neighborhoods, but they too spend much of their free time with other Gujaratis. Almost everyone still speaks Gujarati at home. While they are deeply grateful for the economic opportunities that America of-fers, they firmly reject certain American values and want to hold fast to Indian culture.

As a result, Udah Bhagats spend evenings and weekends at weddings and holiday celebrations, prayer meetings, study sessions, doing charitable work, or trying to recruit new members. Bhagat families conduct these activities within religious organizations that now operate across borders. Rituals, as well as chari-table obligations, have been redefined so they can be fulfilled in the United States but directly supervised by leaders back in India. For example, the Devotional Associates of Yogeshwar or the Swadhyaya movement requires fol-lowers back in Gujarat to dedicate time each month to collective farming and fishing activities; their earnings are then donated to the poor. An example of such charitable work in Boston is families meeting on weekends to assemble circuit boards on sub-contract for a computer company. For the Udah Bhagats, religious life not only reaffirms their homeland ties but also erects clear barriers against aspects of American life they want to avoid. Not all Indians are pleased that Hindu migrants are so religious in America. While some view the faithful as important guardians of the religious flame, others claim that emigrants abroad are the principal underwriters of the recent wave of Hindu nationalism plaguing In-dia, including the Hindu-Muslim riots that took place in Ahmedabad in 2002.

# THE RISE OF TRANSNATIONAL MIGRATION

Not all migrants are transnational migrants, and not all who take part in transnational practices do so all the time. Studies by Alejandro Portes and his colleagues reveal that fewer than 10 percent of the Dominican, Salvadoran, and Colombian migrants they surveyed regularly participated in transnational economic and political activities. But most migrants do have occasional transnational contacts. At some stages in their lives, they are more focused on their country of origin, and at other times more committed to their host nation. Similarly, they climb two different social ladders. Their social status may improve in one country and decline in the other.

Transnational migration is not new. In the early 1900s, some European immigrants also returned to live in their home countries or stayed in America while being active in economic and political affairs at home. But improvements in telecommunications and travel make it cheaper and easier to remain in touch than ever before. Some migrants stay connected to their homelands daily through e-mail or phone calls. They keep their fingers on the pulse of everyday life and weigh in on family affairs in a much more direct way than their earlier counterparts. Instead of threatening the disobedient grandchild with the age-old refrain, "wait until your father comes home," the grandmother says, "wait until we call your mother in Boston."

The U.S. economy welcomes highly-educated, professional workers from abroad, but in contrast to the early 20th century, is less hospitable to low-skilled industrial workers or those not proficient in English. Because of poverty in their country of origin and insecurity in the United States, living across borders has become a financial necessity for many less skilled migrant workers. At the same time, many highly skilled, professional migrants choose to live transnational lives; they have the money and know-how to take advantage of economic and political opportunities in both settings. These days, America tolerates and even celebrates ethnic diversity—indeed, for some people, remaining "ethnic" is part of being a true American—which also makes long-term participation in the homeland and putting down roots in the United States easier.

Nations of origin are also increasingly supportive of long-distance citizenship, especially countries that depend on the remittances and political clout of migrants. Immigrants are no longer forced to choose between their old and new countries as they had to in the past. Economic self-sufficiency remains elusive for small, non-industrialized countries and renders them dependent on foreign currency, much of it generated by migrants. Some national governments actually factor emigrant remittances into their macro-economic policies and use them to prove credit-worthiness. Others, such as the Philippines, actively promote their citizens as good workers to countries around the world. Transnational migrants become a key export and their country of origin's main connection to the world economy. By footing the bill for school and road construction back home, transnational migrants meet goals that weak home governments cannot. The increasingly interdependent global economy requires developing nations to tie themselves more closely to trade partners. Emigrant communities are also potential ambassadors who can foster closer political and economic relations.

# THE AMERICAN DREAM GOES TRANSNATIONAL

Although few immigrants are regularly active in two nations, their efforts, combined with those of immigrants who participate occasionally, add up. They can transform the economy, culture and everyday life of whole regions in their countries of origin. They transform notions about gender relations, democracy, and what governments should and should not do. For instance, many young women in Miraflores, Dominican Republic no longer want to marry men who have not migrated because they want husbands who will share the housework and take care of the children as the men who have been to the United States do. Other community members argue that Dominican politicians should be held accountable just like Bill Clinton was when he was censured for his questionable real estate dealings and extramarital affairs.

Transnational migration is therefore not just about the people who move. Those who stay behind are also changed. The American-born children of migrants are also shaped by ideas, people, goods, and practices from outside—in their case, from the country of origin—that they may identify with during particular periods in their lives. Although the second generation will not be involved with their ancestral homes in the same ways and with the same intensity as their parents, even those who express little interest in their roots know how to activate these connections if and when they decide to do so. Some children of Gujaratis go back to India to find marriage partners and many second-generation Pakistanis begin to study Islam when they have children. Children of Mirafloreños born in the United States participate actively in fund-raising efforts for Miraflores. Even Dominican political parties have established chapters of second-generation supporters in the United States.

Transnational migrants like the Mirafloreños and the Udah Bhagats in Boston challenge both the host and the origin nations' understanding of citizenship, democracy, and economic development. When individuals belong to two countries, even informally, are they protected by two sets of rights and subject to two sets of responsibilities? Which states are ultimately responsible for which aspects of their lives? The Paraguayan government recently tried to intercede on behalf of a dual national sentenced to death in the United States, arguing that capital punishment is illegal in Paraguay. The Mexican government recently issued a special consular ID card to all Mexican emigrants, including those living without formal authorization in the United States. More than 100 cities, 900 police departments, 100 financial institutions, and 13 states accept the cards as proof of identity for obtaining a drivers' license or opening a bank account. These examples illustrate the ways in which countries of origin assume partial responsibility for emigrants and act on their behalf.

Transnational migration also raises questions about how the United States and other host nations should address immigrant poverty. For example, should transnationals qualify for housing assistance in the United States at the same time that they are building houses back home? What about those who cannot fully support themselves here because they continue to support families in their homelands? Transnational migration also challenges policies of the nations of origin. For example, should social welfare and community development programs discriminate between those who are supported by remittances from the United States and those who have no such outside support? Ideally, social programs in the two nations should address issues of common concern in coordination with one another.

There are also larger concerns about the tension between transnational ties and local loyalties. Some outside observers worry when they see both home country and U.S. flags at a political rally. They fear that immigrants' involvement in homeland politics means that they are less loyal to the United States. Assimilation and transnational connections, however, do not have to conflict. The challenge is to find ways to use the resources and skills that migrants acquire in one context to address issues in the other. For example, Portes and his colleagues find that transnational entrepreneurs are more likely to be U.S. citizens, suggesting that becoming full members of their new land helped them run successful businesses in their countries of origin. Similarly, some Latino activists use the same organizations to promote participation in American politics that they use to mobilize people around homeland issues. Some of the associations created to promote Dominican businesses in New York also played a major role in securing the approval of dual citizenship on the island.

These are difficult issues and some of our old solutions no longer work. Community development efforts directed only at Boston will be inadequate if they do not take into account that Miraflores encompasses Boston and the island, and that significant energy and resources are still directed toward Miraflores. Education and health outcomes will suffer if policymakers do not consider the many users who circulate in and out of two medical and school systems. As belonging to two places becomes increasingly common, we need approaches to social issues that not only recognize, but also take advantage of, these transnational connections.

## REFERENCES

Guarnizo, Luis, Alejandro Portes, and William Haller. "Assimilation and Transnationalism: Determinants of Transnational Political Action among Contemporary Migrants." *American Journal of Sociology* 108(2003): 1211–48.

Portes, Alejandro, William Haller, and Luis Guarnizo. "Transnational Entrepreneurs: The Emergence and Determinants of an Alternative Form of Immigrant Economic Adaptation." *American Sociological Review* 67 (2002): 278–298.

## DISCUSSION QUESTIONS

1.  How are twenty-first century immigrants different from previous generations?

2.  Having read this article, do you think immigrants have to give up a former national identity to become assimilated into a new society?

3.  What are some of the challenges as well as the benefits of dual citizenship?

# Student Exercises

1.  Voting is an important sign of citizenship, yet before 2008, a very low percentage of citizens actually voted. Voting among young people continues to be tentative. Interview ten students from other classes that you are taking and ask them whether or not they voted in the last election and why. Then ask what their reasons are for participating or not in this act of citizenship. Having done so, explain why you think voter turnout among the youth is low. Is race a significant factor in youth voting?

2.  Globalization means that people are often moving to work in other nations—places where they might or might not have rights as citizens. Can you imagine making the decision to relocate to another nation? What might make you do so? What rights would you expect in your new home? What would you do if you were not fully accepted? Having imagined your relocation, think about how this exercise might compare (or not) to the experiences of groups examined in this section.

# Section VI

# The Changing Face of America: Immigration

## Elizabeth Higginbotham
## and Margaret L. Andersen

The United States is a nation of immigrants. Virtually everyone living here has an immigrant experience somewhere in their family's history—with the exception of indigenous Native Americans and African Americans whose ancestors were forced into slavery. At times, immigrants are celebrated in America. Other times, they are chastised and defined as outsiders or aliens. Even our national symbols of immigration have sometimes shone; at other times, they have lost their luster. Ellis Island, the point of entry for most immigrants coming to the East Coast, operated from 1892 until 1954. It was in disrepair for decades, but then was refurbished and opened as a museum in 1990. Like Angel Island, the major entry point on the West Coast, these sites are now marketed as tourist destinations. The story of immigration to the various shores is a complicated one, and the legacy of inequality shapes how we think about our borders and how we define ourselves as a nation. This section explores immigration legislation and the experiences of immigrants. At this writing, the United States Congress has not passed new legislation about immigration, but it will likely do so in the future. Together, these articles introduce you to historical issues and some of the concerns that inform these contemporary debates.

Patterns of immigration are global trends—people are pushed and pulled between nations for economic, political, and religious reasons. Once it became independent of England, the United States found immigrants arriving from many different shores. Germans and other Europeans who entered the United States after the War of 1812 found the native White population a little distant, but most of the immigrants were able to settle in rural and opening sections of the country such as Ohio and Illinois, where they hoped to replicate the lives they left in Europe. Although these newcomers were foreign to native-born White Americans, they were mostly farmers and craftspeople whose way of life did not threaten much of the established population, and over time they were considered Americans.

The Irish, the majority of whom settled in cities in the 1820s to 1850s, faced harsh discrimination for the little work available. Immigrant men worked in the harbors and traveled to work on canals and other industrial sites, while women did domestic work. Clustered in urban centers, such as New York and Boston, the Irish eventually found a niche in society, gaining citizenship and playing a key role in urban and state politics. When immigrants arrived from eastern and southern Europe during the development of our industrial economy (1880s to 1920s), the Irish were a group with some degree of political power in some northeastern cities.

These immigrants from different ports were defined by the majority group as racially inferior. Alarmed at the prospect of downgrading the "stock" of White Americans, people took actions that reflected their beliefs about racial hierarchies. Native-born White Americans used their political power to shape who could enter the country and become future citizens.

The U.S. Congress has the right to regulate immigration as part of its mandate to govern foreign affairs (Ngai 2003). These regulations reflect economic and political power as well as pressure from voting citizens, who are often conflicted about the growing presence of "foreign" people. In 1882, Congress passed the Chinese Exclusion Act, which limited Chinese immigrants to being merchants, thus excluding the working class from entry. This legislation marks "the moment when the golden doorway of admission to the United States began to narrow" (Daniels 2004: 3).

Immigrants from southern and eastern Europe participated in expanding industries (e.g., steel, oil, glass, and railroads) that transformed the nation from an agricultural economy to a major industrial power. Their labor was instrumental in changing the nation from a land of independent farmers and craftsmen to a nation where economic power was concentrated in the hands of a few robber barons. Dramatic changes were welcomed by big businesses and growing corporations, but new immigrants were seen as a threat to many native-born White Americans, who were not directly profiting from these economic changes. Limiting immigration was a means of holding onto their established ways. The ethnic and religious differences between new immigrants and native-born White Americans made scapegoating the former easy, as one population was urban and the other rural, with images of these new groups shaped by the press. Native-born White Americans used their political power to advocate for closing our borders. You might think about how this historical pattern is being played out now in contemporary immigration politics and policies.

Historically, industrial employers wanted to keep the borders open so that they would always be able to hire people at low wages, replacing their current workers with immigrants if the former demanded higher wages. National leaders, including presidents, sided with the big industries because such businesses were important to expanding the nation's economic and political power. Over time, regulation became the rule, with immigration inspectors, as part of the Immigration and Naturalization Service (INS), admitting some and rejecting others, depending on the nation's labor needs and relations with other countries.

The Johnson–Reed Act of 1924 limited immigration to 150,000 people per year and mandated a quota system based on the nation of origin. This act ended immigration for most Asians. The 1924 law, with many modifications, remained the framework of U.S. immigration policy until 1965. By establishing national origin quotas, powerful groups ensured that people from northern and western Europe could continue to enter the United States. Thus, because they held citizenship rights, Irish politicians got a higher quota for Ireland, while the quotas for those from southern and eastern European nations were lower than the numbers entering prior to 1920.

In 1965, the Hart–Celler Act changed the old system by offering the same quota to all eligible nations. Many southern and eastern European families were reunited, and previously excluded groups were granted access. By the end of the twentieth century, the majority of immigrants were coming from Latin American, Asia, and the Caribbean. Again our economy was undergoing change, but from a manufacturing-based economy to one focused on service work— both highly skilled service work (such as doctors, engineers, and college professors) and less-skilled, low-wage service work (such as janitors, domestic workers, and fast-food workers). Workers are necessary for these tasks, and issues of citizenship and race have emerged as key to economic and civic opportunities. After decades of limited immigration, new policies and a new economy have made the United States a major destination for immigrants from around the globe and have introduced some new dilemmas.

Mae M. Ngai ("Impossible Subjects: Illegal Aliens and the Making of Modern America") explains how recent laws have created new categories of "illegal aliens" as well as militarizing our borders, particularly that with Mexico. Her historical perspective puts the current debate about the fate of millions of undocumented workers in a global context.

Nancy Foner ("From Ellis Island to JFK: Education in New York's Two Great Waves of Immigration") compares the pursuit of education by earlier European immigrants (1880–1920) with those coming from Latin America, Asia, and the Caribbean to New York City since the 1965 law. Foner clarifies how the economic progress of earlier, uneducated immigrants translated into educational advantages for their children. Immigrants entering the country in a postindustrial economy are now arriving when higher education is essential for economic security. Many parents make sacrifices so that their children gain those credentials, a trend that promotes their children's incorporation into the mainstream.

Labor needs are attracting immigrants to new regions of the country, particularly the Midwest and the South. Charles Hirschman and Douglas S. Massey ("Places and Peoples: The New American Mosaic") identify how these immigrants are responding to the pulls offered by industrial restructuring in the United States and the pushes within their homelands, where there are fewer opportunities than abroad. Many small and medium-sized firms are seeking a flexible workforce in non-unionized jobs that many native-born workers shun, but that immigrants are eager to take. Now, as in the past, it is the need for labor that brings immigrants to our shores.

Part of the changing face of America is the composition of young people. One in every five children in our nation has either one or two immigrant parents. Hispanics are the largest and youngest minority group, making it important to understand the circumstances of these young people. The Pew Research Center ("Between Two Worlds: How Young Latinos Come of Age in America") details the circumstances of young Latinos through a large national survey. There is great diversity within this group, as it includes immigrants, this second generation, and young people with deep roots on this country. The majority of young Latinos are born here; many forge a complex identity as they grow up in the United States. Understanding their situations can help us think about supporting their transition and growth.

## REFERENCES

Daniels, Roger. 2004. *Guarding the Golden Door*. New York: Hill and Wang.

Ngai, Mae M. 2003. *Impossible Subjects: Illegal Aliens and the Making of Modern America*. Princeton, NJ: Princeton University Press.

## FACE THE FACTS: ORIGINS OF THE FOREIGN-BORN POPULATION

**Origins of Foreign-born
U.S. Population, 2008**

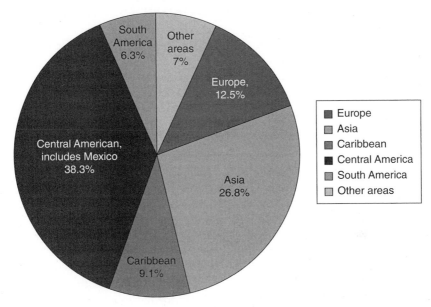

SOURCE: U.S. Census Bureau. 2009. *Statistical Abstract of the United States.* Washington, DC: U.S. Department of Commerce. www.census.gov

***Think about it:*** This pie chart shows you the country of origin of the total U.S. foreign-born population in 2008. Is this population more or less diverse than you thought? What factors might influence immigration from these different regions of the world?

# 24

# Impossible Subjects:
## Illegal Aliens and the Making of Modern America

### MAE M. NGAI

*Ngai reviews the changes in immigration patterns in the United States in recent years and also documents the changes in policy that have become more restrictive and punishing, even while serving the needs of the contemporary labor market. She also assesses alternatives that would have a less punitive effect on immigrants and their families.*

If the Johnson-Reed Act ushered in the most restrictionist era in American immigration law, the Hart-Celler Act, which ended that period, altered and refined but in no way overturned the regime of restriction. Certainly, patterns of immigration changed dramatically in the period after 1965, as the abolition of quotas based on national origin opened the way for increased immigration from the third world. Yet Hart-Celler's continued commitment to numerical restriction, especially its imposition of quotas on Western Hemisphere countries, ensured that illegal immigration would continue and, in fact, increase. During the late twentieth century, illegal immigration became perceived as the central and singularly intractable problem of immigration policy and became a lightning rod in domestic national politics generally. Moreover, legal reform in the area of aliens' rights of due process was decidedly uneven over the past thirty years. Together, these trends have contributed to yet another ethno-racial remapping of the nation.

Notwithstanding the Hart-Celler Act's intentions to keep migration at modest levels, legal immigration into the United States has climbed steadily since 1965. Immigration rose sharply after 1990, when Congress raised the numerical ceiling on immigration by 35 percent in response to the 1980s boom in the U.S. economy and concomitant demands for labor in low-wage sectors and in some high ones as well. By the mid-1990s immigration approached one million a year. Refugee admissions also increased in the mid-and late 1970s from Southeast Asia in the wake of the Vietnam War and in the early 1990s from Russia and former Soviet-bloc countries. Just as important, if not more so, since 1980 Europeans have accounted for only 10 percent of annual legal admissions; Mexico and Caribbean nations account for half the new immigrants and Asia for 40 percent.

SOURCE: Mae M. Ngai. 2004. *Impossible Subjects: Illegal Aliens and the Making of Modern America.* Princeton: Princeton University Press.

Paralleling the increase in legal admission, illegal immigration also increased dramatically after 1965.... Unauthorized migration increased radically in the late 1960s and mid-1970s as a direct result of the imposition of quotas on Western Hemisphere countries, especially Mexico. New illegal-migrant streams also emerged, comprising undocumented migrants displaced from war-torn countries in Central America and from the instabilities of economic transition in China. During the 1990s removals increased 40 percent from the previous decade; by the turn of the twenty-first century the INS was removing some 1.8 million illegal aliens a year. Yet the INS estimates that the illegal population still accretes by 275,000 persons a year.

During the 1980s and 1990s Americans manifested a schizophrenic attitude towards illegal immigration. On the one hand, undocumented workers provided labor for low-wage agricultural, manufacturing, and service industries, as well as in the informal economy of domestic work, housing construction, and other services that support the lifestyle of not just the prosperous but many middle-class Americans. On the other hand, critics worried that the nation had lost control of its borders and was being overrun by undesirable illegal aliens; estimates of the illegal population in the early 1980s ranged from 2 to 8 million. After much contentious debate, Congress passed the Immigration Reform and Control Act of 1986. IRCA provided amnesty for some 2.7 million undocumented immigrants and sought, unsuccessfully, to curb future illegal entries by imposing sanctions against employers who knowingly hire illegal workers and by greater enforcement efforts.

A stunning militarization of the U.S.-Mexico border was accomplished during the 1990s. Congress authorized a doubling of the Border Patrol's force, the erection of fences and walls, and the deployment of all manner of high-tech surveillance on land and by air. Enforcement at the southwest border now costs taxpayers $2 billion a year. The militarization of the border has not stopped illegal entry, but it has made it more difficult and dangerous. The INS's Operation Gatekeeper, initiated in 1994, pushed illegal entries from the San Diego–Tijuana border area to remote desert sections of California and Arizona, which has increased the likelihood that migrants will die of exposure before reaching safety. Others succumb as a result of desperate attempts to enter in the sealed holds of ships, trucks, and boxcars. Many pay thousands of dollars to smugglers or guides to make these dangerous journeys, and those who survive the trip then often remain indentured for years in sweatshops before they can pay their debts. Illegal aliens are at once familiar and invisible to middle-class Americans: their labor is desired but the difficulties of their lives for the most part go unnoticed. The INS now estimates there are at least 5 million undocumented immigrants residing in the United States—of which over half are from Mexico—and debate rages on about what to do about it.

All told, net annual immigration (legal and illegal immigration less emigration) over the last decade averaged about 700,000. At less than one-half of 1 percent of the total U.S. population that is not a terribly large number, but the rate of increase and the shift to third-world immigration have had considerable consequence. The foreign-born now account for nearly 10 percent of the population, which is much greater than in 1970, when they were less than 5 percent, but still less than the historical high of 14 percent before World War I. And, while concentrated in six states, immigrants live and work in every region of the country and are thus a visible presence throughout the United States.

The new demographics have both enhanced the politics of diversity and multiculturalism and provoked nativist sentiment and campaigns, like California's Proposition 187 and the English-only movement, suggesting that Asians and Latinos continue to be constructed as foreign racial others, even as their positions in the American racial landscape have changed. In the last quarter of the twentieth century, Latinos and Latinas and Asian Americans were the fastest growing ethno-racial groups in the United States. The 2000 Census counted 32.8 million Latinos and Latinas in the United States; these diverse communities—Mexican, Cuban, Puerto Rican, Dominican, Salvadoran—collectively account for 12 percent of the total U.S. population. Because Mexicans and Central Americans make up the vast majority of undocumented immigrants today, they experience persistently high levels of poverty and disfranchisement and embody the stereotypical illegal alien. The growth of a dynamic transborder economy and culture in the Southwest has engendered new, hybrid cultures and identities, complicating ideas about national difference and belonging. At the same time, some Latinos and Latinas have achieved a measure of structural assimilation into the mainstream of American society. The business and professional classes (particularly among Cuban Americans and Mexican Americans) have garnered considerable political influence, particularly in Texas and Florida; and Latinos and Latinas such as Ricky Martin, Jennifer Lopez, and Cameron Diaz have broken into the top echelons of the entertainment industry.

The end of Asiatic exclusion, the post-1965 influx of Asian immigrant professionals and, more recently, the arrival of wealthy elites from Hong Kong, South Korea, and Singapore have repositioned Asian Americans as "model minorities." While Asian Americans have in many ways overcome the status of alien citizenship that defined them during exclusion era—through access to naturalized citizenship and occupational and residential mobility—the model minority stereotype elides the existence of large numbers of working-class immigrants, undocumented workers, and refugees. Moreover, the model minority idea reproduces Asian Americans' foreignness. Critics contend it is a new, pernicious form of "yellow peril" that rests on essentialized notions of Asian culture and breeds new forms of race discrimination (occupational segregation and glass ceilings, reverse quotas against Asians in college admissions).

The growth in the size of Asian and Latino and Latina populations has made it manifestly clear that the question of race in the United States is not just about the status of African Americans or black-white relations. Yet, in an important sense, contemporary multiracial politics remain informed by the historically dominant black-white paradigm, and are also implicated in reproducing it. For example, for Afro-Caribbean and African immigrants, foreignness becomes subsumed by racial "blackness," as these immigrants join, variegate, and complicate established African American communities.

Social scientists, journalists, and politicians have used both Asian and Latino and Latina immigrant success narratives to discipline African Americans. Using cultural stereotypes to occlude economic structures, the strategy deploys immigrant exemplars of hard work, thrift, and self-reliance against an alleged "culture of poverty" among native-born minorities (African Americans as well as Puerto Ricans) to explain the persistence of unemployment and poverty among the latter. In a twist to this argument, some liberals blame immigrants for undercutting native-born workers' wages and for displacing African Americans from jobs in the lower strata of the workforce. While economists debate the costs and benefits of immigrant labor to the economy, the more troubling aspect of the new liberal restrictionism is that it posits minority groups in a zero-sum calculus and regards immigrants as outsiders whose claims on society are less deserving than those of citizens.

…What do these trends portend for the central figure of this study, the illegal alien? As in the mid-twentieth century, illegal immigration today results from the confluence of two conditions: macroeconomic structures that push and pull migration from developing countries to low-wage sectors of the United States, and positive domestic law, which sets qualitative terms and quantitative limits on immigration. The first dynamic is also what makes enforcement of the second difficult and is why enforcement alone has never been a solution for illegal immigration.

It may be that illegal immigration will persist as long as the world remains divided into sovereign nation-states and as long as there remains an unequal distribution of wealth among them. Yet, this is not to say that no solutions exist short of open borders, on the one hand, or a totalitarian police state, on the other. We might consider, instead, strategies aimed at altering the push-and-pull dynamics of migration from the developing world to the United States. Trade and investment policies that strengthen the economies of sending nations would lessen pressures on emigration. Raising the numerical ceiling on legal migration, reestablishing a statute of limitations on deportation, enforcing wage and hour standards, and facilitating collective bargaining for workers in agriculture and low-wage industries would counter the reproduction of undocumented workers as an exploited underclass.

Amnesty for undocumented immigrants has diverse political support, including from the presidents of the United States and Mexico, business and organized labor, and human rights activists. Amnesty is a humane gesture that recognizes the claims to legitimate membership that come with settlement. Still, it is a limited reform that addresses only past illegal migration. Some advocates of amnesty (now called "regularization") propose to curb future unlawful entries by channeling would-be illegal migrants into legal, temporary guest-worker programs. But guest-worker programs pose the moral problem of creating a caste of second-class persons that we exploit economically but deny full membership to in the polity. Historical experience also suggests that guest workers do not all quietly return home when their services are no longer needed. Many remain, and in so doing become illegal aliens.

Some legal scholars propose eliminating or vastly lowering restrictions on migration from Mexico and Canada, a plan that has antecedents in the pre-1965 policy of Pan-Americanism. They argue for extending the North American Free Trade Agreement, which has already lowered national barriers in the hemisphere to ease the flow of capital and products, to migration, citing the European Union as a more comprehensive model of integrated regionalism. In fact, some critics believe NAFTA's one-sided emphasis on free trade will lead to greater pressures on illegal migration. Some observers fear that the elimination of tariffs on American agricultural products entering Mexico in 2003 threatens to "crush the ability of millions of Mexican farmers to survive and drive them north."

Indeed, while the hardened nationalism that characterized the world order for so much of the last century has dramatically relaxed in many realms of economic and cultural exchange, it remains in force in other fields, notably politics. Nation-states remain committed to restrictive immigration policies, so that states' regulation of the transnational movement of bodies differs greatly from that of capital, goods, and information. The principle of universal human rights—that there is a moral code higher than positive domestic law—has grown in prominence over the last half-century and has been key to recognizing the claims of refugees and asylum-seekers, particularly in a number of European states. Yet human rights as an international or multinational legal regime carries little authority over sovereign nation-states. As Seyla Benhabib wrote, "There are still no global courts of justice with the jurisdiction to punish sovereign states for the way that they treat refugees, foreigners, and aliens." In fact, notwithstanding the European Convention on Human Rights, the European Union appears to be shifting emphasis to joint border operations against illegal migrants and unwanted asylum seekers.

Thus, even as migration patterns change according to new global conditions, they remain shaped by asymmetrical relations of economic and political power between nation-states. The endurance of "crustacean" immigration policies may not, in fact, be eccentric to twenty-first-century globalization but a constituent element of it, aimed at maintaining the privileged position of the most powerful countries. Transnational migrants today move about in a very different world than that of the mid-twentieth century, navigating faster and more dangerous circuits in pursuit of work and safety, but their journeys reprise past crossings in important ways. They travel with the ghosts of migrants past as they, too, traverse boundaries, negotiate with states over the terms of their inclusion, and alter the future histories of nations and the history of our world.

## DISCUSSION QUESTIONS

1.  In what ways has U.S. immigration policy moved from one of accommodation to punishment? Given Ngai's argument, why would you say this has occurred?

2.  What are the macro-sociological conditions (that is, large-scale changes) that frame the current debates about immigration?

# 25

# From Ellis Island to JFK:

## Education in New York's Two Great Waves of Immigration

NANCY FONER

*Foner compares the experiences of two groups of immigrants—those who entered the United States in the first wave of mass immigration (1880–1920) and those who are contemporary. She shows how the specific economic and social conditions that groups experience at different points in time shape their opportunities and achievements, most notably, in the case she examines, in education.*

New York City is in the midst of a profound transformation as a result of the massive immigration of the last four decades. More than two and a half million immigrants have arrived since 1965, mainly from Latin America, Asia, and the Caribbean, and they are now streaming in at a rate of over 100,000 a year. Immigrants already constitute over a third of the city's population. In the midst of these dramatic changes, commentators and analysts, popular and academic, in the press and in the journals, are comparing the new immigration with the old.

This is not surprising. Few events loom larger in the history of New York City than the wave of immigration that peaked in the first decade of the 20th century. Between 1880 and 1920, close to a million-and-a-half immigrants arrived and settled in the city, so that by 1910, fully 41 percent of all New Yorkers were foreign-born. The immigrants, mostly Eastern European Jews and southern Italians, left an indelible imprint on the city—indeed, a large and influential part of New York's current citizens are their descendants.

An elaborate mythology has grown around immigration of a century ago, and perceptions of that earlier migration deeply color how the newest wave is seen. For many present-day New Yorkers, their Jewish and Italian immigrant forebears have become folk heroes of a sort—and represent a baseline against which current arrivals are compared and, unfortunately, often fail to measure up.

Nowhere is the nostalgia for the past more apparent than when it comes to education. Sentimental notions about Eastern European Jewish immigrants' love

SOURCE: *Brandeis Review* 21(2): 32–37, 2001. Reprinted by permission of the author.

affair with education and their zeal for the life of the mind have become part of our picture of the "world of our fathers." Jews are remembered as the "people of the book" who embraced learning on their climb up the social ladder. These memories set up expectations about what immigrants can and should achieve in the schools. If my grandparents and great-grandparents could succeed in New York City's schools a hundred years ago, without special programs to help them adjust, why—many people say—can't today's immigrants and their children do well when they get so much more assistance?

A comparison of New York's two great waves of immigration shows that inspirational tales about Eastern European Jews' rise through education and their success in New York's schools in the so-called golden immigrant age do not stand up against the hard realities of the time. Today, despite a dramatically different context and significant problems in the schools, many immigrant children are doing remarkably well.

In the years before World War I, most Eastern European Jews did not make the leap from poverty into the middle class through education. Those who made substantial progress up the occupational ladder in this period generally did so through businesses in the garment, fur, shoe, and retail trades and in real estate. It was only in later decades that large numbers of Eastern European Jewish children used secondary and higher education as a means of advancement.

A hundred years ago, most Jewish immigrant children left school with, at best, an eighth grade education; few went to high school, and even fewer graduated. In the first decade of the 20th century, well below five percent of the Russian Jewish children in New York City graduated from high school; less than one percent of Russian Jewish young people of college age ever reached the first year of college. By 1908, the City College of New York (CCNY) had already become a largely Jewish school, but Jewish undergraduates there and at other New York colleges were a select few. Only a tiny number graduated. In 1913, City College's entire graduating class had only 209 students, less than 25 of them Eastern European Jews. The 25 percent of Hunter College graduates who were Eastern European Jews in 1916 amounted to only 58 women. This was at a time when the Jewish population of New York City was almost a million! It was not until the 1930s that there were big graduating classes at City College that contained large numbers of Jews of Russian and Polish origin.

One reason so few Russian Jewish students went to high school or college is that there weren't many high schools or colleges at the time. In 1911, after a decade of expansion, the city had 19 high schools, but the high school student body was still only about a quarter of the size of the four preceding elementary school grades. CCNY and Hunter, the only two public colleges, had about 1,400 students in 1908. A high school degree wasn't necessary for the jobs employing most New Yorkers, and an eighth grade graduate could even get a white-collar job. A business career didn't require four years of college nor did teaching or the law. In any case, extended schooling was a luxury beyond the means of most new immigrant families, who needed their children's contributions to the family income. Even those who managed it rarely saw all their children go to high school.

The elementary school also did a poor job of educating immigrant children so that many were not prepared or motivated to continue on. The schools were severely overcrowded in the wake of the huge immigrant influx and the inability of school construction to keep up with demand. By 1914, enrollments had grown to almost 800,000, more than triple the figure for 1881. Educators of the time joked that teachers should have prior experience in a sardine factory before being hired to work in the New York schools.

There's a nostalgia for the "sink or swim" approach to learning English, but unfortunately, many in the past "sank" rather than "swam." Most non-English speaking children were placed in the lowest grade regardless of their age. (The special "steamer" classes introduced in 1904—in which students were totally immersed in English for a few months—catered to a mere fraction of students needing them, only 1,700 students throughout the city in 1908). Teachers made promotion decisions, and many children who could not do the work were left back. In 1908, over a third of the Russian Jewish elementary school pupils in New York City were two or more years over age for their grade.

If Russian Jewish children in this early period were not the education exemplars often remembered, they did do much better than Italians, the other "greenhorns" at the time. This favorable comparison helps explain why Jewish academic achievements have stood out and received so much attention. Russian Jewish students' progress, however, was fairly similar to that of native white New Yorkers at the elementary school level—and they did less well than native whites in making it to high school and college.

It's difficult to compare immigrants' educational achievements today with those in the past because the context in which education is a path to mobility is so radically different: formal education, and a more extended education, is now more important in getting a job owing to educational upgrading and transformations in the world of work. In the last great immigration wave, high schools were just becoming mass institutions. Today, getting a high school diploma is the norm, achieved by more than 70 percent of New York City's adult population and essential for many low-skilled positions. College is no longer an institution for a tiny elite—in 1990, a quarter of New Yorkers 25 and older had a college degree. Today, college graduates compete for jobs that immigrants with a high school diploma could have obtained a century ago, and a college degree—or more—is required for the growing number of professional, technical, and managerial positions.

How are immigrant children doing today? Admittedly, there are many dropouts and failures—something more serious now when more education is needed to get a decent job. Many, perhaps most, immigrants attend New York City schools where student skill levels are low, dropout rates high, and attendance rates poor. Once again, the surge of immigrants has led to soaring public school enrollments—now over the million mark—and serious overcrowding. New language programs are inadequate to meet the enormous need. Whereas a hundred years ago, New York schools mainly had to cope with Yiddish- and Italian-speaking children, today they confront a bewildering array of languages; a Board of Education count indicates that more than 100 languages are spoken by students from over 200 countries. When high schools were institutions for a minority of the better and more motivated students, violence, crime, and student indifference—and hostility—were not issues the way they are today.

Despite these problems, a substantial number of students are making it, and some are doing exceptionally well. There are new educational opportunities: expanded, and more widely available, higher education; a host of new programs for teaching students English; and even special immigrant schools designed specifically for newcomer children. Perhaps most important, large numbers of immigrant children have highly educated parents, and some immigrant children themselves have previous experience in fine schools in their home country. This has translated into academic success for many newcomers.

Although data on immigrant students in New York City are woefully inadequate—the only Board of Education data on immigrants refer to "recent immigrant students" who have entered as U.S. school system for the first time in the past three years—they show immigrants comparing favorably with other students in several ways. Students who were recent immigrants to the public school system in middle school graduate from high school on time by a slightly greater percentage than their native-born peers. They also have lower dropout rates. Although recent immigrants' median test scores in math, reading, and English are somewhat lower than those of other students, they improved their scores between 1989-1990 and 1990-91 more than the rest of the student body. At the City University of New York, immigrants make up a high, and growing, proportion of the student body: in 1998, 48 percent of CUNY's freshman class were foreign-born.

National studies based on large representative samples show immigrants often outperforming their native-born peers. In a study of eighth graders, children of immigrants—those born abroad and in the United States—earned higher grades and math scores than children of native-born parents even after the effects of race, ethnicity, and parental socioeconomic status were held constant. Another national study, which interviewed more than 21,000 10th and 12th graders in 1980 and followed them over a six-year period, compared immigrants and the native born in the aggregate as well as immigrants and the native born in four different ethnoracial groups (Asian, white, black, and Hispanic). Whichever way they were compared, immigrants were more likely to follow an academic track in high school than their native-born counterparts; once graduated, immigrants were also more likely to enroll in postsecondary education, to attend college, and to stay continuously through four years of college.

Like Jews of an earlier era, today's educational exemplars are Asians; white European immigrants are also doing comparatively well. Most striking is Asian (native as well as foreign born) overrepresentation in New York City's elite public high schools that select students on the basis of notoriously difficult entrance exams. In 1995, an astounding half of the students at the most selective high school of all, Stuyvesant, were Asian; at the Bronx High School of Science, 40 percent were Asian, and at Brooklyn Technical High School, 33 percent. This is at a time when Asians were 10 percent of the city's high school population. So many Korean students now attend Horace Mann School—one of

New York's most competitive private secondary schools—that there is a Korean parents' group there.

Just how important is Asians' culture in accounting for their educational achievements? As among Jews in the past, it plays a role. But I would argue that social-class factors, then as well as now, are more important. And today, race must also be considered.

A major reason Eastern European Jews did so much better academically than Italians in the old days was their occupational head start. Jews were more urban and arrived with higher levels of vocational skills, which gave them a leg up in entering New York's economy. Because the Jewish immigrant population was, from the start, better off economically than the Italian, Jewish parents could afford to keep their children in school more regularly and for longer. The poorer, less skilled Italians were more in need of their children's labor to help in the family. That Jewish children were more likely to have literate parents was also a help; the children themselves often arrived with a reading and writing knowledge of one language, making it easier to learn to read and write English than it was for southern Italian immigrant children, who generally arrived with no such skills.

Today, educational background plays a much larger role in explaining why the children of Asian immigrants are doing so well. Relatively high proportions of Asian students have highly educated parents. Although they often experience downward occupational mobility in New York, highly educated parents have higher educational expectations for their children and provide family environments more conducive to educational attainment. If their children started school in the home country, they typically attended excellent—and rigorous—institutions. Well-educated parents, moreover, are usually more sophisticated about the way the American educational system works and have an easier time, and more confidence in, navigating its complexities—and steering their children into good schools—than those with less education. In large part because so many come from professional and middle-class backgrounds, Asian New Yorkers are also doing fairly well economically. A century ago, economic resources were important because they allowed immigrants to keep their children in school; now they make it possible (or easier) to send children to private schools or to move to areas in the city (and suburbs) with better schools.

As for culture, the high value Jews placed on education—and the fact that southern Italians' heritage made them less oriented to and more skeptical about the value of book learning—helps account for the different educational achievements of the two groups. Today, most immigrant parents, in all groups, arrive with positive attitudes to education and high educational expectations for their children. Asian immigrants have particularly high aspirations for their children, though it's hard to say how much these aspirations are due to the cultural values and resources they bring to America as opposed to social-class advantages. That several national studies show Asian students outperforming all other racial/ethnic groups even after taking into account family income, household composition, and parental education, strongly suggests that culture is a factor.

Among Chinese immigrant parents, for example, hard work and discipline, not innate intelligence, are the keys to educational success. If their children study long hours, parents believe, they can get As, and they put intense pressure on their children to excel. Confucian teaching, it is said, emphasizes discipline, family unity, and obedience to authority, all of which contribute to academic success. Children who do poorly in school bring shame to Chinese families; those who do well bring honor. According to sociologist Min Zhou, Chinese immigrant parents denounce consumption of name-brand clothes and other "too American" luxuries, but do not hesitate to pay for books, after-school programs, Chinese lessons, private tutors, music lessons, and other educationally oriented activities. Chinese and Korean immigrants also have imported after-school institutions that prepare their children for high-school admissions and college-entry exams. In Chinatown, "school after school" has, according to Zhou, become an accepted norm. According to one survey, a fifth of Korean junior and senior high school students in New York City were taking lessons after school, either in a private institution or with a private tutor.

Finally, there is the role of race. At the turn of the 20th century, race was irrelevant in explaining why Jews did better academically than Italians. Both groups were at the bottom of the city's ethnic pecking order, considered to be inferior white races. Today, the way Asian, as opposed to black and Hispanic, immigrants fit into the racial hierarchy makes a difference in the opportunities they can provide their children. Because they are not black, Asian (and white) immigrants have greater freedom in where they can live and, in turn, send their children to schools. Asians have been able to move into heavily white neighborhoods with good schools fairly easily. Moreover, their children are less likely than black or Hispanic immigrants to feel an allegiance with native minorities and be drawn into an oppositional peer culture that emphasizes racial solidarity and opposition to school rules and authorities and sees doing well academically as "acting white."

What, then, in a broad sense, can be learned from this comparison? The remembered past is clearly not the same thing as what actually transpired, and it is wrong to judge today's immigrants by a set of myths rather than actual realities. We place an added burden on the newest arrivals if we expect them to live up to a set of folk heroes and heroines from a mythical golden age of immigration. As New York, and indeed the nation as a whole, continues to be transformed by the current wave of immigration, this is something important to keep in mind.

## DISCUSSION QUESTIONS

1. List two of the major differences that Foner cites comparing immigrants entering New York City at the end of the nineteenth century and those entering now.

2. What factors account for the difference in educational success for Jewish and Italian immigrants to New York early in the twentieth century?

# 26

## Places and Peoples:

### The New American Mosaic

CHARLES HIRSCHMAN AND DOUGLAS S. MASSEY

*Immigrants have discovered the South and Midwest, where they are settling and working in mid-size companies. The authors identify how economic restructuring is pulling immigrants to these regions because companies have redesigned work to cut costs in response to global competition. These new jobs are shunned by many American-born workers, but are attractive to newcomers.*

... In recent years, the "immigration problem," as it has been widely labeled, has been the subject of repeated national commissions, investigative reports, and congressional legislation. Although the apparent goal of American policy has been to cap or reduce immigration, the opposite has occurred. By 2000, there were over 30 million foreign-born persons in the United States, almost one third of whom arrived in the prior decade. Adding together these immigrants and their children (the second generation), more than 60 million people—or one in five Americans—have recent roots in other countries....

The latest surprise has been the shift in the geography of the new immigration. One of the standard findings of research on the post-1965 immigration wave during the 1970s and 1980s was its concentration in the states of New York, California, Texas, Florida, and Illinois, generally within a handful of "gateway" metropolitan areas such as New York, Los Angeles, Houston, Miami, and Chicago. Although different nationalities may have been concentrated in different areas (Puerto Ricans in New York, Cubans in Miami, Mexicans in Los Angeles, etc.) there was a common pattern and interpretation. Once immigrant pioneers had established a beachhead with ethnic neighborhoods and economic niches in certain industries, later immigrants flowed to the same places. Migrants were drawn to immigrant-ethnic communities that could offer assistance to newcomers seeking housing, jobs, and the warmth of familiarity.

The majority of new immigrants still settle in the traditional gateway cities; but ... California and New York became much less dominant in the 1990s and during the early years of the new century than they were during the 1970s and 1980s. Immigrants now settle in small towns as well as large cities and in the

SOURCE: Douglas S. Massey, ed. 2008. *New Faces in New Places: The Changing Geography of American Immigration,* New York: Russell Sage Foundation.

interior as well as on the coasts. Immigrants have discovered the Middle West …
and the South … as well as traditional gateways in the East and West. Given the
virtual absence of immigrants in many regions of the United States up to 1990,
even a small shift away from traditional gateways implied huge relative increases
at new destinations. The absolute numbers of new immigrants arriving in
Georgia, North Carolina, and Nevada may number only in the hundreds of thousands, but in relative terms the growth of immigrant communities in these areas is
frequently off the charts.…

## IMMIGRATION IN AN AGE OF INDUSTRIAL
## RESTRUCTURING

At the individual level, potential migrants are affected by incentives and information. Potential migrants are pushed by hard times, a lack of jobs, or by a shortage
of "good jobs" that provide desired social and economic rewards. Just as potential migrants differ in their skills and needs, what constitutes a sufficient push will
vary between communities and between individuals in the same community.
Landless and small-scale farmers may be pushed off their lands as commercial
markets replace traditional norms of tenancy. At the other end of the spectrum,
college graduates may take flight if they see only dead-end careers with few
rewards and opportunities in the local labor market. People of all classes may
depart in the absence of viable markets for capital and credit, seeking to self-
finance home acquisition with earning from international migration.

Economic pulls attract migrants in ways that complement the variety of push
factors. The promise of wages, even at the lowest levels of compensation, may be
very attractive for poor foreign workers with few choices locally. In professional
and high-tech circles, scholarships, prestige, and opportunities for challenging
careers lure workers to relocate internationally.…

People can only respond to the various pushes and pulls if they are aware of
them, of course; and individuals are not wholly independent actors, but are constrained by information about opportunities in distant locations. Social ties
embedded within migrant networks can provide this information and lower the
costs of migration.… Most migrants follow in the footsteps of friends and family
members who have already made the journey and can offer advice, encouragement, and the funds to subsidize the costs of transportation and settlement.
Migration streams from a specific place of origin to a specific place of destination
reflect the inherent tendency of earlier migrants to assist their relatives and neighbors with temporary housing and the search for employment. The hypothesis of
"cumulative causation" posits that social networks of friends and family broaden
the base of migration so much that other factors—those that originally caused the
migration—become less important over time.…

In modern times, however, "economic development" has generally been most
forceful in promoting long-distance migration, embracing such diverse processes as

the commercialization of agriculture, the development of wage labor markets, the creation of modern consumer tastes, and the loss of traditional forms of social insurance. Once migration develops for these structural reasons, the self-reinforcing nature of social networks and cumulative causation take hold and flows increase and broaden to include other groups far removed from the initial pioneers.

… The big five destination states (New York, California, Illinois, Texas, and Florida) were still attracting most immigrants in the late 1990s, but the proportion of Mexican immigrants going to them has dropped to 60 percent, and less than half of other immigrant streams are now settling in traditional destinations.…

Although smaller in absolute terms than in established areas, the growing number of foreigners is a new phenomenon—at least in the memories of those alive today. Immigrant laborers are creating ethnic niches in local labor markets and schools and churches are struggling to adapt to an upsurge in Spanish-speaking newcomers, as day-laborer sites have created a political storm in some areas. Which immigrants are going to these new destination areas and why?

… Mark Leach and Frank Bean find that immigrants to new destinations are generally heterogeneous in terms of individual traits and characteristics, but the places they go have basic economic facts in common: they tend to be places with well-developed and growing low-skill service sectors, thus pointing toward industrial restructuring as a driving force behind geographic diversification.… Thus, industrial restructuring creates an initial demand for immigrants in new locations and then processes of cumulative causation take over to channel subsequent cohorts of migrants to these new destinations.… The characteristics of immigrants to nonmetropolitan destinations in the South and Midwest were quite different from those going to traditional areas of destination. Especially within counties where the native white population was declining, the new immigrants tended to be younger, more poorly educated, more recently arrived, and more Mexican.…

Thus immigrants appear to be overrepresented in secondary-labor-market jobs that are typically shunned by native-born workers. The presence and expansion of poorly paid jobs that are difficult, dirty, and sometimes dangerous in small towns and rural areas is a common thread in many "new destination" areas.… Where the proportion of immigrants employed in manufacturing declined in metropolitan areas from 1990 to 2000, in nonmetropolitan areas it significantly increased. The native-born, meanwhile, were less likely to work in manufacturing in 2000 than in 1990, regardless of location.

These patterns are most salient in … "offset countries"—those in which immigrant growth offsets a population decline among natives. In these areas, immigrants in general and Mexican immigrants in particular were overrepresented in meatpacking, leather processing, and carpet and rug manufacture. Although these industries were present in 1990, they expanded over the ensuing decade and immigrants appear to have played a major role in their growth in the face of stiff global competition.

It thus appears that the increasing geographic diversity of immigration to the United States is related to broader structural changes in the American economy and to the decreasing attractiveness of certain jobs to native-born workers.…

"Industrial restructuring" is a generic term used ... to describe shifts in the American economy away from large-scale capital-intensive production and a relatively well-paid, unionized, and mostly native workforce toward labor-intensive production and low-paid, non-unionized, foreign workforces. International competition and technological innovation have cut the profit margins of older companies that held virtual monopolies on the manufacture, distribution, and marketing of goods within many industrial sectors. Some American manufacturers were unable to compete with cheaper imports and they shifted production to plants overseas in low-wage countries; but in other cases foreign firms opened more efficient, non-unionized manufacturing plants in the United States, often in rural or small towns with lower prevailing wages.

One of the common features of industrial restructuring is the prevalence of labor subcontracting and the overall informalization of labor relations. Subcontracts to smaller firms allow larger companies to achieve greater flexibility and minimize employment costs, but smaller firms have lower profit margins and are much less likely to offer fringe benefits such as health insurance or retirement programs....

... American industries and employers, facing greater international competition and declining profit margins, have cut costs through subcontracting, deskilling, and decentralizing production to areas with lower wage rates. Although not all industries restructure in the same fashion, they all seek to achive common outcomes that affect workers–lower wages, fewer unions, reduced fringe benefits and easier layoffs. The "new jobs" are not attractive to American-born workers, especially younger workers with credentials or connections to find formal-sector employment. Because immigrants have fewer options and are generally more tolerant of difficult working conditions and job instablity, especially if they lack documents, they fill the gap in an increasingly segmented domestic labor market.

... Social mechanisms allow immigrant workers to respond to changes in economic demand. Just as social networks and institutions of mutual support have led to the concentration of immigrants in traditional gateway cities, immigrant entrepreneurs and middlemen quickly recruit friends, families, and co-ethnics to new destination areas. Each pioneer immigrant community creates the potential for additional immigration through network-driven processes of cumulative causation, and eventually for the creation of satellite settlements in nearby towns where immigrant niches can be reproduced.

## DISCUSSION QUESTIONS

1. Where are immigrants coming from, and why are they settling outside of traditional settlement regions?

2. How does economic restructuring develop new secondary labor market jobs that are attractive to immigrants?

3. Why are native-born Americans likely to reject the work that is attractive to immigrants?

# 27

# Between Two Worlds:

## How Young Latinos Come of Age in America

PEW RESEARCH CENTER

*The category of Latino youth includes those who are born in the United States as well as immigrants from many Latin American nations. Their ethnic identification varies, but they also face many challenges to assimilation, finding themselves moving between two worlds as they grow up in the United States.*

Hispanics are the largest and youngest minority group in the United States. One-in-five schoolchildren is Hispanic. One-in-four newborns is Hispanic. Never before in this country's history has a minority ethnic group made up so large a share of the youngest Americans. By force of numbers alone, the kinds of adults these young Latinos become will help shape the kind of society America becomes in the 21$^{st}$ century.

This report takes an in-depth look at Hispanics who are ages 16 to 25, a phase of life when young people make choices that–for better and worse—set their path to adulthood. For this particular ethnic group, it is also a time when they navigate the intricate, often porous borders between the two cultures they inhabit–American and Latin American.

The report explores the attitudes, values, social behaviors, family characteristics, economic well-being, educational attainment and labor force outcomes of these young Latinos. It is based on a new Pew Hispanic Center telephone survey of a nationally representative sample of 2,012 Latinos, supplemented by the Center's analysis of government demographic, economic, education and health data sets.

The data paint a mixed picture. Young Latinos are satisfied with their lives, optimistic about their futures and place a high value on education, hard work and career success. Yet they are much more likely than other American youths to drop out of school and to become teenage parents. They are more likely than white and Asian youths to live in poverty. And they have high levels of exposure to gangs.

These are attitudes and behaviors that, through history, have often been associated with the immigrant experience. But most Latino youths are *not immigrants*. Two-thirds were born in the United States, many of them descendants of the big, ongoing wave of Latin American immigrants who began coming to this country around 1965.

SOURCE: Pew Research Center, 2009

As might be expected, they do better than foreign-born counterparts on many key economic, social and acculturation indicators analyzed in this report. They are much more proficient in English and are less likely to drop out of high school, live in poverty or become a teen parent.

But on a number of other measures, U.S.-born Latino youths do no better than the foreign born. And on some fronts, they do worse.

For example, native-born Latino youths are about twice as likely as the foreign born to have ties to a gang or to have gotten into a fight or to have carried a weapon in the past year. They are also more likely to be in prison.

The picture becomes even more murky when comparisons are made among youths who are first generation (immigrants themselves), second generation (U.S.-born children of immigrants) and third and higher generation (U.S.-born grandchildren or more far-removed descendants of immigrants).[1]

For example, teen parenthood rates and high school dropout rates are much lower among the second generation than the first, but they appear higher among the third generation than the second. The same is true for poverty rates.

## IDENTITY AND ASSIMILATION

Throughout this nation's history, immigrant assimilation has always meant something more than the sum of the sorts of economic and social measures outlined above. It also has a psychological dimension. Over the course of several generations, the immigrant family typically loosens its sense of identity from the old country and binds it to the new.

It is too soon to tell if this process will play out for today's Hispanic immigrants and their offspring in the same way it did for the European immigrants of the 19th and early 20th centuries. But whatever the ultimate trajectory, it is clear that many of today's Latino youths, be they first or second generation, are straddling two worlds as they adapt to the new homeland.

According to the Pew Hispanic Center's National Survey of Latinos, more than half (52%) of Latinos ages 16 to 25 identify themselves first by their family's country of origin, be it Mexico, Cuba, the Dominican Republic, El Salvador or any of more than a dozen other Spanish-speaking countries. An additional 20% generally use the terms "Hispanic" or "Latino" first when describing themselves. Only about one-in-four (24%) generally use the term "American" first.

Among the U.S.-born children of immigrants, "American" is somewhat more commonly used as a primary term of self-identification. Even so, just 33% of these young second generation Latinos use American first, while 21% refer to themselves first by the terms Hispanic or Latino, and the plurality-41%-refer to themselves first by the country their parents left in order to settle and raise their children in this country.

Only in the third and higher generations do a majority of Hispanic youths (50%) use "American" as their first term of self-description.

## IMMIGRATION IN HISTORICAL PERSPECTIVE

Measured in raw numbers, the modern Latin American-dominated immigration wave is by far the largest in U.S. history. Nearly 40 million immigrants have come to the United States since 1965. About half are from Latin America, a quarter from Asia and the remainder from Europe, Canada, the Middle East and Africa. By contrast, about 14 million immigrants came during the big Northern and Western European immigration wave of the 19th century and about 18 million came during the big Southern and Eastern European-dominated immigration wave of the early 20th century.[2]

However, the population of the United States was much smaller during those earlier waves. When measured against the size of the U.S. population during the period when the immigration occurred, the modern wave's average annual rate of 4.6 new immigrants per 1,000 population falls well below the 7.7 annual rate prevailed in the mid- to late-19th century and the 8.8 rate at the beginning of the 20th century.

All immigration waves produce backlashes of one kind or another, and the latest one is no exception. Illegal immigration, in particular, has become a highly-charged political issue in recent times. It is also a relatively new phenomenon; past immigration waves did not generate large numbers of illegal immigrants because the U.S. imposed fewer restrictions on immigration flow in the past than it does now.

The current wave may differ from earlier waves in other ways as well. More than a few immigration scholars have voiced skepticism that the children and grandchildren of today's Hispanic immigrants will enjoy the same upward mobility experienced by the offspring of European immigrants in previous centuries.

Their reasons vary, and not all are consistent with one another. Some scholars point to structural changes in modern economies that make it more difficult for unskilled laborers to climb into the middle class. Some say the illegal status of so many of today's immigrants is a major obstacle to their upward mobility. Some say the close proximity of today's sending countries and the relative ease of modern global communication reduce the felt need of immigrants and their families to acculturate to their new country. Some say the fatalism of Latin American cultures is a poor fit in a society built on Anglo-Saxon values. Some say that America's growing tolerance for cultural diversity may encourage modern immigrants and their offspring to retain ethnic identities that were seen by yesterday's immigrants as a handicap. (*The melting pot is dead. Long live the salad bowl*). Alternatively, some say that Latinos' brown skin makes assimilation difficult in a country where white remains the racial norm.

It will probably take at least another generation's worth of new facts on the ground to know whether these theories have merit. But it is not too soon to take some snapshots and lay down some markers....

For some in this mixed group, endemic poverty and its attendant social ills have been a part of their families, *barrios* and *colonias* for generations, even centuries. Meantime, others in the third and higher generation have been upwardly mobile in ways consistent with the generational trajectories of European immigrant groups. Because the data we use in this report do not allow us to separate out the different demographic sub-groups within the third and higher generation, the overall numbers we present are averages that often mask large variances within this group.

## NOTES

1.   In this report when we refer to the third and higher generations of Latinos, we are describing a group with diverse family histories vis-à-vis the United States. We estimate that 40% of this group are grandchildren of immigrants. The rest are more far removed from the immigrants in their families. And a small share comes from families that never immigrated at all—their ancestors were living in what was then Mexico when their land became a part of the United States in the 19th century as a result of war, treaty, annexation and/or purchase.

2.   These estimates do not include U.S. residents born in Puerto Rico. However, in the rest of the report, people born in Puerto Rico are included among the foreign born because they are from a Spanish-dominant culture and because on many points their attitudes, views and beliefs are much closer to Hispanics born abroad than to Latinos born in the 50 U.S. states and the District of Columbia.

## DISCUSSION QUESTIONS

1.   How are the life experiences of U.S.-born Latinos different from those of immigrant Latinos?

2.   Latinos are a diverse group from different parts of Latin America. How are different segments likely to identify themselves?

3.   What factors might contribute to the economic well-being and mobility of young Latinos?

# Student Exercises

1.  Is there an immigrant experience in the history of your family? Do you know what it is? If so, when did your family arrive and what were the specific challenges faced by that generation? What pushed them from their original homelands? What were the economic and social conditions in the United States as the time of arrival? If your family history does not include such an experience, why not? How does your family's history relate to that of immigrant groups?

2.  Immigration is transforming the composition of many communities in the United States. Who are the immigrants in your community? Where do they live? What work do they do? Did their children attend school with you? After your own personal assessment, review the newspaper or visual media coverage to understand popular images of these immigrants. Are immigrants presented as a positive influence in the community or a negative one? Do you think the media depictions of the immigrants are related to their nations of origin or their race?

# Section VII

# The Difference It Makes:
## Race, Class, and Gender Inequality
### Elizabeth Higginbotham
### and Margaret L. Andersen

If you were studying race and ethnicity fifty years ago, you would have been only looking at race and ethnicity as important determinants of life changes. Since then scholars have identified how people are not solely members of a racial or ethnic group, but also have a gender and a social class position, along with other social attributes. **Gender** refers to the culturally and socially structured relationship between men and women. Gender is constructed as both a social location and an identity, meaning that what men and women do, how they relate to each other, and how they think about themselves and others is shaped by the society and the interaction of race with gender and social class.

**Social class** refers to the economic circumstances of people and their position in the labor market that will result in different wages and salaries, and whatever material assets they have (or do not). Social class means people have either advantages or disadvantages with regard to life chances and challenges. It includes material and social assets and, like race and gender, is part of a person's identity, even though it is based in social structural conditions. Just as racial hierarchies developed historically, so did social class and gender hierarchies.

These dimensions are fundamental axes of society and are important for understanding how social institutions structure people's lives. By studying the intersection of these different dimensions of advantage and disadvantage, we learn much about the challenges different people face and the resources they receive. Membership in specific racial, gender, and social class groups creates or blocks opportunities. When there are obstacles, as there often are, people have to navigate and resist them as best as they can.

Understanding the intersections of race, class, and gender means that you cannot see all members of a given group as the same. For example, because of the interaction of gender and race, the cultural images that are produced of young African American men differ from those of young African American women. As these young men and women move about society, their reception by others is often based on racial and gendered stereotypes. Furthermore, the interaction of gender and race means that, depending on your social location, different groups, by virtue of their race, gender, and class, will likely travel different streets, attend different schools, and get different jobs. Looking at all of these dimensions—race, class, and gender—is necessary for understanding the specific challenges people face and the resources they have. As you will see in these readings, the complex pattern of advantages and disadvantages is important both at the individual level and at the organizational and institutional levels.

Section VII opens with a work by one of the pioneers in this intersectional approach in the social sciences, Patricia Hill Collins ("Toward a New Vision: Race, Class, and Gender as Categories of Analysis and Connection"). In her classic article, Collins identifies how using these three dimensions of oppression explains the structural basis of domination and subordination. Collins offers an analysis that explores the institutional, symbolic, and individual dimensions of race, class, and gender, arguing that people sit within a *matrix of domination*.

Collins's work cautions us about using a singular lens to understand people's experience, even though in some studies, people rely on only one such dimension. Thus, in women's studies, people sometimes only analyze gender; in race studies, attention is sometimes solely on racial differences; and in social class studies, people may only highlight class hierarchies. In her article, Yen Le Espiritu ("Theorizing Race, Gender, and Class") calls for greater integration of these perspectives. She provides a quick history of these scholarly fields, making it clear that failure of an intersectional analysis makes it difficult to build those coalitions for social change. The racial, gender, and class complexity of the United States means we must move towards this direction of analysis to understand each other and our society.

As an example, fatherhood is a masculine gender role, but how men are able to perform this role is related to race, social class, *and* gender, as well as other sets of opportunities at given points in time. So you often hear about the absence of fathers in many African American families, as if the men themselves had failed. But when fathers are missing, it is not just about individual failings. There are social constraints on people's abilities to parent. Roberta Coles and Charles Green ("The Myth of the Missing Black Father") identify some of the unique challenges faced by African American men in a context of deindustrialization and the consequent lack of employment opportunities for Black men. They also show that, contrary to dominant assumptions, Black men value fatherhood and fulfill parental roles both inside and outside of marriage.

Deindustrialization also produces barriers for young Black women in distressed neighborhoods. In such communities, young Black women have limited choices. Nikki Jones ("Between Good and Ghetto: African American Girls and Inner City Violence") observes how young African American girls come of age in a Philadelphia neighborhood. They have to understand "the code of the street," which Elijah Anderson (1999) sees as a system of accountability that governs public relations. These young girls have to survive within this code, navigating their way through a complex world of competing expectations for being "good" girls.

Women of color must negotiate worlds where race, class, and gender merge. Even success does not end the need to negotiate these locations. Gladys

García-Lopez and Denise A. Segura ("'They Are Testing You All the Time': Negotiating Dual Femininities among Chicana Attorneys") report on their interviews with fifteen Chicana lawyers who face racism and sexism in the legal system, even though they are also aware of their newly found social class status. These Chicana lawyers see their degrees as a means to help their community, but find they have to display "femininity" in order to achieve their goals within their communities and the legal system. Although these women attorneys are very different from the high school students observed by Nikki Jones, both exist with in a web of privilege and disadvantage that is part of American society.

While everyone is located differently within the race, class, and gender hierarchies, these readings highlight how this matrix of domination works.

## REFERENCES

Elijah Anderson. 1999. *Code of the Streets: Decency, Violence, and the Moral Life of the Inner City.* New York: W.W. Norton.

## FACE THE FACTS: THE RACE-GENDER WAGE GAP

**Median Weekly Earnings
Full-time Workers, 2007**

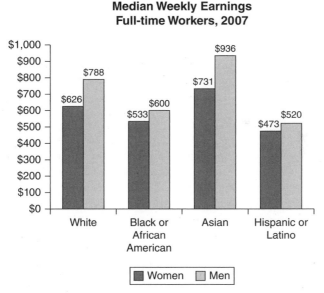

SOURCE: U.S. Department of Labor. 2008. *Employment and Earnings.* Washington, DC: U.S. Department of Commerce.

***Think about it***: You have probably heard that the income gap between women and men is about 73 percent, but, as the chart above shows, when you add race to the picture, the gap between women and men within different racial-ethnic groups changes. Where is the gap the largest? The smallest? Why is this and what does this tell you about the intersection of race, gender, and social class?

# 28

## Toward a New Vision:

### Race, Class, and Gender as Categories of Analysis and Connection

PATRICIA HILL COLLINS

*People are often focused on one dimension of oppression—the one they feel is most important in their lives. Yet we are all positioned with regard to entire sets of advantages and disadvantages. Collins pushes readers to think about new categories that emerge when we explore the connection of institutional, symbolic, and individual dimensions of oppression.*

> The true focus of revolutionary change is never merely the oppressive situations which we seek to escape, but the piece of the oppressor which is planted deep within each of us.
>
> —AUDRE LORDE, *SISTER OUTSIDER*, 123

Audre Lorde's statement raises a troublesome issue for scholars and activists working for social change. While many of us have little difficulty assessing our own victimization within some major system of oppression, whether it be by race, social class, religion, sexual orientation, ethnicity, age or gender, we typically fail to see how our thoughts and actions uphold someone else's subordination. Thus, white feminists routinely point with confidence to their oppression as women but resist seeing how much their white skin privileges them. African-Americans who possess eloquent analyses of racism often persist in viewing poor white women as symbols of white power. The radical left fares little better. "If only people of color and women could see their true class interests," they argue, "class solidarity would eliminate racism and sexism." In essence, each group identifies the type of oppression with which it feels most comfortable as being fundamental and classifies all other types as being of lesser importance.

Oppression is full of such contradictions. Errors in political judgment that we make concerning how we teach our courses, what we tell our children, and which organizations are worthy of our time, talents and financial support flow smoothly from errors in theoretical analysis about the nature of oppression and

SOURCE: In *Race, Sex, & Class*, 1, no. 1, Fall 1993. Reprinted by permission of the author.

activism. Once we realize that there are few pure victims or oppressors, and that each one of us derives varying amounts of penalty and privilege from the multiple systems of oppression that frame our lives, then we will be in a position to see the need for new ways of thought and action.

To get at that "piece of the oppressor which is planted deep within each of us," we need at least two things. First, we need new visions of what oppression is, new categories of analysis that are inclusive of race, class, and gender as distinctive yet interlocking structures of oppression. Adhering to a stance of comparing and ranking oppressions—the proverbial, "I'm more oppressed than you"—locks us all into a dangerous dance of competing for attention, resources, and theoretical supremacy. Instead, I suggest that we examine our different experiences within the more fundamental relationship of domination and subordination. To focus on the particular arrangements that race or class or gender takes in our time and place without seeing these structures as sometimes parallel and sometimes interlocking dimensions of the more fundamental relationship of domination and subordination may temporarily ease our consciences. But while such thinking may lead to short-term social reforms, it is simply inadequate for the task of bringing about long-term social transformation.

While race, class and gender as categories of analysis are essential in helping us understand the structural bases of domination and subordination, new ways of thinking that are not accompanied by new ways of acting offer incomplete prospects for change. To get at that "piece of the oppressor which is planted deep within each of us," we also need to change our daily behavior....

In order to move toward new visions of what oppression is, I think that we need to ask new questions. How are relationships of domination and subordination structured and maintained in the American political economy? How do race, class and gender function as parallel and interlocking systems that shape this basic relationship of domination and subordination? Questions such as these promise to move us away from futile theoretical struggles concerned with ranking oppressions and towards analyses that assume race, class and gender are all present in any given setting, even if one appears more visible and salient than the others. Our task becomes redefined as one of reconceptualizing oppression by uncovering the connections among race, class and gender as categories of analysis.

## 1. THE INSTITUTIONAL DIMENSION
## OF OPPRESSION

Sandra Harding's contention that gender oppression is structured along three main dimensions—the institutional, the symbolic and the individual—offers a useful model for a more comprehensive analysis encompassing race, class and gender oppression (Harding 1986). Systemic relationships of domination and subordination structured through social institutions such as schools, businesses, hospitals, the workplace and government agencies represent the institutional dimension of oppression. Racism, sexism and elitism all have concrete institutional locations. Even though the workings of the institutional dimension of oppression are often obscured with ideologies claiming equality of opportunity, in actuality, race, class and gender place Asian-American women, Native American men, White men, African-American women and other groups in distinct institutional niches with varying degrees of penalty and privilege.

Even though I realize that many … would not share this assumption, let us assume that the institutions of American society discriminate, whether by design or by accident. While many of us are familiar with how race, gender and class operate separately to structure inequality, I want to focus on how these three systems interlock in structuring the institutional dimension of oppression. To get at the interlocking nature of race, class and gender, I want you to think about the antebellum plantation as a guiding metaphor for a variety of American social institutions. Even though slavery is typically analyzed as a racist institution, and occasionally as a class institution, I suggest that slavery was a race, class, gender specific institution. Removing any one piece from our analysis diminishes our understanding of the true nature of relations of domination and subordination under slavery.

Slavery was a profoundly patriarchal institution. It rested on the dual tenets of White male authority and White male property, a joining of the political and the economic within the institution of the family. Heterosexism was assumed and all Whites were expected to marry. Control over affluent White women's sexuality remained key to slavery's survival because property was to be passed on to the legitimate heirs of the slave owner. Ensuring affluent White women's virginity and chastity was deeply intertwined with maintenance of property relations.

Under slavery, we see varying levels of institutional protection given to affluent White women, working class and poor White women and enslaved African women. Poor White women enjoyed few of the protections held out to their upper class sisters. Moreover, the devalued status of Black women was key in keeping all White women in their assigned places. Controlling Black women's fertility was also key to the continuation of slavery, for children born to slave mothers themselves were slaves.

African-American women shared the devalued status of chattel with their husbands, fathers and sons. Racism stripped Blacks as a group of legal rights, education and control over their own persons. African-Americans could be whipped, branded, sold, or killed, not because they were poor, or because they were women, but because they were Black. Racism ensured that Blacks would continue to serve Whites and suffer economic exploitation at the hands of all Whites.

So we have a very interesting chain of command on the plantation—the affluent White master as the reigning patriarch, his White wife helpmate to serve him, help him manage his property and bring up his heirs, his faithful servants whose production and reproduction were tied to the requirements of the capitalist political economy and largely propertyless, working class White men and women watching from afar. In essence, the foundations for the contemporary roles of elite White women, poor Black women, working class White men and a series of other groups can be seen in stark relief in this fundamental American social institution. While Blacks experienced the most harsh treatment under slavery, and thus made slavery clearly visible as a racist institution, race, class and gender interlocked in structuring slavery's systemic organization of domination and subordination.

Even today, the plantation remains a compelling metaphor for institutional oppression. Certainly the actual conditions of oppression are not as severe now as they were then. To argue, as some do, that things have not changed all that much denigrates the achievements of those who struggled for social change before us. But the basic relationships among Black men, Black women, elite White women, elite White men, working class White men and working class White women as groups remain essentially intact....

## 2. THE SYMBOLIC DIMENSION OF OPPRESSION

Widespread, societally sanctioned ideologies used to justify relations of domination and subordination comprise the symbolic dimension of oppression. Central to this process is the use of stereotypical or controlling images of diverse race, class and gender groups. In order to assess the power of this dimension of oppression, I want you to make a list, either on paper or in your head, of "masculine" and "feminine" characteristics. If your list is anything like that compiled by most people, it reflects some variation of the following:

| Masculine | Feminine |
|-----------|----------|
| aggressive | passive |
| leader | follower |
| rational | emotional |
| strong | weak |
| intellectual | physical |

Not only does this list reflect either/or dichotomous thinking and the need to rank both sides of the dichotomy, but ask yourself exactly which men and women you had in mind when compiling these characteristics. This list applies almost exclusively to middle class White men and women. The allegedly "masculine" qualities that you probably listed are only acceptable when exhibited by elite White men, or when used by Black and Hispanic men against each other or against women of color. Aggressive Black and Hispanic men are seen as dangerous, not powerful, and are often penalized when they exhibit any of the allegedly "masculine" characteristics. Working class and poor White men fare slightly better and are also denied the allegedly "masculine" symbols of leadership, intellectual competence, and human rationality. Women of color and working class and poor White women are also not represented on this list, for they have never had the luxury of being "ladies." What appear to be universal categories representing all men and women instead are unmasked as being applicable to only a small group.

It is important to see how the symbolic images applied to different race, class and gender groups interact in maintaining systems of domination and subordination. If I were to ask you to repeat the same assignment, only this time, by making separate lists for Black men, Black women, Hispanic women and Hispanic men, I suspect that your gender symbolism would be quite different. In comparing all of the lists, you might begin to see the interdependence of symbols applied to all groups. For example, the elevated images of White womanhood need devalued images of Black womanhood in order to maintain credibility.

While the above exercise reveals the interlocking nature of race, class and gender in structuring the symbolic dimension of oppression, part of its importance lies in demonstrating how race, class and gender pervade a wide range of what appears to be universal language. Attending to diversity in our scholarship, in our teaching, and in our daily lives provides a new angle of vision on interpretations of reality thought to be natural, normal and "true." Moreover, viewing images of masculinity and femininity as universal gender symbolism, rather than as symbolic images that are race, class and gender specific, renders the experiences of people of color and of nonprivileged White women and men invisible. One way to dehumanize an individual or group is to deny the reality of their experiences. So when we refuse to deal with race or class because they do not appear to be directly relevant to gender, we are actually becoming part of someone else's problem.

Assuming that everyone is affected differently by the same interlocking set of symbolic images allows us to move forward toward new analyses. Women of color and White women have different relations to White male authority and this difference explains the distinct gender symbolism applied to both groups. Black women encounter controlling images such as the mammy, the matriarch, the mule and the whore, that encourage others to reject us as fully human people. Ironically, the negative nature of these images simultaneously encourages us to reject them. In contrast, White women are offered seductive images, those that promise to reward them for supporting the status quo. And yet seductive images can be equally controlling....

Both sets of images stimulate particular political stances. By broadening the analysis beyond the confines of race, we can see the varying levels of rejection and seduction available to each of us due to our race, class and gender identity. Each of us lives with an allotted portion of institutional privilege and penalty, and with varying levels of rejection and seduction inherent in the symbolic images applied to us. This is the context in which we make our choices. Taken together, the institutional and symbolic dimensions of oppression create a structural backdrop against which all of us live our lives.

## 3. THE INDIVIDUAL DIMENSION OF OPPRESSION

Whether we benefit or not, we all live within institutions that reproduce race, class and gender oppression. Even if we never have any contact with members of other race, class and gender groups, we all encounter images of these groups and are exposed to the symbolic meanings attached to those images. On this dimension of oppression, our individual biographies vary tremendously. As a result of our institutional and symbolic statuses, all of our choices become political acts.

Each of us must come to terms with the multiple ways in which race, class and gender as categories of analysis frame our individual biographies. I have lived my entire life as an African-American woman from a working class family and this basic fact has had a profound impact on my personal biography. Imagine how different your life might be if you had been born Black, or White, or poor, or of a different race/class/gender group than the one with which you are most familiar. The institutional treatment you would have received and the symbolic meanings attached to your very existence might differ dramatically from what you now consider to be natural, normal and part of everyday life. You might be the same, but your personal biography might have been quite different.

I believe that each of us carries around the cumulative effect of our lives within multiple structures of oppression. If you want to see how much you have been affected by this whole thing, I ask you one simple question—who are your close friends? Who are the people with whom you can share your hopes, dreams, vulnerabilities, fears and victories? Do they look like you? If they are all the same, circumstance may be the cause. For the first seven years of my life I saw only low income Black people. My friends from those years reflected the composition of my community. But now that I am an adult, can the defense of circumstance explain the patterns of people that I trust as my friends and colleagues? When given other alternatives, if my friends and colleagues reflect the homogeneity of one race, class and gender group, then these categories of analysis have indeed become barriers to connection.

I am not suggesting that people are doomed to follow the paths laid out for them by race, class and gender as categories of analysis. While these three structures certainly frame my opportunity structure, I as an individual always have the choice of accepting things as they are, or trying to change them. As Nikki Giovanni points out, "we've got to live in the real world. If we don't like the world we're living in, change it. And if we can't change it, we change ourselves. We can do something" (Tate 1983, 68). While a piece of the oppressor may be planted deep within each of us, we each have the choice of accepting that piece or challenging it as part of the "true focus of revolutionary change."

## REFERENCES

Harding, Sandra. 1986. *The Science Question in Feminism*. Ithaca, New York: Cornell University Press.

Lorde, Audre. 1984. *Sister Outsider*. Trumansberg, New York: The Crossing Press.

Tate, Claudia, ed. 1983. *Black Women Writers at Work*. New York: Continuum.

## DISCUSSION QUESTIONS

1. What does Collins mean by "the contradictions of oppression"?
2. Can you identify how the interlocking nature of race, class, and gender structure inequality in your college or workplace?
3. What role does the symbolic dimension of oppression play in perpetuating inequalities?

# 29

# Theorizing Race, Gender, and Class

YEN LE ESPIRITU

*Here Yen Le Espiritu discusses how approaches to gender issues can limit how scholars explore social issues and also how they understand different groups. As social scientists look at race and social class as well as gender, they are better able to understand the work experiences of men and women of color as well as gender relations within racial communities.*

In the 1970s, when "second-wave" feminist scholars identified gender as a major social force, they pronounced that traditional scholarship represented the experiences of men as gender neutral, making it unnecessary to deal with women. With men's lives assumed to be the norm, women's experiences were subsumed under those of men, narrowly categorized, or excluded altogether. Responding to the omission of women, the feminist agenda at the time was to fill in gaps, correct sexist biases, and create new topics out of women's experiences. In the next stage, feminist thinking advanced from woman-centered analyses to providing a "gendered" understanding of sociological phenomena—one that traced the significance of gender organization and relations in all institutions and in shaping men's as well as women's lives. The feminist agenda shifted then from advocating the inclusion of women to transforming the basic conceptual and theoretical frameworks of the field.

In a 1985 statement on the treatment of gender in research, the American Sociological Association's Committees on the Status of Women in Sociology (1985) urged members of the profession "to explicitly acknowledge the social category of gender, and gender differences in power, at each step of the research process" (p. 1). But a gendered transformation of knowledge has not been achieved in sociology. In a review essay published in the mid-1980s, Judith Stacey and Barrie Thorne (1985) argued that although feminist scholarship has made important contributions to sociology in terms of uncovering and filling gaps in sociological knowledge, the gender paradigm, which positions gender as a prominent social category creating hierarchies of difference in society, remains a minority position within the discipline. As in anthropology and history, feminist sociology seems to have been "both co-opted and ghettoized, while the discipline as a whole and its dominant paradigms have proceeded relatively unchanged" (Stacey and Thorne, 1985, p. 302). The use of gender as a variable,

SOURCE: From Yen Le Espiritu. 2008. *Asian American Women and Men: Labor, Laws and Love*, 2nd ed, Lanham, MD: Rowman & Littlefield.

conceptualized in terms of sex difference rather than as a central theoretical concept, is a prime example of the co-optation of feminist perspectives (Stacey and Thorne, 1985, p. 308). Six years later, Sylvia Pedraza (1991) reached the same conclusion: within sociology, "a truly gendered understanding of most topics has not been achieved" (p. 305). A 2007 assessment of the field remained bleak: although the quantity of work on gender in leading sociological journals in the twenty-first century has increased, "the critical turn for the discipline, one in which a structurally-oriented gender perspective would inform sociology as a whole, has yet to happen" (Ferree, Khan, and Morimoto, 2007, p. 473).

Although the concept of gender is invaluable, the gender process cannot be understood independently of class and race. Class and gender overlap when the culture of patriarchy enables capitalists to benefit from the exploitation of the labor of both men and women. Because patriarchy mandates that men serve as good financial providers, it obliges them to toil in the exploitative capitalist wage market. Correspondingly, the patriarchal assumption that women are not the breadwinners—and therefore can afford to work for less—allows employers to justify hiring women at lower wages and under poorer working conditions. On the other hand, in however limited a way, wage employment affords women a measure of economic and personal independence, thus strengthening their claims against patriarchal authority. This is just one of the many contradictions that occur in the interstices of the race/gender/class matrix.

Economic oppression in the United States is not only gendered but also racist. Historically, people of color in the United States have encountered institutionalized economic and cultural racism that has restricted their economic mobility. Due to their gender, race, and noncitizen status, immigrant women of color fare the worst because they are seen as being the most desperate for work at any wage. Within this racially based social order, there are no gender relations per se, only gender relations as constructed by and between races. Jane Gaines (1990, p. 198) suggested that insofar as the focus on gender keeps women from seeing other structures of oppression, it functions ideologically in the interests of the dominant group(s). That is, to conceptualize oppression only in terms of male dominance and female subordination is to obscure the centrality of classism, racism, and other forms of inequality in U.S. society (Stacey and Thorne, 1985, p. 311).

Women of color have charged that feminist theory falsely universalizes the category of "woman." As bell hooks (1984) observed, this gender analysis centers on the experiences of white, middle-class women and ignores the way women in different racial groups and social classes experience oppression. For women of color, gender is only part of a larger pattern of unequal social relations. In their daily lives, these women brave not only sexism but also the "entire system of racial and ethnic stratification that defines, stigmatizes, and controls the minority group as a whole" (Healey, 1995, p. 26). These systems of power render irrelevant the public/private distinction for many women of color. As Aida Hurtado (1989) reminded us,

> Women of color have not had the benefit of the economic conditions
> that underlie the public/private distinction.... Welfare programs and

policies have discouraged family life, sterilization programs have re-stricted production rights, government has drafted and armed dispro-portionate numbers of people of color to fight its wars overseas, and locally, police forces in the criminal justice system arrest and incarcerate disproportionate numbers of people of color. There is no such thing as a private sphere for people of color except that which they manage to create and protect in an otherwise hostile environment. (p. 849)

In this hostile environment, some women of color, in contrast to their white counterparts, view unpaid domestic work—having children and maintaining families—more as a form of resistance to racist oppression than as a form of exploitation by men.

Women of color have also protested their marginalization in traditional fem-inist scholarship, charging that they have been added to feminist analysis only "as an afterthought" (Gaines, 1990, p. 201). This tokenism is evident in the manner in which race has been added to, rather than integrated into, traditional feminist scholarship. Gaines (1990, p. 201) pointed out that although feminist anthologies consistently include writings by women of color, the issue of race is conspicu-ously absent from the rest of the volume. This practice suggests that white fem-inists view racism as an issue that affects only people of color and not as a system that organizes and shapes the daily experiences of *all* people. In an anthology of writings by radical women of color, Cherrie Moraga (1981) recorded her reac-tion to "dealing with white women": "I have felt so very dark: dark with anger, with silence, with the feeling of being walked over" (p. xv). Similarly, Bettie Luke Kan referred to racism when she explained why Chinese American women rejected the National Organization of Women: "No matter how hard you fight to reduce the sexism, when it's all done and over with, you still have the racism. Because white women will be racist as easily as their male counterparts. And white women continue to get preferential treatment over women of color" (cited in Yung, 1986, p. 98). Recentering race, then, requires feminist scholars to reshape the basic concepts and theories of their field and to acknowledge that the experiences of white women and women of color are not just different but connected in systematic ways.

Bringing class and race into the study of gender also requires us to explicate the positions that white men *and* white women occupy over men of color. Many white feminists include all males, regardless of color and social class, into their critique of sexist power relations. This argument assumes that *any* male in U.S. society benefits from a patriarchal system designed to maintain the unequal rela-tionship that exists between men and women. Pointing to the multiplicities of men's lives, feminists of color have argued that, depending on their race and class, men experience gender differently. Presenting race and gender as relation-ally constructed, black feminists have referred instead to "racial patriarchy"—a concept that calls attention to the white patriarch/master in American history and his dominance over the black male as well as the black female (Gaines, 1990, p. 202). Providing yet another dimension to the intersections of race and gender, Gaines (1990) pointed out that the "notion of patriarchy is most obtuse when it disregards the position white women occupy over Black men as well as Black women" (p. 202). The failure of feminist scholarship to theorize the historically specific experiences of men of color makes it difficult for women of color to rally to the feminist cause without feeling divided or without being accused of betrayal. As Kimberlee Crenshaw (1989) observed, "Although patriarchy clearly operates within the Black community, ... the racial context in which Black women find themselves makes the creation of a political consciousness that is oppositional to Black men difficult" (p. 162). Along the same lines, King-Kok Cheung (1990) exhorted white scholars to acknowledge that, like female voices, "the voices of many men of color have been historically silenced or dismissed" (p. 246). David Eng (2001) reminded us that Asian Americanists have also neglected the topic of Asian American masculinities: in the last fifteen years, while the field of Asian American studies has produced substantial research on Asian American female identity, mother-daughter relations, and feminism, comparatively less critical attention has been paid to the formation of Asian American male subjectivity (p. 15).

## REFERENCES

American Sociological Association, Committee on the Status of Women in Sociology 1985. *The treatment of gender in research.* Washington, DC.

Cheung, K.-K. 1990. The woman warrior versus the Chinaman pacific: Must a Chinese American critic choose between feminism and heroism? In M. Hirsch and E. F. Keller (eds.), *Conflicts in feminism* (pp. 234–51). New York: Routledge.

Crenshaw, K. 1989. Demarginalizing the intersection of race and sex: A black feminist critique of antidiscrimination doctrine, feminist theory and antiracist politics. In *University of Chicago Legal Forum: Feminism in the law: Theory, practice, and criticism* (pp. 139–67). Chicago: University of Chicago Press.

Eng, D.L. 2001. *Racial castration: Managing masculinity in Asian America.* Durham, NC: Duke University Press.

Ferree, M. M., S. Khan, and S. A. Morimoto 2007. Assessing the feminist revolution: The presence and absence of gender in theory and practice. In C. Calhoun (ed.), *America Sociology: A Centenary* (pp. 438–79). Chicago: University of Chicago Press.

Gaines, J. 1990. White privilege and looking relations: Race and gender in feminist film theory. In P. Erens (ed), *Issues in feminist film criticism* (pp. 197–214). Bloomington: Indiana University Press.

Healey, J. F. 1995. *Race, ethnicity, gender, and class: The sociology of group conflict and change.* Thousand Oaks, CA: Pine Forge.

hooks, b. 1984 *Feminist theory: From margin to center.* Boston: South End.

Hurtado, A. 1989. Relating to privilege: Seduction and rejection in the subordination of white women and women of color. *Signs: Journal of Women in Culture and Society*, 14, (pp. 833–55).

Moraga, C. 1981. Preface. In C. Moraga and G. Anzaidua (eds.), *This bridge called my back: Writings by radical women of color* (pp. xiii–xix). Watertown, MA: Persephone.

Pedraza, S. 1991. Women and migration: The social consequences of gender. *Annual Review of Sociology*, 17, (pp. 303–25).

Stacey, J., and B. Thorne 1985. The missing feminist revolution in sociology. *Social Problems*, 32, (pp. 301–16).

Yung, J. 1986. *Chinese women of America: A pictorial essay*. Seattle: University of Washington Press.

## DISCUSSION QUESTIONS

1.  What do scholars mean when they suggest that the category of woman has been universalized, and why is that a problem?

2.  How does attention to race and class expand our understanding of gender and patriarchy?

3.  How can we see both race and gender expressed in the work options of men and women of different racial groups?

# 30

# The Myth of the Missing Black Father

### ROBERTA L. COLES AND CHARLES GREEN

*Why is there such a high number of mother-only families in the Black commu-
nity when this has not historically been the case? These authors identify how race
and social class barriers are important in shaping Black men's ability to fill an
important gendered role—being a father.*

The black male. A demographic. A sociological construct. A media caricature.
A crime statistic. Aside from rage or lust, he is seldom seen as an emotionally
embodied person. Rarely a father. Indeed, if one judged by popular and academic coverage, one might think the term "black fatherhood" an oxymoron.
In their parenting role, African American men are viewed as verbs but not
nouns; that is; it is frequently assumed that Black men *father* children but seldom
*are* fathers. ...

But this stereotype did not arise from thin air. ... In 2000, only 16 percent
of African American households were married couples with children, the lowest
of all racial groups in America. On the other hand, 19 percent of Black households
were female-headed with children, the highest of all racial groups. From the perspective of children's living arrangements, shown in [Table 1] over 50 percent of
African American children lived in mother-only households in 2004, again the
highest of all racial groups. Although African American teens experienced the
largest decline in births of all racial groups in the 1990s, still in 2000, 68 percent
of all births to African American women were nonmarital, suggesting the pattern
of single-mother parenting may be sustained for some time into the future. This
statistic could easily lead observers to assume that the fathers are absent.

While it would be remiss to argue that there are not many absent black
fathers, absence is only one slice of the fatherhood pie and a smaller slice than
is normally thought. The problem with "absence," as is fairly well established
now, is that it's an ill defined pejorative concept usually denoting nonresidence
with the child, and it is sometimes *assumed* in cases where there is no legal marriage to the mother. More importantly, absence connotes invisibility and nonin-
volvement, which further investigation has proven to be exaggerated (as will
be discussed below). Furthermore, statistics on children's living arrangements
(Table 1) also indicate that nearly 41 percent of black children live with their
fathers, either in a married or cohabiting couple household or with a single dad.

SOURCE: Roberta L. Coles and Charles Green., eds. 2010. *The Myth of the Missing Black
Father.* New York: Columbia University Press.

**TABLE 1    Living Arrangements of Black Children, 2004**

| Living arrangements | Black children |
| --- | --- |
| Two parents | 37.6% |
| Married | 33.9 |
| Unmarried | 3.7 |
| Mother-only | 50.4 |
| Father-only | 3.3 |
| Neither parent | 8.8 |

Source: Kreider (2008).

These African American family-structure trends are reflections of large-scale societal trends—historical, economic, and demographic—that have affected all American families over the past centuries. Transformations of the American society from an agricultural to an industrial economy and, more recently, from an industrial to a service economy entailed adjustments in the timing of marriage, family structure and the dynamics of family life. The transition from an industrial to a service economy has been accompanied by a movement of jobs out of cities; a decline in real wages for men; increased labor-force participation for women; a decline in fertility; postponement of marriage; and increases in divorce, nonmarital births, and single-parent and nonfamily households.

These historical transformations of American society also led to changes in the expected and idealized roles of family members. According to Lamb (1986), during the agricultural era, fathers were expected to be the "moral teachers"; during industrialization, breadwinners and sex-role models; and during the service economy, nurturers. It is doubtful that these idealized roles were as discrete as implied.... It is likely that many men had trouble fulfilling these idealized roles despite the legal buttress of patriarchy, but it was surely difficult for African American men to fulfill these roles in the context of slavery, segregation, and, even today, more modern forms of discrimination. A comparison of the socioeconomic status of black and white fathers illustrates some of the disadvantages black fathers must surmount to fulfill fathering expectations.... In 1999 only 33.4 percent of black fathers had attained at least a college education, compared to 68.5 percent of white fathers. In 1998, 25.5 percent of black fathers were un- or underemployed, while 17.4 percent of white fathers fell into that category. Nearly 23 percent of black fathers' income was half of the poverty threshold, while 15 percent of white fathers had incomes that low.

The historical transformations were experienced across racial groups but not to the same extent. The family forms of all racial groups in America have become more diverse, or at least recognition of the diversity of family structure has increased, but the proportions of family types vary across racial groups. Because African American employment was more highly concentrated in blue-collar jobs, recent economic restructuring had harsher implications for black communities and families. The higher and more concentrated poverty levels and greater income and wealth inequality—both among African Americans and between African Americans and whites—expose African American men, directly and indirectly, to continued lower life expectancy, higher mortality, and, hence, a skewed gender ratio that leaves black women outnumbering black men by the age of eighteen.

All of these societal and family-level trends affect black men's propensity to parent and their styles of parenting in ways we have yet to fully articulate. For instance, Americans in general have responded to these trends by postponing marriage by two to four years over the last few decades, but that trend is quite pronounced among African Americans, to the point that it is estimated that whereas 93 percent of whites born from 1960 through 1964 will eventually marry, only 64 percent of blacks born in the same period ever will.

... Wilson (1987) and others have suggested that black men's underemployment, along with black women's higher educational attainment in relation to black men may decrease both men's and women's desire to marry and may hinder some black men's efforts to be involved fathers. However, other research has found that even college-educated and employed black men have exhibited declines in marriage, and yet additional research points to attitudinal factors with black men desiring marriage less than white and Latino men....

In conjunction with [an] increased amount of research and, in fact, frequently fueling the research, has been a proliferation of public and private programs and grants aimed at creating "responsible fatherhood." While many of the programs have been successful in educating men on how to be qualitatively better fathers, many have aimed primarily either at encouraging fathers to marry the mothers of their children or at securing child support. Marriage and child support are important aspects of family commitment, but marriage is no guarantee of attentive fathering, and garnished child support alone, particularly if it goes to the state and not to the mother and child, is hardly better parenting....

Given the increased focus on fatherhood in scholarly and popular venues, what do we really know about black men and parenting? ... African American fathers, when discussed at all, continue to be addressed predominantly under categories frequently associated with parenting from afar, as nonresident, nonmarital fathers. Even books specifically on black fathers concentrate almost exclusively on nonresident fathers.

... Studies on this ilk of fathers indicate that generally a large portion of nonresident fathers are literally absent from their children's lives or, if in contact, their involvement decreases substantially over time. A number of memoirs by black men and women, sons and daughters of literally absent fathers, attest to the painful experience that this can be for the offspring—both sons and daughters—of these physically or emotionally missing fathers. For instance, writing in his 1999 book *Becoming Dad: Black Men and the Journey to Fatherhood,* award-winning journalist Leonard Pitts wrote of his own father and others:

> He was one thing many other fathers were not: He was there. Present and accounted for every day. Emotionally absent, mind you. But there, at least, in body. I know so many men, so many black men, who cannot

say the same. So many men for whom the absence of father is a wound that never scabbed over.

… Although these anguished experiences are too common, they remain only one part, though often the more visible part, of the larger fatherhood picture. An increasing number of quantitative and qualitative studies find that of men who become fathers through nonmarital births, black men are least likely (when compared to white and Hispanic fathers) to marry or cohabit with the mother. But they were found to have the highest rates (estimates range from 20 percent to over 50 percent) of visitation or provision of some caretaking or in-kind support (more than formal child support). For instance, Carlson and McLanahan's (2002) figures indicated that only 37 percent of black nonmarital fathers were cohabiting with the child (compared to 66 percent of white fathers and 59 percent of Hispanic), but of those who weren't cohabiting, 44 percent of unmarried black fathers were visiting the child, compared to only 17 percent of white and 26 percent of Hispanic fathers. These studies also suggested that black nonresident fathers tend to maintain their level of involvement over time longer than do white and Hispanic nonresident fathers.

Sometimes social, fictive, or "other" fathers step in for or supplement nonresident biological fathers. Little research has been conducted on social fathers, but it is known they come in a wide variety: relatives, such as grandfathers and uncles; friends, romantic partners and new husbands of the mother, cohabiting or not; and community figures, such as teachers, coaches, or community-center staff. Although virtually impossible to capture clearly in census data, it is known that a high proportion of black men act as social fathers of one sort or another, yet few studies exist on this group of dads.

Jarrett, Roy, and Burton's (2002:234) review of qualitative studies of black fathers managed to capture the perspectives of a few low-income social fathers. One sixteen-year-old talked about his fatherlike relationship with the young daughter of a friend.

> Tiffany (a pseudonym) is not my baby, but she needs a father. To be with her, I work in the day care center at school during my lunch hour. I feed her, change her diapers, and play with her. I buy her clothes when I can because I don't make much money. I keep her sometimes. Her mother and her family appreciate what I do and Tiffany loves me too. Every time she sees me she reaches for me and smiles.

Fagan's (1998) study of low-income biological and stepfathers in two-parent homes found that the two types of fathers of black children were equally involved. Similarly, McLanahan and Sandefur (1994) found that, compared to those who live in single-parent homes, black male teens who lived with stepfathers were significantly less likely to drop out of school and black teen females were significantly less likely to become teen mothers. The authors speculated that the income, supervision, and role models that stepfathers provide may help compensate for communities with few resources and social control. Although they are often pictured as childless men, these social fathers may also be some other child's biological father, sometimes a nonresident father himself. Consequently, it is not easy and is certainly misleading to discuss fathers as if they come in discreet, nonoverlapping categories of biological or social.

A smaller amount of research has been conducted on black fathers in two-parent families, which are more likely to also be middle-class families. Allen (1981), looking at wives' reports, found black wives reported a higher level of father involvement in childrearing than did white wives. McAdoo (1988) and Bowman (1993) also concluded that black fathers are more involved than white fathers in childrearing....

Finally, and ironically, most *absent* in the literature on black fatherhood have been those fathers who are most *present*: black, single full-time fathers. About 6 percent of black households are male-headed, with no spouse present; about half of those contain children under eighteen years old. These men also may be biological or adoptive fathers, but little is known about them....

In sum, research on black fathers has been limited in quantity and has narrowly focused on nonmarital, nonresident fathers and only secondarily on dads in married-couple households. This oversight is not merely intentional, for black men are only about 6 percent of the U.S. population and obviously a smaller percent are fathers. ... We want to adjust the public's visual lens from a zoom to a wide angle to view black fathers in a realistic landscape, to illustrate that they are quite varied in their living arrangements, marital status, and styles of parenting.

... We want to consider policies, tried or suggested, that impede or facilitate parenting on the part of black men. We seek to provide a forum in which black fathers in their full range of parenting take center stage. We feel that the timing ... is opportune, with the recent election of the first black president of the United States. Many African Americans are optimistic that President Barack Obama, who experienced the absence of his own father and is expressly committed to furthering involved fatherhood, will be able to significantly weaken the existing stereotype of the black father, both through his own public example and through facilitative policy.

## REFERENCES

Allen, W. 1981. "Mom, Dads, and Boys: Race and Sex Differences in the Socialization of Male Children." In *Black Men*, ed. L. Gary, 99–114. Beverly Hills, Calif.: Sage.

Bowman, P. 1993. "The Impact of Economic Marginality on African-American Husbands and Fathers." In *Family Ethnicity*, ed. H. McAdoo, 120–137. Newbury Park, Calif.: Sage.

Carlson, M. J. and S. S. McLanahan. 2002. "Fragile Families, Father Involvement, and Public Policy" In *Handbook of Father Involvement: Multidisciplinary Perspectives*, ed. Catherine Tamis-LeMonda and Natasha Cabrera, 461–88. Mahwah, N.J.: Lawrence Erlbaum.

Fagan, J. 1998. "Correlates of Low-Income African American and Puerto Rican Fathers' Involvement with Their Children." *Journal of Black Psychology* 24 (3): 351–67.

Jarrett, R. L., K. M. Roy, and L. M. Burton. 2002. "Fathers in the 'Hood': Insights from Qualitative Research on Low-Income African American Men." In *Handbook of Father Involvement: Multidisciplinary Perspectives*, ed. C. Tamis-LeMonda and N. Cabrera, 221–48. New York: Lawrence Erlbaum.

Kreider, R. M. 2008. *Living Arrangements of Children: 2004*. U.S. Department of Commerce, U.S. Census Bureau.

Lamb, M. E. 1986. "The Changing Role of Fathers." In *The Father's Role: Applied Perspectives*, ed. M. E. Lamb, 3–27. New York: Wiley.

McAdoo, J. L. 1988. "The Roles of Black Fathers in the Socialization of Black Children." In *Black Families*, ed. H. P. McAdoo, 257–69. Newbury Park, Calif.: Sage.

McLanahan, S., and G. Sandefur. 1994. *Growing Up with a Single Parent. What Hurts, What Helps*. Cambridge, Mass.: Harvard University Press.

Pitts, L., Jr. 1999. *Becoming Dad: Black Men and the Journey to Fatherhood*. Atlanta: Longstreet.

Wilson, W. J. 1987. *The Truly Disadvantaged: The Inner City, the Underclass, and Public Policy*. Chicago: University of Chicago Press.

## DISCUSSION QUESTIONS

1. What factors are important in affecting Black men's parenting roles?

2. What do the authors mean by "social fathers"? Why do they see such roles as important?

3. How does understanding the intersection of race and social class enhance our understanding of how Black men parent?

# 31

## Between Good and Ghetto

NIKKI JONES

*The everyday experiences of growing up are shaped by race, gender, and social class, as some communities can offer shelter from difficulties, while others require young people to negotiate complex environments. The African American girls observed by Nikki Jones had to be sensitive to the code of the street in their community, even if they were shaping a gender role that was opposed to that code.*

...The need to avoid or overcome dangers throughout their adolescence presents a uniquely gendered challenge for girls who grow up in distressed inner-city neighborhoods. As a system of accountability, gender reflects widely held beliefs, or normative expectations, about the "attitudes and activities appropriate for one's sex category." During interactions and encounters with others, children and adults evaluate themselves and others in light of these normative gender expectations in ways that reinforce or challenge beliefs about the natural qualities of boys and girls, and especially the essential differences between the two. Generally, women and girls who are able to mirror normative expectations of femininity during their interactions with others ..., are evaluated by adults (e.g., family members, teachers, counselors) and by peers as appropriately feminine girls or *good* girls. Meanwhile, girls or women who seem to violate perceived gender boundaries by embracing stereotypically masculine behaviors often are disparagingly categorized as "unnaturally strong" (Collins 2004, 193–199).

The intersection of gender, race, and class further complicates the degree to which girls measure up to gender expectations. African American, inner-city girls in the United States are evaluated not only in light of mainstream gender expectations but also by the standards of Black respectability: the set of expectations governing how Black women and girls ought to behave. These are reflected in images of "the Black lady"—think Claire Huxtable from the popular 1980s sitcom *The Cosby Show*—the middle-class, Black woman who reflects many of the expectations of White, middle-class femininity.

... Inner-city girls who live in distressed urban neighborhoods face a gendered dilemma: they must learn how to effectively manage potential threats of interpersonal violence—in most cases this means that they must work the code as

SOURCE: From Nikki Jones. 2010. *Between Good and Ghetto: African American Girls and Inner-City Violence*. New Brunswick: Rutgers University Press.

boys and men do—at the risk of violating mainstream and local expectations regarding appropriate feminine behavior. This is a uniquely difficult dilemma for
girls, since the gendered expectations surrounding girls' and women's use or control of violence are especially constraining…. Inner-city girls, like most American
girls, feel pressure to be "good," "decent," and "respectable." Yet, like some
inner-city boys, they may also feel pressure to "go for bad" or to establish a
"tough front" in order to deter potential challenges on the street or in the school
setting. They too may believe that "sometimes you do got to fight"—and sometimes they do. In doing so, these girls, and especially those girls who become
deeply invested in crafting a public persona as a tough or violent girl, risk evaluation by peers, adults, and outsiders as "street" or "ghetto."

Among urban and suburban adolescents, "ghetto" is a popular slang term that
is commonly used to categorize a person or behavior as ignorant, stupid, or otherwise morally deficient…. Analytically, the pairs "ghetto" and "good," or "street"
and "decent," are used to represent "two poles of value orientation, two contrast-
ing conceptual categories" that structure the moral order of inner-city life. In
inner-city neighborhoods, the decent/street or good/ghetto distinctions are powerful. Community members use these distinctions as a basis for understanding, interpreting, and predicting their own and others' actions, attitudes, and behaviors,
especially when it comes to interpersonal violence (Anderson 1999, 35)….

The branding of adolescent girls as ghetto is self-perpetuating, alienating the
institutional forces that protect good girls and forcing adolescent girls who work
the code of the street to become increasingly independent. Girls who are
evaluated by adults or peers as ghetto, as opposed to those evaluated as good,
ultimately may have the code as their only protection in the too often violent
inner-city world in which they live. Their efforts to protect themselves put them
at risk of losing access to formal institutional settings like schools or the church,
where girls who mirror normative gender expectations—girls who are perceived
by others as good—can take some refuge. Yet, even for those good girls, this
institutional protection is inadequate—they are aware that they may become targets in school or on the street and they too feel pressure to develop strategies that
will help them successfully navigate their neighborhoods. Thus, inner city girls find
themselves caught in what amounts to a perpetual dilemma, forced by violent circumstances to choose between two options, neither of which offers the level of
security that is generally taken for granted in areas outside of urban poverty….

My conversations with girls about their experiences with violence, along
with my observations of their actions and conversations with others, revealed
that girls astutely worked the code *between* the equal and opposing pressures of
good and ghetto. From this social location, girls are able to challenge and manip-
ulate the constraining social and cultural expectations embedded in gender and
the code, depending on the situation. Elijah Anderson defines the activity of
adapting one's behavior to the set of rules that govern a situation—decent or
street, good or ghetto—as "code switching." Inner-city families and youth,
most of whom strive for decency, put a "special premium" on the ability to
"switch codes and play by the rules of the street," when necessary (1999, 36,
98–106). Of course, this act is complicated for girls whose working of the code

is likely to challenge expectations regarding appropriate feminine behavior. Inner-city girls work the code with the understanding that they are always accountable to these gendered expectations and that gender violations are likely to open them up to a series of public or private sanctions. Girls' lives seemed to be defined by this every day struggle to balance the need to protect themselves with the pressure to meet normative expectations associated with their gender, race, and class positions. Girls' accounts of how they manage these expectations, including how they work the code, defy any simple categorizations or stereotypical evaluations of girls as *either* good or ghetto. Instead, girls' accounts of violent incidents reveal that they embrace, challenge, reinforce, reflect, and contradict normative expectations of femininity and Black respectability *as* they work the code. Girls' accounts of navigating inner-city adolescence are characterized by this fluidity.

...[Girls] who are invested in being perceived by others as good tend to have a limited fighting history. They also typically have a network of family members who are committed to isolating (or at least buffering) them from actual and potential dangers of the street. Nevertheless, they understand how the code organizes social life in troubled neighborhoods. They may be reluctant to engage in physical battles, but they have also learned that "sometimes you got to fight." Even as they work the code, enacting displays of strength or aggression typically expected from boys, these adolescent girls indicate an acceptance of fundamental elements of mainstream femininity and Black respectability. These expectations, in turn, modulate the way these girls handle conflict and violence in neighborhood settings. Their general commitment to presenting themselves in ways that conform to normative expectations is typically rewarded by adults. In the school setting, teachers and counselors are likely to treat girls who conform to gender expectations in ways that reaffirm their public personas as young ladies who are good and appropriately feminine. These girls' peers, and particularly their female peers, however, may observe such girls' appearance, attitude, and behavior—their "presentation of self" (Goffman 1959)—as evidence that these girls think they are better than their peers (i.e., that "they think they are all that"). For their part, these girls often conclude that other girls are "jealous." These competing perceptions often either instigate or are used to justify physical battles between girls. In other words, what protects and insulates them is at the same time their point of entry for violent conflict. This complexity is characteristic and illustrates how girls who may otherwise gravitate toward a set of behaviors or beliefs that are commonly perceived as appropriately feminine and essentially good get caught between good and ghetto.

In contrast to girls who see themselves as good girls, girls who see themselves as fighters tend to view life as an ongoing battle. They seem ready to fight at the slightest of provocation, even seeking out opportunities to prove their reputations by courting conflict or by engaging in the kinds of "campaigns for respect" (Anderson 1999, 68) that are common among inner-city boys. These girls are generally aware that others consider them outsiders, either because of their physical attributes or their behavior. Girls who are known for fighting and winning, ... embrace those elements of the code that directly challenge gendered expectations associated with White, middle-class femininity and Black respectability. These girls may come to embrace this outsider status with confidence and without apology, often to the disappointment or frustration of those who would prefer that they look or act more like good girls. Yet, girl fighters may also be striving for decency in ways that are not obvious to most outsiders. Understanding how girls reconcile the gendered dilemmas that accompany working the code, whether they identify as good girls or fighters, expands and enriches our understanding of African American, inner–city girls' beliefs and behaviors, and deepens our understanding of how the contemporary circumstances of life in distressed inner-city neighborhoods shape their social world.

## REFERENCES

Anderson, Elijah. 1999. *Code of the street: Decency, violence, and the moral life of the inner city.* New York: W. W. Norton.

Collins, Patricia Hill. 2004. *Black sexual politics: African Americans, gender, and the new racism.* New York: Routledge.

Goffman, Erving. 1959. *The presentation of self in everyday life.* New York: Anchor Books.

## DISCUSSION QUESTIONS

1. What are the challenges to meeting normative gender expectations of femininity for Black girls in depressed neighborhoods?

2. Why do young African American women have to understand the code of the streets?

3. How is the symbolic power of being either a "good girl" or a "ghetto girl" expressed in the day-to-day life of young women?

# 32

# "They Are Testing You All the Time":
## Negotiating Dual Femininities among
## Chicana Attorneys

GLADYS GARCÍA-LÓPEZ AND DENISE A. SEGURA

*The pool of Chicana attorneys is growing, but their numbers are still small. This article identifies how these educated professional women have to negotiate between their work and political communities. They are involved in delicate race and gender negotiations as they face barriers to mobility in the workplace, but also have to retain their feminine ways to retain their connections with their ethnic communities.*

> "I think they always want to *see* if you really are a lawyer ...
> they are testing you all the time."
> —YVETTE GARCÍA

A small but growing number of Chicanas are entering highly prestigious and male-dominated legal professions. This relatively recent development raises a number of questions regarding their incorporation and mobility within this field. For example, how are Chicanas received by other lawyers, judges, and clients? How do Chicanas negotiate space in this masculine terrain?...

In this article we present findings from a recent qualitative study of 15 "successful" Chicana attorneys. These Chicana attorneys define success as "making a difference in my community," "empowering people and in turn, myself," and "having a balance between my career and my family." Our analysis of the ways that co-workers, supervisors, and clients "are testing" successful Chicanas' qualifications and legitimacy "all the time" reveals key mechanisms in workplace structure and interaction that reinforce racially gendered boundaries that contradict larger goals of social diversity. We also find that Chicanas who enter the legal profession aspiring to serve their community navigate across racially gendered worksites by developing a number of strategies including negotiating a

SOURCE: *Feminist Studies* 34, nos. 1/2 (Spring/Summer 2008). © 2008 by Gladys García-López and Denise A. Segura.

distinctive Chicana practice and presentation of self, or what we term "dual femininities," as part of the process of becoming "legitimate" lawyers.

Dual femininities are enactments of a culturally gendered ideology and practice produced through a series of negotiations across class, culture, and gender systems. In their professional lives, Chicana attorneys are held accountable to white, middle-class femininities that are becoming more visible within increasing numbers of legal settings. Unlike white women, however, Chicanas are expected to adhere to distinct values and perform culturally gendered behaviors by their Chicano/Latino clients. Dual femininities are thus ideological and dynamic gender practices that are produced contextually as Chicanas negotiate the "shoulds" of culture with the "musts" of the profession....

How Chicana attorneys navigate across these accountability systems to develop strategies for acceptance and legitimacy in both their political communities and worksites is our major question....

Chicanas in law are submerged in an institution where white heterosexual masculinities define success. Successful attorneys must immerse themselves within the profession on a full-time basis, leaving little time for other social obligations, such as parenting, which has been defined as "women's work." The current legal professional structure does not support women's reproductive abilities, choices, and desires....

Women are expected to "do" a relatively narrow form of "white heterosexual femininity" that has gained some acceptance within this profession. As working-class women of color, however, Chicanas "do difference." This difference often makes it difficult if not impossible for them to be accepted as competent attorneys. Furthermore, within their respective racial-ethnic communities, Chicana attorneys are expected to perform "heterosexual Chicana femininity." These femininities are accomplished through culturally gendered practices in their respective workplaces.

...Drawing on the work of Patricia Yancey Martin, who encourages "focusing on practicing femininity/femininities and practicing masculinity/masculinities ... because multiple masculinities and femininities exist, and people practice, and are held accountable to, specific kinds depending on their bodies," we argue that multiple forms of Chicana femininities operate. These dual femininities are physically and ideologically negotiated depending on the context (for example, private versus public-interest firm, courtroom versus community).[1]...

Similar to white women attorneys, Chicanas are evaluated by colleagues and clients within a masculine normative order that typically regards feminine gender displays and ways of being as antithetical to "winning." Jennifer Pierce argues that women lawyers are often caught in a "double bind" between the requirements of "the good woman" and the role of the adversary.[2] Chicanas' double bind involves developing gendered practices that incorporate various Chicana/o cultural values, behaviors, and presentations of self within environments that are slowly accepting some aspects of white femininities as the profession diversifies....

Chicana lawyers are expected to continue their role as intermediaries or brokers between the Chicano/Latino community and the legal system. This expectation extends the well-documented practice of Latino immigrants to utilize the bilingual skills of their children as cultural and linguistic translators. These idealized traits associated with Chicano/Latino culture exist in varying levels among Chicanos and Latinos depending on generation in the United States; neighborhood ethnic composition; region; interaction levels with kin, particularly relatives in Mexico or Latin America; educational levels; and social class. Chicanas who violate these idealized norms and expectations are often viewed as "*vendidas*" or "cultural sell-outs."...

## PHYSICAL AND IDEOLOGICAL NEGOTIATIONS OF DUAL FEMININITY IN THE WORKPLACE

In this research the courtroom and the law firm (both private and public-interest) are key workplace sites. The firm provides the context for Chicanas' development of gender practices and deployment of dual femininities through the cases they handle, their interaction with colleagues and clients, and their presentation of self. The courtroom is a racialized site characterized by defendants of color and white attorneys. In both settings, the mere presence of a Chicana attorney disrupts racial and gendered displays and configurations. ...

Chicanas report that they are often the "token" woman of color attorney in their respective workplaces. Their low visibility magnifies the ascendancy of white male culture and leads them to voice feelings of what several term *rechazo*, or rejection. Laura Esquivél is one of the many women who perceive rejection and a sense of undesirability by male counterparts:

> I feel that we [Chicanas] are not accepted as attorneys, litigators, and thinkers, or as contributing to the legal society, but rather we're just kind of like there for their viewing [laughter]. I don't think that's imagined either. There are certain people who think like that.

A workplace organization that puts Chicanas "on display" for "their viewing" is a mechanism that "others" Chicanas.... Chicanas, unlike white women, are often viewed as "foreigners" by the state and within the system that regulates the state—the law. Their "othering" can be described as part of their "perceived difference":[3] they do not simulate the prototypical white male attorney or, in most cases, idealized white femininities.

"Othering" is also reinforced by the misidentification process. Misidentification reflects the hierarchal structure of the legal profession along racial-ethnic, gendered, heterosexual, and class lines.... Ana claims she is confused with courtroom support staff as a result of "looking young, and especially because I'm not that tall." Irma describes how her body evokes a negative assessment of her legal qualifications from judges.

> You know, a judge will get a paper and it will say, "Irma Juárez, attorney-at-law, Compton, California," and he will probably assume that I went to some correspondence school instead of Harvard Law.

There are definitely some mixed notions of who I am and sometimes I feel that they think that I am just this short, young-looking girl....

Misidentification serves as an insidious reminder to Chicanas that they are outsiders. For Chicanas, misidentification results not only from body signification, but what they *sound* like as well. Research indicates that women tend to have softer voices vis-à-vis men, which are a significant drawback in the legal profession and points to an additional layer of complexity for Chicanas with Spanish-accented voices....

Seen and unseen, Linda [Guardia], like all of the Chicana attorneys with Spanish accents, assert that their voices heighten misidentification and lead to inquiries and doubts about their professional competence. Oftentimes, speaking with an accent in the United States is equated to "thinking with an accent."...

Misidentification does, however, negatively impact Chicana lawyering. Olivia states:

> When people confuse me with something else, that tells me that they don't think that I am capable or that I can't do it. All Latinas encounter situations where you walk into a courtroom and they think you are the interpreter or they think you're the defendant. My feelings no longer get hurt when that happens. What happens is that I hope that this person learned a lesson today....

## THE COMMUNITY

Chicanas' enactments of dual femininities negotiate across the accountability systems of the profession and their interpretations of the culturally gendered expectations of the Chicano/Latino community. Chicana lawyers in this study assert they are held accountable to being a "Chicana" lawyer and enact appropriate forms of Chicana (heterosexual) femininity. Chicana lawyers are expected to be proficient in Chicano/Latino cultural capital (for example, knowing how to act, speak, and present oneself in ways that demonstrate affinity with other Chicanos/ Latinos). They are expected to use their education to improve the overall conditions of their respective communities. Women are expected to prioritize marriage, motherhood, and their extended families. Chicana lawyers are expected to be family spokespersons and to resolve any public discord that family members may be undergoing. These expectations have been directed at other highly educated Chicanas as well.

Chicanas in this study report they were drawn to the legal profession not only as a form of acquiring upward mobility, but also as a way of producing positive social change for their communities. "I feel a unique responsibility to be accessible to the Latino community," states Ana Cantú. Laura Esquivél prizes her ability to "look at a code and know how to interpret it and lead my clients in the right direction" as "empowering." She states that these actions fulfill her purpose and need to combat "oppression against Latinos."

Chicanas' strong sense of accountability to their communities and their sense of how this should be done differs from that of the prototypical white attorney. That is, Chicanas are far less interested in asserting positional aspects of their profession or "doing dominance" with their clients. As Marissa Muñoz states, "I make sure my clients know that I don't think I'm better than somebody else." Rather, Chicana attorneys in this study emphasize relationship building that utilizes femininities constructed in dialogue with Chicano/Latino culture and the culture of the profession to connect individuals to larger community agendas, for example, housing, immigration, and underemployment. ...

Non-Chicano/Latino lawyers may speak Spanish, but they cannot speak to clients from the standpoint of a Chicana from a working-class urban *barrio*. According to Linda Guardia, the fact that she is "Latina," and the majority of her clients are of this same background, "makes it an easy connection, an easy relationship." Olivia's narrative offers yet another example of a unique Chicana relationship-building model based on race-ethnicity, class, language, and culture.

> I think it definitely helps that they see me walking into their home and I feel comfortable in their home. I don't find it weird that there are three beds in the living room, and I'm not going to refuse a glass of water that they give me. There's something more of a comfort level when you see someone like you because you have no idea what's in store [as far as the legal system]. I was born here but you have a very good idea of the sacrifices that these people make....

Chicana femininities enact an aura of feeling comfortable in Chicano/Latino and Mexican immigrant homes.... Olivia notes that non-Chicano/Latino lawyers also can be effective; however, the way they engage in building relations is qualitatively distinct....

## CONCLUSION

Across the firm, courtroom, and the community, the narratives of the fifteen Chicana attorneys in this study demonstrate an ongoing set of negotiations between the "shoulds" of Chicana femininities and the "musts" of the legal profession. Chicanas are held accountable to working-class, culturally gendered ways of being that emphasize respect, daughterly deference, family advocacy, and extended cultural brokering. They bring these funds of knowledge to environments that traditionally have valued adversarial masculinities where relation building is instrumental rather than affirming. In this environment femininities tend to be accepted when they contribute to winning cases. Negotiating across these competing expectations produces what we have termed Chicana "dual femininities."

Chicanas are aware that what is considered "appropriate" femininity in the legal profession is typically that which is performed by white, middle-class female attorneys in the firm. As women, Chicana attorneys are provided with examples of successful white, female attorneys whose presentation of self and lawyering styles are available for review. Workplace organization is critical to the options

Chicanas choose. In the courtroom, Chicanas are confronted with furnishings and infrastructure designed to display male bodies and diminish the voice of women, particularly short-statured women, as are many Chicanas. Although they may stand to the side of the podium to be heard, their Spanish-accented voices exacerbate their "othering." Misidentification as support staff and defendants is another form of "othering" that requires Chicana attorneys to craft professional presentations of self. In the firm, Chicanas' misidentification for janitorial and support staff despite their professional attire speaks to their "difference" from the workplace norm. These differences are compounded daily in interaction with partners and clients who "are testing" Chicana attorneys "all the time," but from different standpoints. Some Chicanas, particularly those with fairer skin tones and nonindigenous phenotypes might come close to enacting preferred forms of "white heterosexual femininities," but most either could not or would not abide by such norms.

... Chicanas make decisions about the appropriate gender display and femininities to enact in work and community settings. With twin goals of establishing legitimacy as lawyers and to serve communities historically disenfranchised by legal chicanery, Chicana attorneys map the terrain of the courtroom and the firm to insert new gender practices that affirm their cultural affinities. Chicana dual femininities are strategies women enact to bridge the gap between professional norms and cultural expectations.

## NOTES

1. Patricia Yancey Martin, " 'Said and Done' vs. 'Said and Doing': Gendering Practices, Practicing Gender at Work," *Gender & Society* 17, no. 3 (2003): 355.

2. See Jennifer Pierce, *Gender Trials: Emotional Lives in Contemporary Law Firms* (Berkeley: University of California Press, 1995): 20.

3. Gloria Anzaldúa, *Borderlands/La Frontera* (San Francisco: Aunt Lute Books, 1999): 60–61.

## DISCUSSION QUESTIONS

1. Law is a male-dominated profession, but White women and Chicanas face different challenges to being taken seriously as competent professionals. Why do their experiences differ?

2. As raised working-class women, how do Chicana attorneys define success as a professional?

3. What are the barriers to Chicana attorneys achieving their professional and community goals?

# Student Exercises

1. What does race mean in your life? What does social class mean in your life? What does gender mean in your life? Following the model of these articles, how would you describe where you sit with regard to a matrix of domination? What does your location mean in terms of advantages and disadvantages with regard to the challenges of securing housing, pursuing your education, and negotiating your family's strategies for raising you? How is your location linked to specific resources (these can be material resources or social networks) that you use to resolve the challenges you face? In class, break into groups in which students have had different experiences, and share your insights.

2. We often learn perceptions of others via the media. How is your race, class, and gender group presented in the media? Investigate cultural images of your group. Look for media about people like you. Identify ten specific images—these can be on the Internet or in print advertisements, television shows, movies, and so on. The media helps to present a public face for people in your unique groups. How does that public presentation influence how you are received by people as you go about your daily life? How are the media images of you different from those of other groups? Do you think that media images hamper or support your movements in public spaces?

# PART III

# Race and Social Institutions

# Section VIII

# Race and the Workplace

Elizabeth Higginbotham
and Margaret L. Andersen

The history of discrimination still plagues us. The United States has made progress toward reducing inequality by changing key laws that have blocked people's access to education and work opportunities. However, even after the passage of key civil rights legislation and important Supreme Court decisions, we still have **institutionalized racism**—that is, power and privilege based on race result in discrimination in most areas of social life. Discrimination plays a role in shaping different sets of opportunities for people in the country. Without interventions, discrimination will continue to promote inequalities among citizens and new immigrants.

Establishing coalitions to work for racial justice means we have to understand institutionalized racism, especially as it operates today. But segregation and isolation mean there are significant barriers that hinder that understanding. Members of privileged groups know little about the lives of others beyond what the media project, and many members of minority groups know little about each other and might be distrustful of more powerful people in the society.

Part Three explores the segregated nature of many contemporary social institutions and what this means for various segments of the population. We begin that exploration with an examination of race in the workplace. Labor exploitation has historically established a racialized labor system, although in more recent decades, much has been done to dismantle it. The Civil Rights Act of 1964 outlawed discrimination based on race, creed (that is, religion), national origin, and sex whether in employment, education, or public accommodation (such as restaurants, buses, and trains). Because of this act and other anti-discrimination policies, there is now clear progress as some racial minorities have been able to enter well-paid industrial and service jobs, for example, as bus drivers and police officers. Access to educational spheres has enabled some to enter the ranks of professional and managerial occupations. However, racial disparities remain in educational and occupational attainment. Many of these disparities are linked to a legacy of exclusion in the past. In their groundbreaking book *Black Wealth/White Wealth*, Melvin Oliver and Thomas Shapiro (2006) argue that exclusion from decent work and other social investments were historic barriers for racial minorities, and the effects of those historic barriers persist over time. Because opportunities were denied to racial minorities for decades, even when jobs and opportunities were finally opened in the mid-1960s, racial minorities were still vulnerable to and continue to have a difficult path to advancement. Oliver and Shapiro call this phenomenon the **sedimentation of inequality.**

Efforts to end historic inequalities do not take place in a vacuum. There were significant improvements in the 1960s, but since the 1970s, the U.S. political economy has accelerated the shift from one based on industrial jobs to one based on service jobs, a process called **economic restructuring.** Technological changes mean that we need fewer unskilled and semi-skilled workers in our industries. Economic restructuring produces jobs at the upper end of the workforce that require high levels of skill. Without higher education and training, workers are limited to low-wage employment, often in the growing personal service market. Although economic restructuring is a relatively new pattern, it builds upon the exclusion of the past, leaving racial minority workers vulnerable to new forms of job exclusion and blocked mobility.

William Julius Wilson ("Toward a Framework for Understanding Forces that Contribute to or Reinforce Racial Inequality") examines economic and political forces that can appear to be race neutral, but indirectly contribute to racial inequality. Technological innovations have revolutionized the workplace and global trade has shifted job locations, making it difficult for African Americans with limited educational attainment to find work.

How do people navigate this shifting economic environment? Whether one can find a good job is, as the popular adage goes, often a matter of who you know. Deirdre A. Royster ("Race and the Invisible Hand: How White Networks Exclude Black Men from Blue-Collar Jobs") shows that racism trumps the "invisible hand" of the market by limiting the networks and connections that Black men have, leaving them less likely than White men to find employment, even in working-class jobs. Her research provides a rich account of the processes by which discrimination operates against Black men, as well as other groups who do not have access to White networks.

Immigrants, too, confront a labor market marked by patterns of segregation and inequality. Pierrette Hondagneu-Sotelo ("Families on the Frontier") examines how immigrant women, especially those from Latin America, enter the racialized and gendered niche of service work in private homes, hotels, nursing homes, and hospitals. Hondagneu-Sotelo notes that earlier, men would cross borders for industrial and agricultural work to support their families in home countries, but now it is most frequently women who cross borders and send their wages back home to support families. This arrangement shifts the nature of parenting for these women.

Angela Stuesse ("Race, Migration, and Labor Control") has studied the poultry factories of Mississippi, where production has expanded and employers have found new strategies to control the workforce. Employers have recruited immigrants from Latin America to join the local African Americans in these plants, while offering few benefits and neglecting workers' rights. At the same time, the increased diversity of the workforce creates challenges for union organizers, who see unionization as a way to confront the exploitative practices of employers and protect all workers.

Together, the articles in this section show how racial systems continue to be embedded in the workplace. Understanding the racial structures in the workplace and the problems they pose for people who are seeking work can help us think about the nature of interventions for this century. Can we build a workforce that is not based on exclusion and exploitation? This is the challenge for the future.

## REFERENCE

Oliver, Melvin, and Thomas Shapiro. 2006. *Black Wealth/White Wealth: A New Perspective on Racial Inequality*. 2nd edition. New York: Routledge.

## FACE THE FACTS: MEDIAN WEEKLY EARNINGS OF FULL-TIME WORKERS (2007)

### Face the Facts:
### Median Household Income by Race, 2008

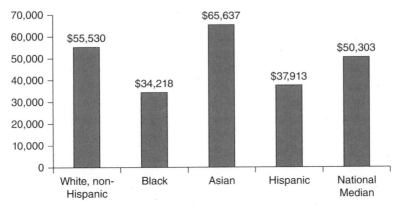

SOURCE: U.S. DeNavas-Walt, Carmen, Bernadette D. Proctor, and Jessica C. Smith, 2009. *Income, Poverty, and Health Insurance Coverage in the United States: 2008*. Washington, DC: U.S. Department of Commerce. www.census.gov

***Think about it:*** Look carefully at the chart above and think about what the median incomes reported here suggest about what different groups can afford. If you had the income of any two of these groups, how might your lifestyle be different? (It will help if you break down the annual income into monthly income; remember that this is pre-tax income).

# 33

# Toward a Framework for Understanding Forces that Contribute to or Reinforce Racial Inequality

WILLIAM JULIUS WILSON

*We still have individuals who discriminate in hiring and promotion. These are social acts, but we need to examine social processes that might indirectly affect racial inequality. Wilson highlights how economic and political changes, such as the shifts in the nature and location of jobs and government responses, are important in shaping the employment options for African Americans who have not overcome a legacy of exclusion in education and employment.*

## UNDERSTANDING THE IMPACT OF STRUCTURAL FORCES

Two types of structural forces contribute directly to racial group outcomes such as differences in poverty and employment rate: social acts and social processes. "Social acts" refers to the behavior of individuals within society. Examples of social acts are stereotyping; stigmatization; discrimination in hiring, job promotions, housing, and admission to educational institutions—as well as exclusion from unions, employers' associations, and clubs—when any of these are the act of an individual or group exercising power over others.

"Social processes" refers to the "machinery" of society that exists to promote ongoing relations among members of the larger group. Examples of social processes that contribute directly to racial group outcomes include laws, policies, and institutional practices that exclude people on the basis of race or ethnicity. These range from explicit arrangements such as Jim Crow segregation laws and voting restrictions to more subtle institutional processes, such as school tracking that purports to be academic but often reproduces traditional segregation, racial profiling by police that purports to be about public safety but focuses solely on minorities, and redlining by banks that purports to be about sound fiscal policy

SOURCE: *Race and Social Problems* (2009) Vol. 1, pp. 3–11.

but results in the exclusion of people of color from home ownership. In all of these cases, ideologies about group differences are embedded in organizational arrangements.

However, many social observers who are sensitive to and often outraged by the direct forces of racism, such as discrimination and segregation, have paid far less attention to those political and economic forces that *indirectly* contribute to racial inequality. I have in mind political actions that have an impact on racial group outcomes, even though they are not explicitly designed or publicly discussed as matters involving race, as well as impersonal economic forces that reinforce longstanding forms of racial inequality. These structural forces are classified as indirect because they are mediated by the racial groups' position in the system of social stratification (the extent to which the members of a group occupy positions of power, influence, privilege, and prestige). In other words, economic changes and political decisions may have a greater adverse impact on some groups than on others simply because the former are more vulnerable as a consequence of their position in the social stratification system....

Take, for instance, impersonal economic forces, which sharply increased joblessness and declining real wages among many poor African Americans in the last several decades. As with all other Americans, the economic fate of African Americans is inextricably connected with the structure and functioning of a much broader, globally influenced modern economy. In recent years, the growth and spread of new technologies and the growing internationalization of economic activity have changed the relative demand for different types of workers. The wedding of emerging technologies and international competition has eroded the basic institutions of the mass production system and eradicated related jobs in manufacturing in the United States. In the last several decades, almost all of the improvements in productivity have been associated with technology and human capital, thereby drastically reducing the importance of physical capital and natural resources. The changes in technology that are producing new jobs are making many others obsolete.

Although these trends tend to benefit highly educated or highly skilled workers, they have contributed to the growing threat of job displacement and eroding wages for unskilled workers. This development is particularly problematic for African Americans who have a much higher proportion of workers in low-skilled jobs than whites....

The workplace has been revolutionized by technological changes that range from mechanical development like robotics, to advances in information technology like computers and the internet. While even educated workers are struggling to keep pace with technological changes, lower-skilled workers with less education are falling behind with the increased use of information-based technologies and computers and face the growing threat of job displacement in certain industries. To illustrate, in 1962 the employment-to-population ratio—the percentage of adults who are employed—was 52.5% for those with less than a high school diploma, but by 1990 it had plummeted to 37.0%. By 2006 it rebounded slightly to 43.2%, possibly because of the influx of low-skilled Latino immigrants in low-wage service-sector jobs.

In the new global economy, highly educated, well-trained men and women are in demand, as illustrated most dramatically in the sharp differences in employment experiences among men. Compared to men with lower levels of education, college-educated men spend more time working, not less. The shift in the demand for labor is especially devastating for those low-skilled workers whose incorporation into the mainstream economy has been marginal or recent. Even before the economic restructuring of the nation's economy, low-skilled African Americans were at the end of the employment line, often the last to be hired and the first to be let go.

The computer revolution is a major reason for the shift in the demand for skilled workers. Even "unskilled" jobs such as fast food service require employees to work with computerized systems, even though they are not considered skilled workers. Whereas only one-quarter of U.S. workers directly used a computer on their jobs in 1984, by 2003 that figure had risen to more than half (56.1%) the workforce....

The shift in the United States away from low-skilled workers can also be related to the growing internationalization of economic activity, including increased trade with countries that have large numbers of low-skilled, low-wage workers. Two developments facilitated the growth in global economic activity: (1) advances in information and communication technologies, which enabled companies to shift work to areas around the world where wages for unskilled work are much lower than in the "first world;" and (2) the expansion of free trade, which reduced the price of imports and raised the output of export industries. But increasing imports that compete with labor-intensive industries (e.g., apparel, textile, toys, footwear, and some manufacturing) hurts unskilled labor.

Since the late 1960s international trade has accounted for an increasing share of the U.S. economy, and, beginning in the early 1980s, imports of manufactured goods from developing countries have soared. According to economic theory, the expansion of trade with countries that have a large proportion of relatively unskilled labor will result in downward pressure on the wages of low-skilled Americans because of the lower prices of the goods those foreign workers produce. Because of the concentration of low-skilled black workers in vulnerable labor-intensive industries (e.g., 40% of textile workers are African American even though blacks are only about 13% of the general population; this overrepresentation is typical in many low-skill industries), developments in international trade are likely to further exacerbate their declining labor market experiences.

Note that the sharp decline in the relative demand for low-skilled labor has had a more adverse effect on blacks than on whites in the United States because a substantially larger proportion of African Americans are unskilled. Indeed, the disproportionate percentage of unskilled African Americans is one of the legacies of historic racial subjugation. Black mobility in the economy was severely impeded by job discrimination as well as failing segregated public schools, where per-capita expenditures to educate African American children were far below amounts provided for white public schools. While the more educated and highly trained African Americans, like their counterparts among other racial groups, have very likely benefited from the shifts in labor demand, those with lesser skills have suffered....

The economic situation for many African Americans has now been further weakened because they tend not only to reside in communities that have higher jobless rates and lower employment growth—for example, places like Detroit or Philadelphia—but also they lack access to areas of higher employment growth....

The growing suburbanization of jobs means that labor markets today are mainly regional, and long commutes in automobiles are common among blue-collar as well as white-collar workers. For those who cannot afford to own, operate, and insure a private automobile, the commute between inner-city neighborhoods and suburban job locations becomes a Herculean task. For example, Boston welfare recipients found that only 14% of the entry-level jobs in the fast-growth areas of the Boston metropolitan region could be accessed via public transit in less than 1 hour. And in the Atlanta metropolitan area, fewer than one-half of the entry level jobs are located within a quarter mile of a public transit systems. To make matters worse, many inner-city residents lack information about suburban job opportunities. In the segregated inner-city ghettos, the breakdown of the informal job information network magnifies the problems of *job spatial mismatch*—the notion that work and people are located in two different places.

Although racial discrimination and segregation exacerbate the labor-market problems of the low-skilled African Americans, many of these problems are currently driven by shifts in the economy. Between 1947 and the early 1970s, all income groups in America experienced economic advancement. In fact, poor families enjoyed higher growth in annual real income than did other families. In the early 1970s, however, this pattern began to change. American families in higher income groups, especially those in the top 20%, continued to enjoy steady income gains (adjusted for inflation), while those in the lowest 40% experienced declining or stagnating incomes. This growing disparity in income, which continued through the mid-1990s, was related to a slowdown in productivity growth and the resulting downward pressure on wages....

More than any other group, low-skilled workers depend upon a strong economy, particularly a sustained tight labor market—that is, one in which there are ample jobs for all applicants. In a slack labor market—a labor market with high unemployment—employers can afford to be more selective in recruiting and granting promotions. With fewer jobs to award, they can inflate job requirements, pursuing workers with college degrees, for example, in jobs that have traditionally been associated with high school-level education. In such an economic climate, discrimination rises and disadvantaged minorities, especially those with low levels of literacy, suffer disproportionately.

Conversely, in a tight labor market job vacancies are numerous, unemployment is of short duration, and wages are higher. Moreover, in a tight labor market the labor force expands because increased job opportunities not only reduce unemployment but also draw in workers who previously dropped out of the labor force altogether during a slack labor market period. Thus, in a tight labor market the status of all workers—including disadvantaged minorities—improves....

Undoubtedly, if the robust economy could have been extended for several more years, rather than coming to an abrupt halt in 2001, joblessness and concentrated poverty in inner cities would have declined even more. Nonetheless, many people concerned about poverty and rising inequality have noted that productivity and economic growth are only part of the picture.

Thanks to the Clinton-era economic boom, in the latter 1990s there were signs that America's rising economic inequality that began in the early 1970s was finally in remission. Nonetheless, worrisome questions were raised by many observers at that time: Will this new economy eventually produce the sort of progress that prevailed in the two and a half decades prior to 1970—a pattern in which a rising tide did indeed lift all boats? Or would the government's social and economic policies prevent us from duplicating this prolonged pattern of broadly equal economic gains? In other words, the future of ordinary families, especially poor working families, depends a great deal on how the government decides to react to changes in the economy, and often this reaction has a profound effect on racial outcomes....

During Bill Clinton's 8 years in office, redistribution measures were taken to increase the minimum wage. But the George W. Bush administration halted increases in the minimum wage for nearly a decade, until the Democrats regained control of Congress in 2006 and voted to again increase the minimum wage in 2007. All of these political acts contributed to the decline in real wages experienced by the working poor. Because people of color are disproportionately represented among the working poor, these political acts have reinforced their position in the bottom rungs of the racial stratification ladder. In short, in terms of structural factors that contribute to racial inequality, there are indeed nonracial political forces that definitely have to be taken into account.

## DISCUSSION QUESTIONS

1.  Why does Wilson look beyond social acts of discrimination to explore the continuation of racial inequality?

2.  Why are many African Americans unable to successfully compete in the new economy?

3.  How can you apply Wilson's argument about the indirect effect of economic and political forces to thinking about the economy today and the prospects for different groups?

# 34

# Race and the Invisible Hand:

## How White Networks Exclude Black Men from Blue-Collar Jobs

DEIRDRE A. ROYSTER

*The "invisible hand" refers to a meritocracy, where who you know shapes employment options. By comparing the experiences of Black and White working-class men who graduated from the same vocational high school, Royster shows how race can shape Black men's access to job networks—making who you know a critical factor in their economic well-being.*

In the late 1970s, [African American sociologist William Julius Wilson] published an extremely influential book, *The Declining Significance of Race*. In this book, Wilson argued that race was becoming less and less important in predicting the economic possibilities for well-educated African Americans. In other words, the black-led Civil Rights movement had been successful in removing many of the barriers that made it difficult, if not impossible, for well-trained blacks to gain access to appropriate educational and occupational opportunities. Wilson argued that this new pattern of much greater (but not perfect) access was unprecedented in the racial stratification system in the United States and that it would result in significant and lasting gains for African American families with significant educational attainment.

Recent research on the black middle class has only partially supported Wilson's optimistic prognosis. While blacks did experience significant educational and occupational gains during the 1970s, their upward trajectory appears to have tapered off in the 1980s and 1990s. Moreover, some blacks have found themselves tracked into minority-oriented community relations positions within the professional and managerial occupational sphere. Even more troubling are data indicating that the proportion of blacks who attend and graduate from college appears to be shrinking, with the inevitable result that fewer blacks will have the credentials and skills necessary to get the better jobs in the growing technical and professional occupational categories. Despite real concerns about the stability

SOURCE: From Deirdre A. Royster, *Race and the Invisible Hand: How White Networks Exclude Black Men from Blue-Collar Jobs* (Berkeley, CA: University of California Press, 2003), pp. 18–23, 58–59, 180–189. Copyright © 2003 by The Regents of the University of California. Reprinted by permission.

of the black middle class and some glitches in the workings of the professional labor market, no one doubts that a substantial portion of the black population now enjoys access to middle-class opportunities and amenities—including decent homes, educational facilities, public services, and most importantly, jobs—commensurate with their substantial education and job experience, or in economic terms, their endowment of human capital.

While other scholars were investigating his theories about the black middle class, Wilson became distressed about the pessimistic prospects of blacks who were both poorly educated and increasingly isolated in urban ghettos with high rates of poverty and unemployment. His main concern was that changing labor demands that increase opportunities for highly skilled workers have the potential of making unskilled black labor obsolete. According to Wilson's next two books, *The Truly Disadvantaged* and *When Work Disappears,* this group's inability to gain access to mobility-enhancing educational opportunities is exacerbated by the further problems of a deficiency of useful employment contacts, lack of reliable transportation, crowded and substandard housing options, a growing sense of frustration, and an image among urban employers that blacks are undesirable workers, not to mention the loss of manufacturing and other blue-collar jobs. These factors, and a host of others, contribute to the extraordinarily difficult and unique problems faced by the poorest inner-city blacks in attempting to advance economically. Wilson and hundreds of other scholars—even those who disagree with certain aspects of his thesis—argue that this group needs special assistance in order to overcome the obstacles they face.

If Wilson intended the *Declining Significance of Race* thesis (and its underclass corollary) to apply mainly to well educated blacks and ghetto residents, then Wilson only explained the life chances of at most 30 to 40 percent of the black population.[1] The rest of the black population neither resides in socially and geographically isolated ghettoes, nor holds significant human capital, in the form of college degrees or professional work experience. Looking at five-year cohorts beginning at the turn of the century, Mare found that the cohort born between 1946 and 1950 reached a record high when 13 percent of its members managed to earn bachelor's degrees. Recent cohorts born during the Civil Rights era (1960s) have not reached the 13 percent high mark set by the first cohort to benefit from Civil Rights era victories. As a result, today the total percentage of African Americans age 25 and over who have four or more years of college is just under 14 percent.[2] According to demographer Reynolds Farley, while college attendance rates for white males (age 18–24) have rebounded from dips in the 1970s back to about 40 percent, black male rates have remained constant at about 30 percent since the 1960s. Figures like these suggest that Civil Rights era "victories" have not resulted in increasing percentages of blacks gaining access to college training. Instead, most blacks today attempt to establish careers with only modest educational credentials, just as earlier cohorts did. Thus the vast majority of blacks are neither extremely poor nor particularly well educated; most blacks would be considered lower-middle- or working-class and modestly educated. That is, most blacks (75 percent) lack bachelor's degrees but hold high school diplomas or GEDs; most blacks (92 percent) are working rather than unemployed; and most (79 percent) work at jobs that are lower-white- or blue-collar rather than professional.³ Given that modestly educated blacks make up the bulk of the black population, it is surprising that more attention has not been devoted to explicating the factors that influence their life chances.

Wilson's focus on the extremes within the black population, though understandable, points to a troubling underspecification in his thesis: it is unclear whether Wilson sees individuals with modest educational credentials—high school diplomas, GEDs, associate's degrees, or some college or other post-secondary training, but not the bachelor's degree—as cobeneficiaries of civil rights victories alongside more affluent blacks. The logic of his thesis implies that as long as they do not reside in socially and geographically isolated communities filled with poor and unemployed residents, from which industrial jobs have departed, then modestly educated blacks, like highly educated blacks, ought to do about as well as their white counterparts....

Because he argues that *past* racial discrimination created the ghetto poor, or underclass, while macro-economic changes—and not current racial discrimination—explain their current economic plight, Wilson's perspective implies that white attempts to exclude blacks are probably of little significance today. In addition, Wilson offers a geographic, rather than racial, explanation for whites' labor market advantages when he argues that because most poor whites live outside urban centers, they do not suffer the same sort of structural dislocation or labor obsolescence as black ghetto residents. If Wilson's reasoning holds, there is no reason to expect parity among the poorest blacks and whites in the United States without significant government intervention. Despite a conspicuous silence regarding the prospects for parity among modestly educated blacks and whites, Wilson's corpus of research and theory offers the most race and class integrative market approach available. First, Wilson specifies how supply and demand mechanisms work differently for blacks depending on their class status. Specifically, he argues that there is now a permanent and thriving pool (labor supply) of educationally competitive middle-class blacks, while simultaneously arguing that changes in the job structure in inner cities have disrupted the employment opportunities (labor demand) for poorer blacks. Second, Wilson argues that contemporary racial disparity results, by and large, from the structural difficulties faced by poor blacks rather than racial privileges enjoyed by (or racial discrimination practiced by) poor or more affluent whites. One of the questions guiding this study is whether the life chances of modestly educated whites and blacks are becoming more similar, as with blacks and whites who are well-educated, or more divergent, as with blacks and whites on the bottom....

The fifty young men interviewed for this study may have been, in some ways, atypical. For example, none of them had dropped out of school, and all were extremely polite and articulate. I suspect that these men were among the easiest to contact because of high residential stability and well-maintained friendship networks. Their phone numbers had remained the same or they had kept in touch with friends since graduating from high school two to three years earlier. Because of these factors, I may have tapped into a sample of men who were more likely to be success stories than most would have been. Researchers call this sampling dilemma creaming, because the sample may reflect those who were most likely to rise to the top or be seen as the cream of the crop, rather than those of average or mixed potential. In this study, however, it may have been an advantage to have "creamed," since I wanted to compare black and white men with as much potential for success as possible. Moreover, men who are personable and who have stable residences and friendship networks might be most able to tap into institutional and personal contacts in their job searches—one of my main research queries....

While I don't think there were idiosyncratic differences among the black and white men I studied or between the men I found and their same-race peers that I didn't find, some of the positive attributes of my sample suggest that my findings may generalize only to young men who generally play by the rules. Of course, it isn't all that clear what proportion of young working-class males (black or white) try to play by the rules—maybe the vast majority try to do so. Nor is it clear by what full set of criteria my subjects did, in fact, play by the rules. My sample includes men who had brushes with the law as well as some who might be considered "goody two-shoes." In other words I'm not sure that the specific men with whom I spoke are atypical among working-class men, but I am willing to acknowledge that they may be. Perhaps what is most important to remember about the sample, who seemed to me to be pretty ordinary, "All-American" men, is my contention that this set of men ought to have similar levels of success in the blue-collar labor market—if, that is, we have finally reached a time when race doesn't matter....

One narrative, the achievement ideology, asserts that formal training, demonstrated ability, and appropriate personal traits will assure employment access and career mobility. The second narrative, the contacts ideology, emphasizes personal ties and affiliations as a mechanism for employment referrals, access, and mobility....The achievement ideology has persistently dominated American understanding of occupational success, even though everyone, it seems, is willing to admit that "who you know" is at least as important as "what you know" in gaining access to opportunities in American society. All of the men in this study, for example, said that contacts were very important in establishing young men like themselves in careers. One offered a more nuanced explanation: "It's not [just] who you know, it's how they know you." That is, it is not simply knowing the right people that matters; it is sharing the right sort of bonds with the right people that influences what those people would be willing to do to assist you.

The black and white men in this study had more achievements in common than contacts. They were trained in the same school, in many of the same trades, and by the same instructors. They had formal access to the same job listing services and work-study programs. Instructors and students alike agreed, and records confirm, that in terms of vocational skills and performance, the blacks and whites in this sample were among the stronger students....

Black Glendale graduates trail behind their white peers. They are employed less often in the skilled trades, especially within the fields for which they have been trained; they earn less per hour; they hold lower status positions; they receive fewer promotions; and they experience more and longer periods of unemployment. No set of educational, skill, performance, or personal characteristics unique to either the black or white students differentiates them in ways that would explain the unequal outcomes.... Only their racial status and the way it situates them in racially exclusive networks during the school-work transition process adequately explain their divergent paths from seemingly equal beginnings.

In this study of one variant of the "who you know" versus "what you know" conundrum, it is manifestly and perpetually evident that racial dynamics are a key arbiter of employment outcome. Yet challenging the power of the achievement ideology in American society requires a careful exposition of how factors such as race throw a wrench into the presumption of meritocracy. In addition, the contacts ideology must be uniquely construed to take into account the significance of racially determined patterns of affiliation within a class, in this case the working class.

## RACE, AN ARBITER OF EMPLOYMENT NETWORKS

Researchers have long argued that black males lack access to the types of personal contacts that white males appear to have in abundance.[4] I would argue that it's more than not having the right contacts. In terms of social networks, black men are at a disadvantage in terms of configuration, content, and operation, a disadvantage that is exacerbated in sectors with long traditions of racial exclusion, such as the blue-collar trades. Even when blacks and whites have access to some of the same connections, as in this study, care must be taken to examine exactly what transpires. For example, I noted that black and white males were assisted differently by the same white male teachers. If I had only asked students whether they considered their shop teachers contacts on whom they could rely, equal numbers of black and white males would have answered affirmatively. But this would have told us nothing about *how* white male teachers *chose to know and help* their black male students. The teachers chose to verbally encourage black students, while providing more active assistance to white students. I discovered a munificent flow of various forms of assistance, including vacancy information, referrals, direct job recruitment, formal and informal training, vouching behaviors, and leniency in supervision. For white students, this practice, which repeated neighborhood and community patterns within school walls, served to convert institutional ties (as teachers) into personal ones (as friends) that are intended to and do endure well beyond high school....

Even without teachers consciously discriminating, significant employment information and assistance remained racially privatized within this public school context. In that white male teachers provided a parallel or shadow transition system for white students that was not equally available to black students, segregated networks still governed the school-work transition at Glendale even though classrooms had long been desegregated.

The implications for black men are devastating. Despite having unprecedented access to the same preparatory institution as their white peers, black males could not effectively use the institutional connection to establish successful trade entry. Moreover, segregation in multiple social arenas, beyond schools, all but precluded the possibility of network overlaps among working-class black and white men. As a result, black men sought employment using a truncated, resource-impoverished network consisting of strong ties to other blacks (family, friends, and school officials) who like themselves lacked efficacious ties to employment.

Beyond school, matters were even worse. Without being aware of it, white males' descriptions of their experiences revealed a pattern of intergenerational intraracial assistance networks among young and older white men that assured even the worst young troublemaker a solid place within the blue-collar fold. The white men I studied were not in any way rugged individualists; rather they survived and thrived in rich, racially exclusive networks.

For the white men, neighborhood taverns, restaurants, and bars served as informal job placement centers where busboys were recruited to union apprentice programs, pizza delivery boys learned to be refrigeration specialists, and dishwashers studying drafting could work alongside master electricians then switch back to drafting if they wished. I learned of opportunities that kept coming, even when young men weren't particularly deserving. One young man had been able to hold onto his job after verbally abusing his boss. Another got a job installing burglar alarms after meeting the vice president of the company at a cookout—without ever having to reveal his prison record, which included a conviction for burglary.

Again and again, the white men I spoke with described opportunities that had landed in their laps, not as the result of outstanding achievements or personal characteristics, but rather as the result of the assistance of older white neighbors, brothers, family friends, teachers, uncles, fathers, and sometimes mothers, aunts, and girlfriends (and their families), all of whom overlooked the men's flaws. It never seemed to matter that the men were not A students, that they occasionally got into legal trouble, that they lied about work experiences from time to time, or that they engaged in horseplay on the job. All of this was expected, brushed off as typical "boys will be boys" behavior, and it was sometimes the source of laughter at the dining room table. In other words, there were no significant costs for white men associated with being young and inexperienced, somewhat immature, and undisciplined.

The sympathetic pleasure I felt at hearing stories of easy survival among working-class white men in an era of deindustrialization was only offset by the depressing stories I heard from the twenty-five black men. Their early employment experiences were dismal in comparison, providing a stark and disturbing contrast. Whereas white men can be thought of as the second-chance kids, black men's opportunities were so fragile that most could not have recovered from even the relatively insignificant mishaps that white men reported in passing.

Black men were rarely able to stay in the trades they studied, and they were far less likely than white men to start in one trade and later switch to a different one, landing on their feet. Once out of the skilled trade sphere, they sank to the low-skill service sector, usually retail or food services. The black men had nu-merous experiences of discrimination at the hands of older white male supervi-sors, who did not offer to help them and frequently denigrated them, using familiar racial epithets. The young black men I spoke with also had to be careful when using older black social contacts. More than one man indicated to me that, when being interviewed by a white person, the wisest course of action is to be-have as if you don't know anyone who works at the plant, even if a current worker told you about the opening. These young black men, who had been on the labor market between two and three years, were becoming discouraged. While they had not yet left the labor force altogether, many (with the help of parents) had invested time and resources in training programs or college courses that they and their families hoped would open up new opportunities in or be-yond the blue-collar skilled labor market. Many of the men had begun to lose the skills they had learned in high school; others, particularly those who'd had a spell or two of unemployment, showed signs of depression.

My systematic examination of the experiences of these fifty matched young men leads me to conclude that the blue-collar labor market does not function as a market in the classic sense. No pool of workers presents itself, offering sets of skills and work values that determine who gets matched with the most and least desirable opportunities. Rather, older men who recruit, hire, and fire young workers choose those with whom they are comfortable or familiar. Visible hands trump the "invisible hand"—and norms of racial exclusivity passed down from generation to generation in American cities continue to inhibit black men's entry into the better skilled jobs in the blue-collar sector.

Claims of meritocratic sorting in the blue-collar sector are simply false; equally false are claims that young black men are inadequately educated, inher-ently hostile, or too uninterested in hard work or skill mastery to be desirable workers. These sorts of claims seek to locate working-class black men's employ-ment difficulties in the men's alleged deficits—bad attitudes, shiftlessness, poor skills—rather than in the structures and procedures of worker selection that are typically under the direct control of older white men whose preferences, by cus-tom, do not reflect meritocratic criteria.

Few, if any, political pressures, laws, or policies provide sufficient incentives or sanctions to prevent such employers from arbitrarily excluding black workers or hiring them only for menial jobs for which they are vastly overqualified. More-over, in recent years, affirmative action policies that required that government contracts occasionally be awarded to black-owned firms or white-owned firms that consistently hire black workers have come under attack—eroding the paltry incentives for inclusion set forth during the Civil Rights era, nearly forty years ago. Indeed, there is far less pressure today than in the past for white-owned firms to hire black working men. And given persistent patterns of segregation—equivalent to an American apartheid, according to leading sociologists—there remain few incentives for white men to adopt young black men into informal, neighborhood-generated networks. As a result, occupational apartheid reigns in the sector that has always held the greatest potential for upward mobility, or just basic security, for modestly educated Americans.

## IDEOLOGY AND THE DEFENSE OF RACIALIZED EMPLOYMENT NETWORKS

The public perception of the causes of black men's labor difficulties—namely, that the men themselves are to be blamed—contrasts with my findings. And my research is consistent with that of hundreds of social scientists who have demonstrated state-supported and informal patterns of racial exclusion in housing, education, labor markets, and even investment opportunities. Racism continues to limit the life chances of modestly educated black men....

## WHITE PRIVILEGE, BLACK ACCOMMODATION

How, then, do black males, if they wish to earn a living in the surviving trades, negotiate training and employment opportunities in which networks of gatekeepers remain committed to maintaining white privilege? The present research suggests that the options are few: either accommodation to the parameters of a racialized system or failure in establishing a successful trade career. The interviews revealed that forms of black accommodation begin early, as when young men avoided training in trades of interest because they were known to hold little promise for integration and advancement. For those who made such discoveries later, accommodation took the form of disengaging from specific trades, such as electrical construction, and pursuing whatever jobs became available. For some, the disengaging process involved the claim that they were never really committed to the original trade field, but I suspect that such claims merely served to soften the blow of almost inevitable career failure. For the determined, accommodation required suppressing anger at racially motivated insults and biased employment decisions in the majority-white trade settings. If this strategy wore thin, two difficult accommodations remained. The first involved finding a work setting—not necessarily within one's trade—in which the workplace culture was, if not actively receptive to black inclusion, at least neutral. The second involved finding ways to work in white-dominated fields without having to work beside whites.

A word needs to be said about a particularly troubling accommodative behavior adopted by the black men: not actively and persistently pursuing offers of assistance. It is not clear to what extent the black men were fully cognizant of the extent and potency of whites' informal networks or of the cultural norms governing their operation. But, while it is evident that the older white men who were network gatekeepers did not extend the same access and support to the black men, the black men may also have been less proactive in pursuing older white men who might have assisted them.

Generally, the white men in the study appear to have more actively followed up on offers of assistance. And although their careers developed much more smoothly than those of the black men, they were certainly not without the difficulties of not being hired, workplace dissatisfaction, competing vocational interests, and unemployment. Nevertheless, they returned, sometimes repeatedly, to contacts for further assistance. Certainly, demonstrating proactivity toward a typically racially exclusive white network would be especially problematic for black men.

Undoubtedly, black men's exclusion from white personal settings where easy informal contact is facilitated, like neighborhoods and family, contributes to black men's reluctance to pursue whites for assistance. In addition, black men's lack of personal familiarity with normative expectations among whites probably hampers their efforts to imitate their white peers' more forward network behaviors. Furthermore, any efforts by blacks to engage in such behaviors might not be similarly regarded as appropriate, and might instead be interpreted as aggressive, "uppity," or indicative of a feeling of entitlement. Finally, black men's early experiences of racial exclusion, bias, and hostility in the school and the workplace inform not only their assessment of employment prospects, but also their actual employment strategies. Given these complicated contingencies, perhaps the somewhat hesitant responses of black men are, on the whole, not unreasonable.

## CONCLUSION

Black men have paid a great price for exclusion from blue-collar trades and the networks that supply those trades, but they have not paid it alone. The pain of black men's unemployment and underemployment spreads across black communities in a ripple effect. Less able to contribute financially to the care of children and parents, or to combine resources with black women or assist other men with work entry and "learning the ropes" on the job, black men withdraw from the support structures that they need and that they are needed to support emotionally as well as economically. The enduring power of segregated networks in the blue-collar trades is as responsible as segregated neighborhoods for the existence of extremely poor and isolated black communities and of the disproportionately black and male prison population—in fact, more so. While many black families live in stable communities that are mostly, if not entirely, black, the inability to find remunerative jobs that do not require expensive college training makes living decently anywhere extremely difficult. And the loss of manufacturing jobs cannot account for black men's underemployment in the remaining blue-collar fields—especially construction, auto mechanics, plumbing, computer repair, and carpentry....

My findings demonstrate that, without governmental initiatives that provide strong incentives for inclusion, white tradesmen will have no reason to open their networks to men of color. As a result, the work trajectories of white and black men who start out on an equal footing will continue to diverge into skilled and unskilled work paths because of business-as-usual patterns of exclusion. Although there are few precedents for intervening in the private sector, there are strong precedents for intervening in the public sector, where the tax dollars of majority and minority citizens must not be redistributed in ways that condone customs of exclusion....

Without the government taking a lead, the young black men I studied—who played by the rules—are unlikely to ever reach their potential as skilled workers or to take their places as blue-collar entrepreneurs, as so many of their white peers are poised to do. This tragedy could have been averted. My hope is that it will be averted in the next generation.

## NOTES

1.  Haywood Horton, Beverlyn Lundy, Cedric Herring, and Melvin E. Thomas, "Lost in the Storm: The Sociology of the Black Working Class, 1850 to 1990," *American Sociological Review* 65, no.1 (2000): 128–137.

2.  Robert Mare, "Changes in Educational Attainment and School Enrollment," in *State of the Union*, ed. Reynolds Farley (New York: Russell Sage Foundation, 1995); Nancy Folbre, *The Field Guide to the U.S. Economy* (New York: New Press, 1999).

3.  Ibid.

4.  Richard Freeman and Harry J. Holzer, *The Black Youth Employment Crisis* (Chicago: University of Chicago Press, 1986); William Julius Wilson, *The Truly Disadvantaged* (Chicago: University of Chicago Press, 1987); Paul Osterman, *Getting Started: The Youth Labor Market* (Cambridge: Massachusetts Institute of Technology Press, 1980).

## DISCUSSION QUESTIONS

1.  Why is economic advancement difficult for inner-city Black men with modest levels of education?

2.  What are the differences in the way networks operated for Black men and White men in Royster's study? What are the implications of those outcomes for their futures?

3.  Rather than just let market forces work as they will, why does Royster think government intervention is important?

# 35

## Families on the Frontier:

### From Braceros in the Fields to Braceras in the Home

PIERRETTE HONDAGNEU-SOTELO

*The increased employment of women, especially White women, in the United States has created a growing demand for paid domestic work. Hondagneu-Sotelo's study of immigrant domestic workers shows how inequality between nations means that women fill this need. Women (especially those from the Philippines, the Caribbean, Central America, and Mexico) have had to move for such work, often becoming separated from their own families.*

Why are thousands of Central American and Mexican immigrant women living and working in California and other parts of the United States while their children and other family members remain in their countries of origin?... I argue that U.S. labor demand, immigration restrictions, and cultural transformations have encouraged the emergence of new transnational family forms among Central American and Mexican immigrant women. Postindustrial economies bring with them a labor demand for immigrant workers that is differently gendered from that typical of industrial or industrializing societies. In all post-industrial nations, we see an increase in demand for jobs dedicated to social reproduction, jobs typically coded as "women's jobs." In many of these countries, such jobs are filled by immigrant women from developing nations. Many of these women, because of occupational constraints—and, in some cases, specific restrictionist contract labor policies—must live and work apart from their families.

My discussion focuses on private paid domestic work, a job that in California is nearly always performed by Central American and Mexican immigrant women. Not formally negotiated labor contracts, but rather informal occupational constraints, as well as legal status, mandate the long-term spatial and temporal separation of these women from their families and children. For many Central American and Mexican women who work in the United States, new international divisions of social reproductive labor have brought about transnational family forms and new meanings of family and motherhood. In this respect, the United States has entered a new era of dependency on braceras. Consequently, many Mexican,

SOURCE: From *Latinos: Remaking America*, ed. by Marcelo Suarez-Orozco and Mariela M. Paez (Berkeley, CA: University of California Press, 2002), pp. 259–266. Copyright © 2002 by The Regents of the University of California. Reprinted by permission.

Salvadoran, and Guatemalan immigrant families look quite different from the images suggested by Latino familism.

This [article] is informed by an occupational study I conducted of over two hundred Mexican and Central American women who do paid domestic work in private homes in Los Angeles (Hondagneu-Sotelo 2001). Here, I focus not on the work but on the migration and family arrangements conditioned by the way paid domestic work is organized today in the United States. I begin by noting the ways in which demand for Mexican—and increasingly Central American—immigrant labor shifted in the twentieth century from a gendered labor demand favoring men to one characterized by robust labor demand for women in a diversity of jobs, including those devoted to commodified social reproduction. Commodified social reproduction refers to the purchase of all kinds of services needed for daily human upkeep, such as cleaning and caring work. The way these jobs are organized often mandates transnational family forms....

I ... note the parallels between family migration patterns prompted by the Bracero Program and long-term male sojourning, when many women sought to follow their husbands to the United States, and the situation today, when many children and youths are apparently traveling north unaccompanied by adults, in hopes of being reunited with their mothers. In the earlier era, men were recruited and wives struggled to migrate; in a minority of cases, Mexican immigrant husbands working in the United States brought their wives against the latters' will. Today, women are recruited for work, and increasingly, their children migrate north some ten to fifteen years after their mothers. Just as Mexican immigrant husbands and wives did not necessarily agree on migration strategies in the earlier era, we see conflicts among today's immigrant mothers in the United States and the children with whom they are being reunited. In this regard, we might suggest that the contention of family power in migration has shifted from gender to generation....

## GENDERED LABOR DEMAND AND SOCIAL REPRODUCTION

Throughout the United States, a plethora of occupations today increasingly rely on the work performed by Latina and Asian immigrant women. Among these are jobs in downgraded manufacturing, jobs in retail, and a broad spectrum of service jobs in hotels, restaurants, hospitals, convalescent homes, office buildings, and private residences. In some cases, such as in the janitorial industry and in light manufacturing, jobs have been re-gendered and re-racialized so that jobs previously held by U.S.-born white or black men are now increasingly held by Latina immigrant women. Jobs in nursing and paid domestic work have long been regarded as "women's jobs," seen as natural outgrowths of essential notions of women as care providers. In the late twentieth-century United States, however, these jobs have entered the global marketplace, and immigrant women from developing nations around the globe are increasingly represented in them.

In major metropolitan centers around the country, Filipina and Indian immigrant women make up a sizable proportion of HMO nursing staffs—a result due in no small part to deliberate recruitment efforts. Caribbean, Mexican, and Central American women increasingly predominate in low-wage service jobs, including paid domestic work.

This diverse gendered labor demand is quite a departure from patterns that prevailed in the western region of the United States only a few decades ago. The relatively dramatic transition from the explicit demand for Mexican and Asian immigrant *male* workers to demand that today includes women has its roots in a changing political economy. From the late nineteenth century until 1964, the period during which various contract labor programs were in place, the economies of the Southwest and the West relied on primary extractive industries. As is well known, Mexican, Chinese, Japanese, and Filipino immigrant workers, primarily men, were recruited for jobs in agriculture, mining, and railroads. These migrant workers were recruited and incorporated in ways that mandated their long-term separation from their families of origin.

As the twentieth century turned into the twenty-first, the United States was once again a nation of immigration. This time, however, immigrant labor is not involved in primary, extractive industry. Agribusiness continues to be a financial leader in the state of California, relying primarily on Mexican immigrant labor and increasingly on indigenous workers from Mexico, but only a fraction of Mexican immigrant workers are employed in agriculture. Labor demand is now extremely heterogeneous and is structurally embedded in the economy of California (Cornelius 1998). In the current period, which some commentators have termed "postindustrial," business and financial services, computer and other high-technology firms, and trade and retail prevail alongside manufacturing, construction, hotels, restaurants, and agriculture as the principal sources of demand for immigrant labor in the western region of the United States.

As the demand for immigrant women's labor has increased, more and more Mexican and (especially) Central American women have left their families and young children behind to seek employment in the United States. Women who work in the United States in order to maintain their families in their countries of origin constitute members of new transnational families, and because these arrangements are choices that the women make in the context of very limited options, they resemble apartheid-like exclusions. These women work in one nation-state but raise their children in another. Strikingly, no formalized temporary contract labor program mandates these separations. Rather, this pattern is related to the contemporary arrangements of social reproduction in the United States.

## WHY THE EXPANSION IN PAID DOMESTIC WORK?

Who could have foreseen that as the twentieth century turned into the twenty-first, paid domestic work would become a growth occupation? Only a few decades ago, observers confidently predicted that this job would soon become obsolete, replaced by such labor-saving household devices as automatic dish-washers, disposable diapers, and microwave ovens and by consumer goods and services purchased outside the home, such as fast food and dry cleaning (Coser 1974). Instead, paid domestic work has expanded. Why?

The exponential growth in paid domestic work is due in large part to the increased employment of women, especially married women with children, to the underdeveloped nature of child care centers in the United States, and to patterns of U.S. income inequality and global inequalities. National and global trends have fueled this growing demand for paid domestic services. Increasing global competition and new communications technologies have led to work speedups in all sorts of jobs, and the much bemoaned "time bind" has hit professionals and managers particularly hard (Hochschild 1997). Meanwhile, normative middle-class ideals of child rearing have been elaborated (consider the proliferation of soccer, music lessons, and tutors). At the other end of the age spectrum, greater longevity among the elderly has prompted new demands for care work.

Several commentators, most notably Saskia Sassen, have commented on the expansion of jobs in personal services in the late twentieth century. Sassen located this trend in the rise of new "global cities," cities that serve as business and managerial command points in a new system of intricately connected nodes of global corporations. Unlike New York City, Los Angeles is not home to a slew of Fortune 500 companies, but in the 1990s it exhibited remarkable economic dynamism. Entrepreneurial endeavors proliferated and continued to drive the creation of jobs in business services, such as insurance, real estate, public relations, and so on. These industries, together with the high-tech and entertainment industries in Los Angeles, spawned many high-income managerial and professional jobs, and the occupants of these high-income positions require many personal services that are performed by low-wage immigrant workers. Sassen provides the quintessentially "New York" examples of dog walkers and cooks who prepare gourmet take-out food for penthouse dwellers. The Los Angeles counterparts might include gardeners and car valets, jobs filled primarily by Mexican and Central American immigrant men, and nannies and house cleaners, jobs filled by Mexican and Central American immigrant women. In fact, the numbers of domestic workers in private homes counted by the Bureau of the Census doubled from 1980 to 1990 (Waldinger 1996).

I favor an analysis that does not speak in terms of "personal services," which seems to imply services that are somehow private, individual rather than social, and are superfluous to the way society is organized. A feminist concept that was originally introduced to valorize the nonremunerated household work of women, *social reproduction* or alternately, *reproductive labor,* might be more usefully employed. Replacing *personal services* with *social reproduction* shifts the focus by underlining the objective of the work, the societal functions, and the impact on immigrant workers and their own families.

Social reproduction consists of those activities that are necessary to maintain human life, daily and intergenerationally. This includes how we take care of ourselves, our children and elderly, and our homes. Social reproduction encompasses the purchasing and preparation of food, shelter, and clothing; the routine daily upkeep of these, such as cooking, cleaning and laundering; the emotional care and support of children and adults; and the maintenance of family and community ties. The way a society organizes social reproduction has far-reaching consequences not only for individuals and families but also for macrohistorical processes (Laslett and Brenner 1989).

Many components of social reproduction have become commodified and outsourced in all kinds of new ways. Today, for example, not only can you purchase fast-food meals, but you can also purchase, through the Internet, the home delivery of customized lists of grocery items. Whereas mothers were once available to buy and wrap Christmas presents, pick up dry cleaning, shop for groceries and wait around for the plumber, today new businesses have sprung up to meet these demands—for a fee.

In this new milieu, private paid domestic work is just one example of the commodification of social reproduction. Of course, domestic workers and servants of all kinds have been cleaning and cooking for others and caring for other people's children for centuries, but there is today an increasing proliferation of these services among various class sectors and a new flexibility in how these services are purchased.

## GLOBAL TRENDS IN PAID DOMESTIC WORK

Just as paid domestic work has expanded in the United States, so too it appears to have grown in many other postindustrial societies, in the "newly industrialized countries" (NICs) of Asia, in the oil-rich nations of the Middle East, in Canada, and in parts of Europe. In paid domestic work around the globe, Caribbean, Mexican, Central American, Peruvian, Sri Lankan, Indonesian, Eastern European, and Filipina women—the latter in disproportionately large numbers—predominate. Worldwide, paid domestic work continues its long legacy as a racialized and gendered occupation, but today, divisions of nation and citizenship are increasingly salient.

The inequality of nations is a key factor in the globalization of contemporary paid domestic work. This has led to three outcomes: (1) Around the globe, paid domestic work is increasingly performed by women who leave their own nations, their communities, and often their families of origin to do the work. (2) The occupation draws not only women from the poor socioeconomic classes, but also women who hail from nations that colonialism has made much poorer than those countries where they go to do domestic work. This explains why it is not unusual to find college-educated women from the middle class working in other countries as private domestic workers. (3) Largely because of the long, uninterrupted schedules of service required, domestic workers are not allowed to migrate as members of families.

Nations that "import" domestic workers from other countries do so using vastly different methods. Some countries have developed highly regulated, government-operated, contract labor programs that have institutionalized both the recruitment and the bonded servitude of migrant domestic workers. Canada and Hong Kong provide paradigmatic examples of this approach. Since 1981 the Canadian federal government has formally recruited thousands of women to work as live-in nannies/housekeepers for Canadian families. Most of these women came from Third World countries in the 1990s (the majority came from the Philippines, in the 1980s from the Caribbean), and once in Canada, they must remain in live-in domestic service for two years, until they obtain their landed immigrant status, the equivalent of the U.S. "green card." This reflects, as Bakan and Stasiulis (1997) have noted, a type of indentured servitude and a decline in the citizenship rights of foreign domestic workers, one that coincides with the racialization of the occupation. When Canadians recruited white British women for domestic work in the 1940s, they did so under far less controlling mechanisms than those applied to Caribbean and Filipina domestic workers. Today, foreign domestic workers in Canada may not quit their jobs or collectively organize to improve the conditions under which they work.

Similarly, since 1973 Hong Kong has relied on the formal recruitment of domestic workers, mostly Filipinas, to work on a full-time, live-in basis for Chinese families. Of the 150,000 foreign domestic workers in Hong Kong in 1995, 130,000 hailed from the Philippines, and smaller numbers were drawn from Thailand, Indonesia, India, Sri Lanka, and Nepal (Constable 1997, p. 3). Just as it is now rare to find African American women employed in private domestic work in Los Angeles, so too have Chinese women vanished from the occupation in Hong Kong. As Nicole Constable reveals in her detailed study, Filipina domestic workers in Hong Kong are controlled and disciplined by official employment agencies, employers, and strict government policies. Filipinas and other foreign domestic workers recruited to Hong Kong find themselves working primarily in live-in jobs and bound by two-year contracts that stipulate lists of job rules, regulations for bodily display and discipline (no lipstick, nail polish, or long hair, submission to pregnancy tests, etc.), task timetables, and the policing of personal privacy. Taiwan has adopted a similarly formal and restrictive government policy to regulate the incorporation of Filipina domestic workers (Lan 2000).

In this global context, the United States remains distinctive, because it takes more of a laissez-faire approach to the incorporation of immigrant women into paid domestic work. No formal government system or policy exists to legally contract foreign domestic workers in the United States. Although in the past, private employers in the United States were able to "sponsor" individual immigrant women who were working as domestics for their "green cards" using labor certification (sometimes these employers personally recruited them while vacationing or working in foreign countries), this route is unusual in Los Angeles today. Obtaining legal status through labor certification requires documentation that there is a shortage of labor to perform a particular, specialized occupation. In Los Angeles and in many parts of the country today, a shortage of domestic workers is increasingly difficult to prove. And it is apparently unnecessary, because the significant demand for domestic workers in the United States is largely filled not through formal channels of foreign recruitment but through informal recruitment from the growing number of Caribbean and Latina immigrant women who are *already* legally or illegally living in the United States. The Immigration and Naturalization Service, the federal agency charged with enforcement of illegal-migration laws, has historically served the interests of domestic employers and winked at the employment of undocumented immigrant women in private homes.

As we compare the hyperregulated employment systems in Hong Kong and Canada with the more laissez-faire system for domestic work in the United States, we find that although the methods of recruitment and hiring and the roles of the state in these processes are quite different, the consequences are similar. Both systems require the incorporation as workers of migrant women who can be separated from their families.

The requirements of live-in domestic jobs, in particular, virtually mandate this. Many immigrant women who work in live-in jobs find that they must be "on call" during all waking hours and often throughout the night, so there is no clear line between working and nonworking hours. The line between job space and private space is similarly blurred, and rules and regulations may extend around the clock. Some employers restrict the ability of their live-in employees to receive phone calls, entertain friends, attend evening ESL classes, or see boyfriends during the workweek. Other employers do not impose these sorts of restrictions, but because their homes are located in remote hillsides, suburban enclaves, or gated communities, live-in nannies/housekeepers are effectively restricted from participating in anything resembling social life, family life of their own, or public culture.

These domestic workers—the Filipinas working in Hong Kong or Taiwan, the Caribbean women working on the East Coast, and the Central American and Mexican immigrant women working in California constitute the new "braceras." They are literally "pairs of arms," disembodied and dislocated from their families and communities of origin, and yet they are not temporary sojourners.

## REFERENCES

Bakan, Abigail B., and Daiva Stasiulis. 1997. "Foreign Domestic Worker Policy in Canada and the Social Boundaries of Modern Citizenship." In Abigail B. Bakan and Daiva Stasiulis (eds.), *Not One of the Family: Foreign Domestic Workers in Canada,* Toronto: University of Toronto Press, pp. 29–52.

Constable, Nicole. 1997. *Maid to Order in Hong Kong: Stories of Filipina Workers.* Ithaca and London: Cornell University Press.

Cornelius, Wayne. 1998. "The Structural Embeddedness of Demand for Mexican Immigrant Labor: New Evidence from California." In Marcelo M. Suárez-Orozco (ed.), *Crossings: Mexican Immigration in Interdisciplinary Perspectives.* Cambridge, MA: Harvard University, David Rockefeller Center for Latin American Studies, pp. 113–44.

Coser, Lewis. 1974. "Servants: The Obsolescence of an Occupational Role." *Social Forces* 52:31–40.

Hochschild, Arlie. 1997. *The Time Bind: When Work Becomes Home and Home Becomes Work*. New York: Metropolitan Books, Henry Holt.

Hondagneu-Sotelo, Pierrette. 1994. *Doméstica: Immigrant Workers and Their Employers*. Berkeley: University of California Press.

Lan, Pei-chia. 2000. "Global Divisions, Local Identities: Filipina Migrant Domestic Workers and Taiwanese Employers." Dissertation, Northwestern University.

Laslett, Barbara, and Johanna Brenner. 1989. "Gender and Social Reproduction: Historical Perspectives," *Annual Review of Sociology* 15:381–404.

Waldinger, Roger, and Mehdi Bozorgmehr. 1996. "The Making of a Multicultural Metropolis." In Roger Waldinger and Mehdi Bozorgmehr (eds.), *Ethnic Los Angeles*. New York: Russell Sage Foundation, pp. 3–37.

## DISCUSSION QUESTIONS

1. What does Hondagneu-Sotelo mean by the term *transnational families*?

2. Define social reproductive work and explain why Hondagneu-Sotelo thinks this work is performed by immigrant women.

3. Social reproductive work is part of the industrial societies of the United States, Canada, and Hong Kong, but what are the different national strategies for securing workers?

# 36

# Race, Migration, and Labor Control

### ANGELA C. STUESSE

*Neoliberal policies call for little state interference with market forces. However, if government does not support workers' rights, they are vulnerable to exploitation. Research on the poultry industry reveals the practices of both management, in controlling workers, and union organizers, who work to unite African Americans with the diverse population of Latinos in Mississippi's high-tech but low-wage industry.*

## ... RACE, MIGRATION, AND THE TRANSFORMATION OF THE RURAL SOUTH

The poultry-processing industry in the United States, located predominantly in the South, has gone through a radical transformation in recent decades. Today Americans eat almost twice as much chicken per capita (89.1 pounds annually) than they did in 1980 (48.0 pounds), and as consumption skyrocketed the industry began massive recruitment of foreign-born labor. Whereas traditionally local whites and (later) African Americans supplied the industry's labor power, in many areas today Latino migrants constitute the majority of workers. As of 2000, Latinos represented 29 percent of all meat-processing workers, in comparison to only 9 percent twenty years prior, and 82 percent of these "Hispanic" laborers are foreign born. In fact, over 50 percent of the nation's quarter-million poultry workers are now immigrants....

Although immigration is not new to the South or even to Mississippi, the intensity and breadth of recent transnational migrant flows is novel. The historically rooted Black-white racial binary continues to frame social relations in this region, and the recent arrival of Latinos to rural areas complicates traditional hierarchies....

Mississippi is both the poorest state in the nation and one of the world's leading producers of chicken, selling more than $2.2 billion in poultry products annually. It is also the most recent southern state to feel the effects of the poultry industry's recruitment of transnational labor, as busloads of Latinos began arriving

SOURCE: *Latino Immigrants and the Transformation of the U.S. South*, edited by Mary E. Odem and Elaime Lacy. Athens GA: Univ. of Georgia Press. 2009.

only ten years ago, in the mid–1990s. Scott County is the principal poultry-producing area in Mississippi, with eleven processing plants there and in surrounding counties as of 2005. Scott County is also home to the state's greatest concentration of Latinos, a demographic shift driven by a family–owned poultry plant that began recruiting workers through its institutionalized Hispanic Project in 1993…. Scott County's Latino population is exceptionally diverse, representing over a dozen countries. The largest groups come from Mexico (almost exclusively from the new sending regions of southeastern Mexico), Guatemala (predominantly the department of San Marcos), Argentina, and Peru. The diversity of Mississippi's foreign born goes beyond questions of nationality, as ethnic, racial, linguistic, gender, class, and educational differences also create divisions and tensions within and between migrant groups.

Although obstacles such as pervasive poverty, institutionalized racism, legislation that favors corporations, and the undocumented legal status of many in the workforce discourage workers from claiming and exercising their rights in the contemporary South, Mississippi is an important case study because it possesses a rich history of community organizing, particularly during, but not limited to, the African American freedom struggle of the 1950s and 1960s. More recently, there has been limited but significant labor-organizing activity within the state's poultry-producing region. Ongoing poultry worker justice efforts there include those of the Mississippi Immigrants' Rights Alliance (MIRA) and MPOWER (formerly the Mississippi Poultry Workers' Center), both of which have identified the importance of cultivating relationships between workers of different backgrounds, specifically among African Americans and new immigrants, in order to achieve greater power and voice within the industry.…

## NEOLIBERAL GLOBALIZATION AND THE RESTRUCTURING OF THE POULTRY INDUSTRY

… In this transnational present, the hypermobility of capital and labor provide new opportunities for capitalist exploitation and regulation of low-income communities and individuals. Like classic liberalism, the neoliberalism of the current moment suggests that the state should interfere as little as possible with the market, allowing its "invisible hand" to guide economic, political, and social relationships. However, unlike the liberalism of the earlier twentieth century, today's economic, cultural and political logic is fueled by the transnational processes described above, suggesting that the term "neoliberal globalization" may more accurately describe the dynamics currently at play.…

The U.S. poultry industry is a critical site for studying the changing effects of neoliberal globalization on local subjectivities because its innovative labor control practices are increasingly embraced as a model by other industries aiming to boost profits in the economy of advanced capitalism. Neoliberal globalization has played a fundamental role in the restructuring of industry from a Fordist regime to a post–Fordist model of "flexible accumulation," in which corporate strategies such as outsourcing, contracting, part-time employment, and recruitment of migrant workers allow for greater capital accumulation.... Whereas in the past an individual might spend the majority of his or her working years with one company, gradually accruing seniority, benefits, and company loyalty over time, today's poultry industry displays little concern for worker retention.

Native and immigrant workers alike complain of a myriad of unjust practices in the poultry industry, including unpaid wages, denial of bathroom breaks, dangerous conditions that cause chronic injuries and illnesses, unauthorized paycheck deductions, abuse by plant supervisors and management, deceptive use of labor contractors, discrimination, and sexual harassment. Jobs have been "deskilled" and production has been accelerated through massive technological advances, so that the average worker now repeats the same monotonous—and often dangerous—movement up to 30,000 times per day. As a direct result, repetitive-motion injuries now plague the workforce. Plants are often out of compliance with federal safety and health regulations, and the government agency charged with oversight of these laws, the Occupational Safety and Health Administration (OSHA), is appallingly underresourced and, consequently, largely ineffective. Management frequently discourages workers from seeing doctors or filing workers' compensation claims for on-the-job injuries. In addition, in a recent national survey the U.S. Department of Labor found violations of federal minimum wage laws in 100 percent of poultry plants, while the industry's corporate earnings have risen more than 300 percent since 1987. It is not surprising then that annual turnover of workers is as high as 300 percent annually in some locations. Aside from their claim to being the only major employer in many rural towns, poultry companies give their workers virtually no incentive to stay....

The industry's increasing reliance on the most marginal of workers—recent immigrants—demonstrates its shrewd understanding of the workings of our globalized, neoliberal present.

## CHALLENGES TO ORGANIZING MISSISSIPPI'S POULTRY WORKERS ACROSS DIFFERENCE

Bobby Robertson, an African American former poultry worker and leader at one union local that represents poultry workers in processing plants in Central Mississippi, has witnessed the rapid Latinization of his surroundings.... In the mid-1990s, his coworkers began to organize for their workplace rights and sought to find union representation. Robertson joined in the campaign, became an active union member, and eventually became business manager of the union local. He recalls, after a long uphill battle, when the plants began to heavily recruit immigrant workers—an industry tactic that he says displaced Black workers and significantly weakened the union's membership and bargaining power.... Over time Robertson eventually acknowledged that he and his mostly African American coworkers could do very little to keep new migrants from arriving. He recognized that if the labor movement in Central Mississippi were to survive, it would have to embrace new strategies of organizing to defend the rights of *all* poultry workers.

The task Robertson set for himself was challenging, not only because of the lack of local understanding about immigration and immigrants, but also because he was unable to communicate with these new potential union members. When he acknowledged that he "needed somebody who could speak Mexican," his union's international office responded by sending a bilingual organizer to work with him for a few weeks. The knowledge gained from being able to communicate with immigrant workers,… convinced Robertson that his organizing efforts must work to bridge differences of race, language, and national origin. Robertson's union and other workers' rights organizations in Mississippi have taken up this objective in recent years and have begun looking for ways to increase worker unity across difference.

Supporting local leaders like Robertson in the struggle to bring together immigrant and nonimmigrant poultry workers in defense of their rights are MIRA and MPOWER. MIRA is a statewide coalition of immigrant and civil rights advocates that works closely with progressive elected officials to encourage the legislature and other state institutions to adopt immigrant-friendly policies. MIRA's founders emphasize the importance of bringing organized labor and progressive churches— always a crucial partnership for organizing in the South—together in the struggle for social justice, and the organization maintains strong ideological links between its work today and the efforts of Mississippi's civil rights workers of the 1960s. African Americans, particularly in the state legislature, have been key participants in MIRA's campaigns and are central to the struggle for immigrant rights in Mississippi. MPOWER is "a collaboration among poultry workers of diverse backgrounds, civil rights and immigrants' rights organizations, religious leaders, labor unions, employment justice groups, and other community partners." It "aims to increase workers' and advocates' abilities to ensure equity and justice on the job and in our communities by developing leadership among workers, strengthening [their] capacity to organize collectively, enhancing our access to knowledge, skills and resources, [and] building relationships across differences of race, culture, gender, language and religious affiliation." The goals of bridging differences in ideology, strategy, and identity in the fight for worker justice are explicit in MPOWER's mission statement and exhibited through the intentional relationship and leadership building that has taken place there in recent years.

## DISCUSSION QUESTIONS

1. How have Mississippi poultry processing plants met the challenge of the increased demand for their product?

2. What does Stuesse mean by "neoliberal policies," and who do these policies benefit?

3. What are the challenges to organizing the diverse work force in poultry processing plants?

# Student Exercises

1.  Think about your first experience of finding employment. Did you start in the informal economy, perhaps babysitting or doing yard work, or in the formal economy, perhaps at a fast food restaurant or a retail job at the mall? At what age did you start to work? How much money did you earn and what did you do with it? What role did your community or your family network play in your securing of employment? Having thought about this, share your experiences with the whole class, or in groups of eight or ten, so that there will be some diversity of experiences heard. How do you think your own work experience is related to your race, gender, social class, and residential location? What lessons can you learn from this exercise about how race, gender, social class, residential location, and other social factors are related to a young person's job search?

2.  Who works here? You may not always stop to notice the workers in the settings where you shop or eat. However, there could be social hierarchies in who works in a location and the tasks they perform. The next time you are in either a retail store or a restaurant, take notice of the staff in the establishment. If it is a retail store, who is at the counter, who is walking the floor, who works behind the scenes organizing the commodities, and who is the manager? If it is a restaurant, who greets you when you enter, who is your server, who busses the tables, and who are the other people in the front of the restaurant? See if you can identify who is in the back of the house: who is cooking, washing dishes, and performing essential chores in this restaurant? Make note of the gender, race or ethnicity, and age of the people doing various tasks. Even if you live in a homogeneous area, there might be significance in the gender and age of workers. In your class, talk with your classmates about your findings. Are your observations similar, or do people see different workers in urban areas, small towns, suburban malls, strip malls, or other locations? How can you relate your findings to economic restructuring and the location of work and workers?

# Section IX

# Shaping Lives and Love:

## Race, Families, and Communities

### Elizabeth Higginbotham
### and Margaret L. Andersen

In his groundbreaking work on the differences between immigrant and colonized minorities, Robert Blauner (1972) identifies how the lack of economic and political rights have made African Americans, Asian Americans, Native Americans, and Latino groups vulnerable to many cultural assaults, including their ability to function as families. Families' functioning is linked to their resources and to state support for the varied assistance that helps with socialization and support, such as public schools, health care, and public safety. Race influences the opportunities for individuals to form families, provide for those families, and live together as families. The national discussion might be about family values, but the resources available to families are structured by race, social class, and citizenship status.

Many people marry within their own racial or ethnic groups, but in a diverse society people can be attracted to individuals from different backgrounds. Race has historically been a factor in people's abilities to form the families of their choice. Today we might think that one of the signs of a free society is that people are able to freely associate with others—as peers, friends, neighbors, lovers, marriage partners, or in any other relationship. Therefore, it is hard to believe that for much of history people have not had that opportunity. Racism has invaded the most intimate areas of people's lives. For many years, thirty states had laws that prohibited White people from marrying someone of a different race. These **anti-miscegenation laws** prohibited so-called race mixing. For example, the state of California passed a law in 1880 prohibiting any White person from marrying a "negro, mulatto, or Mongolian." This law was designed to prevent marriages between White people and Chinese immigrants (Takaki 1989). Specific laws against intermarriage varied from state to state. All southern states prohibited White people from marrying Negroes, while some western states, like California, were also anti-Asian. Laws did not prohibit non-White groups from marrying each other. Thus in Mississippi the Chinese could and did marry Negroes, although neither group was allowed to marry White people.

In order to enforce anti-miscegenation laws, states had to devise ways to define race. States varied in this practice as well. "Alabama and Arkansas defined anyone with one-drop of 'Negro' blood as Black; Florida had a one-eighth rule"; and other states varied in their racial definitions (Lopez 1996: 118). If someone wanted to marry a person of another race, they had to do so in a state that did not prohibit the union. However, they risked having their marriage denied if they moved to a state where such arrangements were illegal.

Mildred Jeter (a Black woman) and Richard Loving (a White man), who had been married in the District of Columbia in 1958, had this experience. When they returned to their home state of Virginia, they were indicted and charged with violating Virginia's law banning interracial marriage. They were convicted and sentenced to one year in jail—a term they never served because the judge suspended the sentence on the condition that they leave Virginia. They challenged this law when they returned to Virginia five years later to appeal the decision through the courts. Their case, *Loving v. Virginia,* went all the way to the U.S. Supreme Court, which declared such laws unconstitutional. In 1967 the Supreme Court decided the case based on the argument that laws against intermarriage violated the 14th Amendment, which states: "No State shall make or enforce any laws which shall abridge the privileges or immunities of citizens of the United States; nor shall any State deprive any person of life, liberty, or property, without due process of law; nor deny to any person within its jurisdiction the equal protection of the laws" (U.S. Constitution, Amendment 14, Section 1).

In this section, we explore how racism influences both the challenges that families face and changes in marriage patterns. "Shaping Lives and Love: Race, Families, and Communities opens with a discussion of the impact of racism on Black families and communities". Joe R. Feagin and Karyn D. McKinney ("The Family and Community Costs of Racism") explore the real harm of racism to people and the places they live. Yet people of color actively resist such psychic and physical assaults, working as they can to build families and communities for social support and, at times, overt resistance.

Race, as well as other dimensions of inequality like social class, influences not only a family's resources, but how others in the society perceive them. Many families of color have been historically neglected by government agencies, and current trends indicate that agencies intervene in their lives in negative ways—a burden that more privileged families seldom face. Dorothy Roberts ("Child Welfare as a Racial Justice Issue") argues that the greater likelihood of children of color to be in foster care is evidence of racial inequality within state policies. As the clients in the child welfare systems around the nation changed from being a majority of White children to a majority of children of color, the remedy also changed—from providing services and keeping children at home to removing them from homes and placing them in foster care. Negative ideas about people of color still found in the media and other segments of the society foster the type of mistreatment that Roberts addresses.

Many people enter relations across racial and ethnic lines with admiration for the qualities in each other, but racism can also shape how we see the people we care about. Kumiko Nemoto's essay ("Interracial Relationships: Discourses and Images") examines the history of sexual relationships between American men and Asian women that have their roots in wars and the presence of military bases in Asia. While early legislation prohibited interracial marriage, many Asian women entered the United States after World War II and the Korean War as war brides who were celebrated for their alleged hyper-femininity. We also find more interracial marriage with the increasing number of Asian immigrants since the 1965 Immigration Act. Recently, we have seen a shift to a more modern image of professional Asian women who represent a model minority who can readily assimilate to this society.

Recently, there has been an increase in the numbers of couples marrying outside their race or ethnicity, as about 15 percent of new marriages are interracial or interethnic (Passel, Wang, and Taylor 2010). Zhenchao Qian's essay ("Breaking the Last Taboo: Interracial Marriage in America") discusses how public opinion about interracial dating and marriage has changed—and how it has not. Trends in cross-race interactions are also part of major racial division between families.

Increasingly, the United States is made of people with mixed racial and ethnic heritages. Diversity will likely continue in that direction, particularly as the walls of segregation are shattered and people have more face-to-face interaction with people of different backgrounds. Dismantling all forms of segregation is important because, with integration, people are less dependent on media images for information about people. As people meet, they can connect across shared interests and passions.

## REFERENCES

Blauner, Robert. 1972. *Racial Oppression in America*. New York: Harper & Row.

Lopez, Ian F. Haney. 1996. *White by Law: The Legal Construction of Race*. New York: New York University Press.

Passel, Jeffery, Wendy Wang, and Paul Taylor. 2010. "Marrying Out: One-in-Seven New U.S. Marriage is Interracial or Interethnic," Pew Research Center Publication.

Takaki, Ronald. 1989. *Strangers from a Different Shore: A History of Asian Americans*. New York: Penguin.

## FACE THE FACTS: MARITAL STATUS OF THE U.S. POPULATION

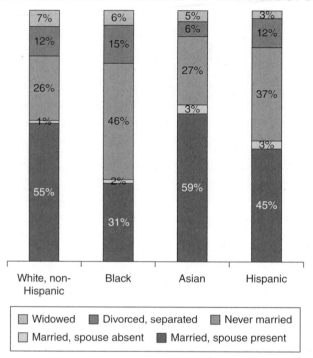

SOURCE: U.S. Census Bureau. 2009. *America's Families and Living Arrangements 2009*. Washington, DC: U.S. Department of Commerce. www.census.gov.

***Think about it***: Diversity in families is a fact of life in the United States. What factors might influence the differences in marital status of the different groups that you see here?

# 37

# The Family and Community Costs of Racism

### JOE R. FEAGIN AND KARYN D. MCKINNEY

*Racism takes a toll on people of color in multiple ways. Feagin and McKinney show how, even in the face of racist beliefs and actions that harm Black families and communities, those same institutions provide a defense against the harm that racism produces.*

Most African Americans see their families and communities as primary defenses against the daily assaults of racism. It is in these families that most black children first learn how to cope with racism. Additionally, the local black community often operates as extended kin. As they become adults, most black Americans seek support from others in their communities when they face increasing numbers of racist incidents in a variety of societal settings, including places of employment. As our respondents often note, the damage of a racially hostile workplace does not end at the workplace door. A black individual's experience with racial animosity and mistreatment is personally painful at the moment it happens, and also can have a cumulative and negative impact on other individuals, on one's family, and on one's community.

African American families and communities are negatively affected in many ways by continuing white-on-black racism. The black family has faced physical, ideological, and material assaults from whites for nearly four centuries. Over these centuries, family and community have been closely linked for African Americans, and for that reason the various white assaults on black families often have a significant impact on the larger black communities....

## THE LONG-TERM IDEOLOGICAL ASSAULT
## ON THE BLACK FAMILY

Most social science researchers view African American families as distinctive in certain characteristics when compared to similar white families. What is disputed is whether these differences are problematic. Rather than looking at African

SOURCE: From Joe E. Feagin and Karyn D. McKinney. 2003, *The Many Costs of Racism*. Lanham, MD: Rowman and Littlefield.

American families as simply *different* in certain ways from white families, most of the early literature and much recent literature addressing certain family forms found in black communities has described these family forms as more or less problematical or pathological....

Perhaps the best-known example of this pathological-family viewpoint is the 1965 report by Daniel Patrick Moynihan, *The Negro Family: The Case for National Action.* In this widely cited and influential government policy initiative, Moynihan writes about an allegedly distinctive pathology of black communities: "Obviously, not every instance of social pathology can be traced to the weakness of family structure.... Nonetheless, at the center of the tangle of pathology is the weakness of the family structure. Once or twice removed, it will be found to be the principal source of most of the aberrant, inadequate, or antisocial behavior, that did not establish, but now serves to perpetuate the cycle of poverty and deprivation."[1] In this report, white society is all but absolved of contemporary responsibility for poverty, deprivation, and family stresses in black communities. This labeling of urban black communities as a "tangle of pathology" is common-place and has been copied to the present day. According to many advocates of this pathology, the typical black family is matriarchal in character, has weak kinship ties, has "illegitimate" children, lacks mainstream values, and is present-oriented rather than future-oriented. This group of factors is sometimes referred to as a "culture of poverty."[2] Thus defined, it can seem inevitable and somehow right to whites that African Americans are impoverished....

A primary problem with the "culture of poverty" notion is that it confuses cause and result. While admitting that slavery and racism *began* the problems for the black family and community, those who adhere to this perspective believe that certain contemporary black family and community problems persist primarily because of an unhealthy black subculture. They do not recognize or admit that the original cause of stress on the black family, institutionalized racial oppression, is *still* the major and pervasive cause of much everyday stress for black families and communities today—and that the *results* of that stress include some of the adaptive characteristics often described in negative terms, or that are distorted in many analyses as supposed deviations from white family norms.

## CONFRONTING AND COPING WITH RACISM: STRONG AND ADAPTIVE FAMILIES

Contrary to the common view of the pathology advocates, the African American family has long been characterized by strong kinship bonds and interaction. During slavery, one of the primary reasons that enslaved African Americans ran away was to return to their families. In fact, many who had escaped to freedom in the North again risked their lives to return South in the attempt to free their families. After more than two centuries of colonial and U.S. slavery, during Reconstruction many African American families that had been separated by slaveholders began to regroup and re-create themselves. Although most of these newly freed black parents were struggling to feed their children, they often added to their families' innumerable "fictive kin"—the extended family and non-kin children who were orphaned after slavery. Similar family restructuring, and the care of extended family and "fictive kin," were seen again during the Great Depression of the 1930s. These patterns of care and survival can still be seen in many African American families and communities today.

For several decades now, numerous scholars have shown that African American families and communities are remarkably adaptive and strong....

Contrary to the "culture of poverty" thesis and commonplace white stereotyping, most members of black families exhibit a strong orientation toward work. Many studies show that whites and blacks are similarly oriented in their attitudes toward work. Our data show that rather than working less hard than whites, most African Americans realize that they often must work *harder* than whites to achieve the same results. For example, a sheriff's deputy in one focus group put it this way: "And that's the same thing ... we were talking about on the energy. Burning so much energy trying to educate these people, that we qualify, you know? And I always said if you see a black doctor and a white doctor standing side by side, equal in status, that black man is *twice* as [good], because he had to work harder ... in every profession."

Another strength of the black family is its adaptability in regard to family roles. Contrary to the popular notion that the black family is usually "matriarchal," even when there is a husband present, scholars have found that African American families are often more egalitarian than are white families. Because they are not as patriarchal as the traditional white family, black families thus appear to many (especially white) outsiders as part of a "matriarchal" family system. Evidence of the adaptability of black families can also be seen when young women become mothers, in that the larger family is usually supportive. African American families are twice as likely as whites to take the children of premarital births into the homes of family or friends of the parents, and are *seven times less* likely than whites to put up these children for adoption.

Another adaptation to unfavorable circumstances is a strong achievement orientation.... Education is valued highly in the black family and community as one way to help overcome poverty and racism.

Another distinctive strength of African American families is their often strong religious orientation. This religious orientation contributes to cohesive black families, and has been used as a mechanism of survival for black communities from slavery through the 1960s civil rights movement to the present day....

## BLACK FAMILIES: THE CHALLENGE OF DIFFICULT
## ECONOMIC CONDITIONS

Some of the most difficult problems faced by black families stem from outside economic forces. Some of these economic forces have, unintentionally, a differential impact on African Americans, while other economic problems are created by government or private sector policies that intentionally, or at least knowingly, have a negative impact on African Americans. In other work, Feagin has distinguished between direct and indirect institutionalized discrimination. Direct institutionalized discrimination is that discrimination intentionally built into organizations and institutions by whites in order to have a significant negative impact on people in subordinate racial groups, especially African Americans. Examples include the intentional exclusion, blatant or subtle, of African Americans from traditionally white jobs and neighborhoods, discrimination that can still be found in some sectors of U.S. society. Indirect institutionalized discrimination involves the differential and negative treatment by whites of subordinate racial groups without that treatment being intentional in the present moment. This is oppressive and damaging discrimination, nonetheless, because it carries into the present the impact of the often blatant discrimination of the past. For example, intentional discrimination in the past often creates limited resources for the descendants of those so discriminated against, and current generations of groups once severely subordinated have less inherited wealth and other socioeconomic resources than do current generations of whites. This means that, today, the average African American typically does not have the same economic and educational resources and, thus, opportunities as the average white American. For example,…the average black family today has about 60 percent of the income of the average white family—and only about a tenth of the average white family's wealth.…

In our focus groups and in informal discussions with a number of other African American husbands and wives, we have found that many people explain in some detail how the stress of discrimination at work places an added burden on their relationships with their spouses. When husbands or wives are under great pressure and stress from the racist incidents in their predominantly white workplaces, their energy for, or willingness to interact with, their spouses when they come home is affected. They may wish to be secluded from family relationships, and perhaps just watch television for a long time in order to unwind for a while from the stress of dealing with whites in the workplace or other social settings.… Many African Americans are aware of how this impacts the family, and they try to develop counter-strategies. In other cases, even these veterans of contending with racism do not realize fully just how that racism has negatively affected their intimate family relationships. In these cases, white racism has yet other negative, often energy-draining, effects on African American families.

For the family, another obvious consequence of workplace discrimination occurs when an important breadwinner is fired unfairly from his or her job because of discrimination, and then is not able to provide adequately for the family. Thus, one black man quoted in an earlier research study had worked for ten years as a school employee whose job it was to deal with the community. Although he was hired for his interpersonal skills and life experiences, after ten years on the job his employer decided to put some formal screening credentials into place for such community contact positions. Under this newly imposed standard, although he had personally moved up into the middle-class, his working-class demeanor and speech were no longer valued, and he was fired. At the time of his interview, he was preparing to sue to win back his job and was working a temporary job that paid a very low wage. He had also separated from his wife. In his interview, he spoke of the costs of this experience on his children:

> I honestly believed and felt that I was put on this earth to help to work with kids, I really did. And I was good, I was damn good at my job to have, let's say, less education than probably anybody at that school.…
> I *am* angry. I'm very angry. Because what they did, they didn't do because I was not performing my job. They did [it] because I was a black man and I spoke out on what I believe and felt. If I hadda been one of their little henchmen to say "yes, no"…I would still be there.… They don't know what the hell they have put me through. I know I could have a job if it wasn't for them folks down there. They have shut off my livelihood. They don't know the suffering, not that I am absorbing, but my kids.… I don't know how many times I didn't have the money to pay the rent. They got food on their table when they come in, in the evening. Sometimes I don't·have food here on my table for my kids.

Damage is done to African American families because of workplace mistreatment along racial or racial-class lines. This man cannot feed his children adequately. He also notes that he has had to take his daughter out of college, thereby jeopardizing her ability to be economically successful in the future. And he has separated from his wife. His family has been torn apart by the racist actions of whites in his workplace. We see here the powerful domino effects on families of racist actions in yet another U.S. workplace.

## THE IMPACT OF RACISM: THE BLACK COMMUNITY

… The spin-off effects of animosity and mistreatment in employment settings can be seen in yet other areas of the lives of these African Americans. Another respondent in a focus group sadly noted the negative impact on participation in church activities:

> I have withdrawn from some of the things I was involved with at church that were very important to me, like dealing with the kids at church. Or we had an outreach ministry where we would go out into the low-income housing and we would share about our services, … and I was just so drained, like [another respondent] said, if we are all so drained, and we stop doing that, then we lose our connection. But I, physically, by the time I got home at the end of the day, I was just so tired. I didn't even feel like giving back to my community; I didn't feel like doing anything. And so I withdrew from church activities, to the point where I just really was not contributing anything. And it was pulling all that energy; I was exhausted from dealing with what I had to at work. And then whatever little bit was left went to my family, so there was nothing there to give.

The considerable impact of workplace racism is graphically described here, for even church activities become a challenge for this person. What energy there is left after struggles at work with racism is reserved for the family. These economically successful African Americans can be important role models in their local communities, but only if they have the energy to participate actively in churches and other community organizations. These accounts of withdrawal from, or lack of energy for, community activities are worth considering in the light of the many accomplishments of African Americans in organizing to improve both their own communities, in a variety of churches and other important local organizations, and also in working together in national organizations to improve the larger U.S. society. In spite of the great strain, pain, and energy loss stemming from individual and institutional racism, the majority of African Americans manage to strive, endure, and succeed in raising families and building communities. Being tired from the daily struggle against racism in the workplace, schools, and public accommodations has not stopped many African Americans from continuing to organize, a point we will accent in the conclusion that follows.

## CONCLUSION: CREATING COMMUNITIES
## OF RESISTANCE

Dealing with the family and community impact of racism is a constant challenge for African Americans. The question of how to develop effective community strategies for development or change has been a recurring topic of concern and analysis for African American intellectuals and other leaders....

It is evident in the words of numerous respondents that interpersonal relationships, including family relationships, are sometimes jeopardized because one or both partners does not have much energy left at the end of a racially stressful day. Yet most seem to manage to overcome these huge challenges most of the time. Indeed, from their discussions of the energy-draining aspects of discrimination, one might wonder how African Americans have developed vibrant community organizations and successful resistance movements over nearly four centuries now. Most African Americans persevere through the barriers and manage to overcome the constant "rain" of antiblack racism enough to stay centered in their life struggles and, remarkably, thrive....

These organized efforts have not only liberated African Americans from legal segregation, but also the *country as a whole* from these enormous barriers to human progress. All Americans are the current beneficiaries of organized African American resistance to racial oppression. There is greater freedom in the U.S. today than there would have been without the civil rights movement of the 1960s. This is as true today as it has been in the past. Racism is a destructive force that ultimately affects all residents of this country. As the English poet John Donne said long ago, "Any man's death diminishes me, because I am involved in mankind; and therefore never send to know for whom the bell tolls; it tolls for thee."

## NOTES

1. Daniel P. Moynihan, *The Negro Family: The Case for National Action* (Washington, D.C.: U.S. Department of Labor, 1965), p. 30.

2. This term was first coined by Oscar Lewis to refer to Mexicans in Oscar Lewis, *The Children of Sanchez* (New York: Random House, 1961). He later applied the term to Puerto Ricans. Lewis saw the "culture of poverty," at least in part, as a response to oppressive circumstances. It was Moynihan and others who later applied the term to black Americans and used it in more of a victim-blaming manner.

## DISCUSSION QUESTIONS

1. What does it mean to say that the common view of African American families has been one of "pathology?" How do Feagin and McKinney evaluate this assumption?

2. In what specific ways do African American families adapt to the racism they encounter?

# 38

# Child Welfare as a Racial Justice Issue

DOROTHY ROBERTS

*There are huge racial disparities in the placement of children into foster care, with the vast majority of such children coming from African American families. Roberts, a legal scholar, argues that this reflects the increased surveillance of Black families by the state and racist assumptions that Black families are somehow inadequate.*

My main intellectual and activist project over the last five years has been to explain child welfare policy as an issue of racial justice. Child welfare was once viewed as a social issue, but by the 1970s the main mission of public child welfare departments had become protecting children against maltreatment inflicted by pathological parents. Child welfare decision making has an atomistic focus, that zooms in on the situation of individual children and their families. If child welfare is discussed as a matter of rights at all, it is usually framed as a contest between children's rights and parents' rights, falsely assuming that the interests of parents and children in the child welfare system are always in opposition to each other and that the system treats all parents and children equally badly.

Strangely, criticisms of the child welfare system are not placed among the burning social justice issues of our day. I say strangely because anyone who is familiar with the child welfare system in the nation's large cities knows that it is basically an apartheid institution. Spend a day at any urban dependency court and you will see a starkly segregated operation. If you came with no preconceptions about the purpose of the child welfare system, you would have to conclude that it is an institution designed to monitor, regulate, and punish poor minority families, especially Black families.

The number of Black children in state custody—those in foster care as well as those in juvenile detention, prisons, and other state institutions—is alarming.... Black children make up about two-fifths of the foster care population, although they represented less than one-fifth of the nation's children (Administration for Children and Families 2003). The color of child welfare is most apparent in big cities where there are sizeable Black and foster care populations. In Chicago, for example, almost all of the children in foster care are Black (Pardo 1999). The racial imbalance in New York City's foster care population is also mind-boggling: out of 42,000 children in the system at the end of 1997, only

SOURCE: From Dorothy Roberts, "Child Welfare as a Racial Justice Issue," 2002. Reprinted by permission of the author.

about 1000 were white (Guggenheim 2000). Black children in New York were 10 times as likely as White children to be in state protective custody. Although the total numbers are smaller, this racial disproportionality extends to cities and states where Black children are less visible.

State agencies treat child maltreatment in Black homes in an especially aggressive fashion. They are far more likely to place Black children than other children who come to their attention in foster care instead offering their families less traumatic assistance. A national study of child protective services by the U.S. Department of Health and Human Services reported that "[m]inority children, and in particular African American children, are more likely to be in foster care placement than to receive in-home services, even when they have the same problems and characteristics as white children" (U.S. Dept. HHS 1997). Most white children who enter the system are permitted to stay with their families, avoiding the emotional damage and physical risks of foster care placement, while most Black children are taken away from theirs. Foster care is the main "service" state agencies provide to Black children. And once removed from their homes, Black children remain in foster care longer, are moved more often, receive fewer services, and are less likely to be either returned home or adopted than any other children (Courtney & Wong 1996; Jones 1997).

In some cases, protecting children requires immediately removing them from their homes. But the public often overlooks the costs to children of separating them from their families. In 2001, Judge Jack Weinstein of the Eastern District of New York issued a blistering condemnation of New York City's Administration for Children's Services for automatically removing children from mothers who were victims of domestic violence (*Nicholson v. Scoppetta* 2001). Judge Weinstein's decision is especially noteworthy for its rare judicial recognition of the harm inflicted on children by unnecessarily taking them from their parents. "It hardly needs to be added that the exact language of the Thirteenth Amendment covers protection of the children's rights," Judge Weinstein wrote. "They are continually forcibly removed from their abused mothers without a court adjudication and placed in a forced state custody in either state or privately run institutions for long periods of time. They are disciplined by those not their parents. This is a form of slavery."

A new politics of child welfare threatens to intensify state supervision of Black children. In the last several years, federal and state policy has shifted away from preserving families toward "freeing" children in foster care for adoption by terminating parental rights. The Adoption and Safe Families Act, passed in 1997, imposes an expedited time frame for state agencies to file petitions to terminate parental rights and give states a financial bonus for increasing the number of adoptions of children in foster care—$4,000 per child, $6,000 if the child is classified as having special needs. The campaign to increase adoptions has hinged on the denigration of foster children's parents, the speedy destruction of their family bonds, and the rejection of family preservation as an important goal of child welfare practice. Adoption is increasingly presented not as an option for a minority of foster children who cannot be reunited with their parents, but as the preferred outcome for all children in foster care....

Welfare reform, by throwing many families deeper into poverty, heightens the risk that the most vulnerable children will be placed in foster care. A front page story in the *New York Times,* entitled "Side Effect of Welfare Law: the No-Parent Family," reported a study of census data in all 50 states that found that the rate of Black children in cities living without their parents has more than doubled as a result of welfare reform—an estimated 200,000 more Black children separated from their parents (Bernstein 2002).

In addition, tougher treatment of juvenile offenders, imposed most harshly on African American youth, is increasing the numbers incarcerated in juvenile detention facilities and adult prisons. These political trends are converging to address the deprivation of poor Black children by placing more of them in one form of state custody or another. Child welfare policy conforms to the current political climate, which embraces private solutions—and when those fail, punitive responses—to the seemingly intractable plight of America's isolated and impoverished inner cities. As welfare reform reduces the welfare rolls by promoting marriage and imposing sanctions for failing to find work, child welfare policy reduces the foster care rolls by terminating parental rights and promoting adoption of the "legal orphans" it creates.

The color of America's child welfare system undeniably shows that race matters to state interventions in families. So why have scholars and policymakers been slow to describe the disproportionate involvement of Black families in the system as a racial injustice? Let me suggest three related explanations.

First, there is profound confusion about the reasons for the system's racial disparity. The existing social science literature contains theories that attribute the racial disparity both to differences in the well being of children and to differences in the system's treatment of children. In other words, there is disagreement over whether the disparity stems from societal conditions outside the system, such as higher poverty rates among nonwhite children, or from racially biased practices *within* the system. Many experts believe that this distinction in causes—societal forces vs. child welfare practices—makes a crucial difference in how we should address the racial disparities (Courtney et al. 1998). If the cause of the system's racial imbalance is social and economic inequality, some say, we can't blame the system (Bartholet 1999).

This is related to a second reason for failing to see a racial justice issue—the concern for children's rights. If the child welfare system is simply reflecting inequities in children's living conditions, we would expect—we would even want—the state to intervene more often to protect Black children from the greater harm that they face. Indeed, wouldn't the government violate Black children's rights if it *failed* to intervene more often to protect them?...

A final stumbling block is the official understanding of racial discrimination. Under current civil rights jurisprudence, the racial disparity in the child welfare system may not constitute racial discrimination without a showing of racial motivation. The system is racist only if Black children are pulled out of their homes by bigoted caseworkers or as part of a deliberate government scheme to subjugate Black people. Any other explanation—such as higher rates of Black family poverty or unwed motherhood—negates the significance of race....

But these views of the child welfare system, children's rights, and racial discrimination fail to take into account the political dimension of child welfare policy. To begin with, which harms to children are detected, identified as abuse or neglect, and punished is determined by inequities based on race, class, and gender. The U.S. child welfare system is and always has been designed to deal with the problems of poor families (Pelton 1989). The child welfare system hides the systemic reasons for poor families' hardships by attributing them to parental deficits and pathologies that require therapeutic remedies rather than social change. The harms caused to children by uncaring, substance-abusing, mentally unstable, absentee parents in middle-class and affluent families usually go unheeded. Although these children from privileged homes might spend years in psychotherapy, it is unlikely they will spend any time in foster care. Most child maltreatment charges are for neglect and involve poor parents whose behavior was a consequence of economic desperation as much as lack of caring for their children.

The racial disparity in the child welfare system also reflects a political choice about how to address child neglect. It is no accident that child welfare philosophy became increasingly coercive as Black children made up a greater and greater share of the caseloads. In the past several decades, the number of children receiving child welfare services has declined dramatically, while the foster care population has skyrocketed (U.S. Dept. HHS 1997). As the child welfare system began to serve fewer white children and more Black children, state and federal governments spent more money on out-of-home care and less on in-home services. This mirrors perfectly the metamorphosis of welfare once the welfare rights movement succeeded in making AFDC available to Black families in the 1960s. As welfare became increasingly associated with Black mothers, it became increasingly burdened with behavior modification rules and work requirements until the federal entitlement was abolished altogether in 1996. Both systems responded to their growing Black clientele by reducing their services to families while intensifying their punitive functions.

The child welfare system's reliance on a disruptive, coercive, and punitive approach also inflicts a political harm. American constitutional jurisprudence defines the harm caused by unwarranted state interference in families in terms of individual rights. Wrongfully removing children from the custody of their parents violates parents' due process right to liberty. The earliest cases interpreting the due process clause to protect citizens against government interference in their substantive liberty involved parental rights. But these explanations of harm do not account for the particular injury inflicted by racially disparate state intervention. The over-representation of Black children in the child welfare system, especially foster care, represents massive state supervision and dissolution of families. This interference with families helps to maintain the disadvantaged status of Black people in the United States. The child welfare system not only inflicts general harms disproportionately on Black families. It also inflicts a particular harm—a racial harm—on Black people as a group....

State supervision of families is antithetical to the role families are supposed to play in a democracy, as critical components of civil society. Families are a principal form of "oppositional enclaves" that are essential to citizens' free participation in democratic institutions, to use Harvard political theorist Jane Mansbridge's term (Mansbridge 1996, 58). Family and community disintegration weakens Blacks' collective ability to overcome institutionalized discrimination and to work toward greater political and economic strength. Family integrity is crucial to group welfare and identity because of the role parents and other relatives play in transmitting survival skills, values, and self-esteem to the next generation. Placing large numbers of children in state custody interferes with the group's ability to form healthy and productive connections among its members. The system's racial disparity also reinforces the quintessential racist stereotype: that Black people are incapable of governing themselves and need state supervision.

The impact of state disruption and supervision of families is intensified when it is concentrated in inner-city neighborhoods. In 1998, one out of every ten children in Central Harlem had been taken from their parents and placed in foster care (Center for an Urban Future 1998, 6). In Chicago, almost all child protection cases are clustered in a few zip code areas, which are almost exclusively African American. The spatial concentration of child welfare supervision creates an environment in which state custody of children is a realistic expectation, if not the norm....

The racial disparity in the foster care population should cause us to reconsider the state's current response to child maltreatment. The price of present policies that rely on child removal rather than family support falls unjustly on Black families and communities. In part because of narrow conceptions of racial discrimination and children's rights, judges, politicians, and the public have a hard time seeing this as a racial injustice. I propose that we figure out better ways of measuring and explaining this type of systemic, community-wide racial harm.

## REFERENCES

Administration for Children and Families, U.S. Department of Health & Human Services. 2000. *Child Maltreatment 1998: Reports from the States to the National Child Abuse and Neglect Data System*. Washington, D.C.: U.S. Government Printing Office.

———2003. *The AFCARS Report: Preliminary FY 2001 Estimates as of March 2003*. http://www.acf.hhs.gov/programs/cb/publications/afcars/report8.htm.

Bernstein, Nina. 2002, July 29. "Side Effect of Welfare Law: The No-Parent Family." *New York Times*, p. 1.

Bartholet, Elizabeth. 1999. *Nobody's Children*. Boston: Beacon Press.

Center for an Urban Future, 1998. "Race, Bias, and Power in Child Welfare." *Child Welfare Watch*. Spring/Summer: 1.

Courtney, Mark E., et al. 1998. "Race and Child Welfare Services: Past Research and Future Directions." *Child Welfare* 75:99.

Courtney, Mark E. and Wong, Vin-Ling Irene. 1996. "Comparing the Timing of Exits from Substitute Care." *Child & Youth Services Review* 18:307.

Guggenheim, Martin. 2000. "Somebody's Children: Sustaining the Family's Place in Child Welfare Policy." *Harvard Law Review* 113: 1716–1750.

Jones, Loring P. 1997. "Social Class, Ethnicity, and Child Welfare." *Journal of Multicultural Social Work* 6:123.

Males, Mike and Dan Macallair. 2000. *The Color of Justice: An Analysis of Juvenile Adult Transfers in California.* San Francisco: Justice Policy Institute.

Mansbridge, Jane. 1996. "Using Power/Fighting Power: The Polity." In *Democracy and Difference: Contesting the Boundaries of the Political,* edited by Seyla Benhabib. Princeton: Princeton University Press.

Nicholson v. Scoppetta. 2001. 202 F.R.D. 377. United States District Court, Eastern District of New York.

Pardo, Natalie. 1999. "Losing Their Children." *Chicago Reporter* 28:1.

Pelton, LeRoy H. 1989. *For Reasons of Poverty: A Critical Analysis of the Public Child Welfare System in the United States.* New York: Praeger.

U.S. Department of Health and Human Services. 1997. *National Study of Protective, Preventive, and Reunification Services Delivered to Children and Their Families.* Washington, D.C.: U.S. Government Printing Office.

## DISCUSSION QUESTIONS

1. What does Roberts mean by an apartheid institution? How does this apply to the child welfare systems in major cities?

2. How can children become legal orphans?

3. How does the impact of welfare policies on individual families influence the collective well-being of the Black community?

# 39

# Interracial Relationships:

## Discourses and Images

KUMIKO NEMOTO

*This author asks questions about the sexual relationships between White American and Asian couples in order to place such interactions within a historical context and explore different visions of femininity. Initially subjected to anti-miscegenation laws in many states, the marriage of Asian women and White men became more common after World War II due to war brides and, more recently, increased immigration from Asia.*

... From the early history of the United States, Asian American–white interracial relationships were shaped by American imperialism in Asia, national security issues, immigration regulations, and domestic racial politics. The United States' involvement in three wars—World War II, the Korean War, and the Vietnam War—created a large influx of military brides from these countries. Existing studies indicate that many of these relationships reflect colonial/postcolonial gender dynamics, wherein a woman's legal status and life is entangled with her subordination to the American patriarchal order. In the post-Vietnam period, the image of Asian women as docile and sexually available became pervasive as a result of the U.S. military presence in Vietnam, the development of the sex industry there, and the mail-order bride industry in Southeast Asia.

With the rise of the Asian economy, especially in the 1970s and 1980s, multiculturalism and the idea of the model minority have reshaped and transformed the dominant image of Asian American women from that of the docile wife to that of the economically mobile woman. The concept of the model minority puts an interesting twist on the issue of gender. Feminism among white women, which has generated cultural anxiety among American men, has interested these men in Asian American women, who are marked not only as possessing economic mobility, but also as possessing traditional femininity....

SOURCE: From Kumiko Nemoto, *Racing Romance: Love, Power and Desire among Asian American/ White Couples.* New Brunswick, NJ: Rutgers University Press 2009.

# BRIDES FROM ASIA

Although the image of Asian military brides became widely popular after World War II, Asian American-white interracial relationships were already being influenced by American immigration policy (like America's colonization of the Philippines and the Chinese Exclusion Act) in the nineteenth and early twentieth centuries. In the 1800s, Chinese female immigrants were suspected by the U.S. government of being a demoralizing threat to whites. In the late nineteenth century, due to concerns generated by the growing Asian population and the resulting fear of miscegenation (seen as the "Yellow Peril"), various federal legislatures passed exclusion laws and antimiscegenation laws....

Two distinctive images of Asian American women, as military brides and as prostitutes, derived from the long history of American military presence in Asian countries, where Americans established a stereotype of Asian American-white interracial relationships. As clear from reception of the opera *Madame Butterfly* (1904) and the Hollywood movie *Shanghai Express* (1932), stereotypes of Asian American women as submissive, sexual, and/or treacherous were already popular in America at the beginning of the twentieth century. When a large influx of military brides entered the United States in the 1950s as a result of the United States' wars with Japan and Korea, sexual encounters and marriage between white men and Asian women became far more visible....

Relationships between Asian American women and white men were often depicted, through images of military brides either as submissive wives or as foreign prostitutes, in either case rescued by America's paternal discretion and international power. The subordination of military brides in the colonial/postcolonial gender order translated into national racial politics, in which white men emphasized the overfeminized characteristics of Asian American women to maintain the racial privileges of whites and patriarchal ethos of family and nation....

Military wives were thought of, not only as loyal and docile wives, but often as tied to military prostitution. Encounters between Asian women and American men often occurred on the U.S. military bases in Okinawa, South Korea, Vietnam, and the Philippines in which local women catered to the American soldiers as sex workers.

Popular culture can reproduce and reinforce hegemonic narratives of race and gender. The military brides may have struggled with racism, isolation, and low-wage jobs in the United States, but Hollywood films of the 1950s and 1960s, such as *Japanese War Brides* (1952) and *Sayonara* (1957), portrayed Asian American-white intimacy as evidence of white America's acceptance of Asian brides (who were exemplars of submissiveness, domesticity, and quiet endurance) and as the precursor of the multicultural and multiracial household of the future. Hollywood movies of this era also depicted romantic love between Asian American women and white men as a white-knight assimilationist love story—a white man is depicted as the ideal knight for an Asian woman, who is

the Cinderella figure, to attain material prosperity, spiritual transcendence, freedom, and salvation—and thus such narratives perpetuate her subordinate position
to white men....

Colonial narratives with white-master and native-maid/whore themes remain transnationally circulated in postcolonial Asia, and they have affected
Asian American–white relationships far beyond the contexts of the American
military in Asia. As Ling notes, the popular narrative sells "the wild sexual
adventures of white men with seductive, available 'native women'" in Asia
(playing into the mail-order bride industry in Asian countries), wherein an
Asian woman appears with "her below-the-waist black hair, projectile-like
breasts, tight dress and sexual insouciance" and aims to "trap a high-salaried
white expat."[1] Transnational imagery of Asia mirroring the Western and
American gaze returns to, and recirculates within, the United States.... Asian
American women have been represented as hypersexual individuals, measured
against the white female norm in the United States, and these stereotypes
have been consumed and reinforced in the context of colonial/postcolonial
gender orders.

## THE MODEL MINORITY AND ASIAN
## AMERICAN FEMININITY

... The large influx of middle-class Asian immigrants after the 1965 Immigration
Act, the rise of multiculturalism, and the growing impact of Asian economies
augmented the visibility of highly educated middle-class Asian Americans who
were no longer subordinate laborers for whites but rival agents of whites. The
model minority discourse emerged with such demographic, cultural, and
economic change in the United States. It also transformed the image of Asian
American women from that of the embodiment of colonial femininity to that
of the upwardly mobile middle-class woman.

Susan Koshy argues that images of Asian American women as hypersexual,
combined with the model minority myth, has played a critical role in the popularity of Asian American women since the 1970s. Koshy claims that long-
existing images of these women's supposed subservience and domesticity, along
with the images of them as hypersexual, have enhanced Asian American
women's "sexual capital" and their status as a "sexual model minority."[2] She
notes that the femininity of Asian American women has become popular precisely because it is perceived as a substitute or replacement for traditional white
femininity, perceived as receding due to the rise of feminism among white
women; thus, images of Asian American women as "traditionally" feminine
became a cultural and sexual remedy for white men's masculinity crises in the
face of white feminism....

The transformation of Asian femininity, from the subordination of the colonial era to its more recent status as an indication of sexual capital, also explains

how femininity in present-day America is not merely about docility or "emphasized femininity" per se, but is tied to a particular physical appearance and body physique. As Patricia Hill Collins notes, "evaluations of femininity are fairly clear-cut. [S]kin color, body type, hair texture, and facial features become important dimensions of femininity."[3] Asian American women have been able to attain the ideal "feminine" body because "the Asian beauty of reputedly slender and petite doll-like bodies was extolled in terms of soft skin and silky hair, and contrasted with images of repudiated and coarser African hair and skin." Such cultural imagery of gender marks the physical appearance of Asian Americans with racialized notions of hyperfemininity and hypersexuality.

With the rise of the model minority discourse, professional Asian American women have become more visible in the media. In the 1980s, the appearance of the successful female newscaster Connie Chung "embod[ied] a new bent on racist representations of Asian American as the 'model minority.'"[4] ...The emergence of upwardly mobile Asian American women certainly marked the progression of the cultural signifiers of Asian American femininity, yet these markers remained subordinated to white men.

Like the model minority discourse, Asian American–white interracial relationships served to mitigate tension between multiculturalism and anti-immigration sentiments in the 1980s and 1990s. Interracial relationships were often presented as the multiculturalist solution that confirmed the hegemonic orders of American national identity....

... The image of interracial marriage is built on heterosexuality and the patriotic ideals of family and nation; thus it consists of hegemonic orders of gender and race; interracial relationships have become acceptable or even "revolutionary" that is, when they embody hegemonic orders of gender, nation, and family. It is in precisely such a context that Asian American women have been designated as the sexual model minority for the reconstruction of American's masculine and heterosexual national identity....

The "new" or transgressive images of Asian American women seem, then, to at once reinforce and contest familiar stereotypes. For example, America's model minority myth, which, as Palumbo-Liu notes, is a "fetishized ethnic dilemma" that promotes "healing" through self-affirmation,[5] reproduces the image of Asian American women as both independent and submissive subjects both globally and locally; the popular reception of the Academy Award-winning movie *Memoirs of a Geisha* (2005) exemplifies Hollywood's still popular depiction of Asian women as submissive and hard-working—exemplars of the sexual model minority. Yet the emergence of relatively diverse popular images of Asian American women, including many characters who are strong and independent and not constrained within the boundaries of domesticity, is easily explained by the important roles of East and South Asia in the global economy, as well as by the accomplishments of Asian Americans in the United States in terms of income and education. However, it is highly questionable to what extent the economic mobility and strength of Asian American women depicted in these media images challenge or destabilize patriarchal logics of gender, race, and family.

## NOTES

1.  L. H. M. Ling, "Sex Machine: Global Hypermasculinities and Images of the Asian Women in Modernity," *positions: east asia cultural critique* 7, no. 2 (1999): 294.

2.  Susan Koshy, *Sexual Naturalization: Asian Americans and Miscegenation* (Stanford, CA: Stanford University Press, 2004) p. 135.

3.  Patricia Hill Collins, *Black Sexual Politics: African Americans, Gender, and the New Racism* (New York: Routledge, 2004), 194–195.

4.  Gina Marchetti, *Romance and the "Yellow Peril": Race, Sex, and Discursive Strategies in Hollywood Fiction* (Berkeley and Los Angeles: University of California Press, 1994) p. 216.

5.  David Palumbo-Liu, *Asia/America: Historical Crossings of a Racial Frontier* (Stanford, CA: Stanford University Press, 1999); p. 396.

## DISCUSSION QUESTIONS

1.  How does the history of the United States military in Asia influence images of Asian women and their relationships with American husbands?

2.  What roles do the media play in presenting an image of Asian women?

3.  How do contemporary images of Asian women contrast with those prior to 1965?

# 40

# Breaking the Last Taboo:
## Interracial Marriage in America

ZHENCHAO QIAN

*Although interracial marriage is becoming more common, it is still quite rare. Qian's essay reviews the extent of and attitudes toward interracial dating and marriage and how these vary among different groups in the U.S. population.*

**G**uess *Who's Coming to Dinner,* a movie about a white couple's reaction when their daughter falls in love with a black man, caused a public stir in 1967. That the African-American character was a successful doctor did little to lower the anxieties of white audiences. Now, almost four decades later, the public hardly reacts at all to interracial relationships. Both Hollywood movies and TV shows, including *Die Another Day, Made in America, ER, The West Wing,* and *Friends,* regularly portray interracial romance.

What has changed? In the same year that Sidney Poitier startled Spencer Tracy and Katherine Hepburn, the Supreme Court ruled, in *Loving v. Virginia,* that laws forbidding people of different races to marry were unconstitutional. The civil rights movement helped remove other blatant legal barriers to the integration of racial minorities and fostered the growth of minority middle classes. As racial minorities advanced, public opinion against interracial marriage declined, and rates of interracial marriage grew rapidly.

Between 1970 and 2000, black-white marriages grew more than fivefold from 65 to 363 thousand, and marriages between whites and members of other races grew almost fivefold from 233 thousand to 1.1 million. Proportionately, interracial marriages remain rare, but their rates increased from less than 1 percent of all marriages in 1970 to nearly 3 percent in 2000. This trend shows that the "social distance" between racial groups has narrowed significantly, although not nearly as much as the social distance between religious groups. Interfaith marriages have become common in recent generations. That marriages across racial boundaries remain much rarer than cross-religion marriages reflects the greater prominence of race in America. While the interracial marriage taboo seems to be gradually breaking down, at least for certain groups, intermarriage

SOURCE: From *Contexts* 4(4): 33–37, 2005. Copyright © American Sociological Association.
Reprinted by permission of The University of California Press.

in the United States will not soon match the level of intermarriage that European immigrant groups have achieved over the past century.

## PUBLIC ATTITUDES

Americans have become generally more accepting of other races in recent decades, probably as a result of receiving more education and meeting more people of other races. Americans increasingly work and go to school with people from many groups. As racial gaps in income narrow, more members of racial minorities can afford to live in neighborhoods that had previously been white. Neighbors have opportunities to reduce stereotypes and establish friendships. Tolerance also grows as generations pass; elderly people with racist attitudes die and are replaced by younger, more tolerant people. The general softening of racial antagonisms has also improved attitudes toward interracial marriage.

In 1958, a national survey asked Americans for the first time about their opinions of interracial marriage. Only 4 percent of whites approved of intermarriage with blacks. Almost 40 years later, in 1997, 67 percent of whites approved of such intermarriages. Blacks have been much more accepting; by 1997, 83 percent approved of intermarriage. Whites' support for interracial marriage—which may to some extent only reflect respondents' sense of what they should tell interviewers—lags far behind their support of interracial schools (96 percent), housing (86 percent), and jobs (97 percent). Many white Americans apparently remain uneasy about interracial intimacy generally, and most disapprove of interracial relationships in their own families. Still, such relationships are on the increase.

## INTERRACIAL DATING

According to a recent survey reported by George Yancey, more than one-half of African-, Hispanic-, and Asian-American adults have dated someone from a different racial group, and even more of those who have lived in integrated neighborhoods or attended integrated schools have done so. Most dates, of course, are casual and do not lead to serious commitments, and this is especially true for interracial dating. Analyzing data from the National Longitudinal Study of Adolescent Health, Kara Joyner and Grace Kao find that 71 percent of white adolescents with white boyfriends or girlfriends have introduced them to their families, but only 57 percent of those with nonwhite friends have done so. Similarly, 63 percent of black adolescents with black boyfriends or girlfriends have introduced them to their families, but only 52 percent of those with non-black friends have done so. Data from another national survey show similar patterns for young adults aged 18–29 (61 percent versus 51 percent introducing for whites, and 70 percent versus 47 percent for blacks).

While resistance to interracial relationships in principle has generally declined, opposition remains high among the families of those so involved. Interracial couples express concern about potential crises when their families become aware of such relationships. Their parents, especially white parents, worry about what those outside the family might think and fear that their reputations in the community will suffer.

Maria Root notes that parents actively discourage interracial romance, often pointing to other peoples' prejudice—not their own—and expressing concern for their child's well being: "Marriage is hard enough; why make it more difficult?"

The dating and the parental reservations reveal a generation gap: Young men and women today are more open to interracial relationships than their parents are. This gap may be due simply to youthful experimentation; youngsters tend to push boundaries. As people age, they gradually learn to conform. Kara Joyner and Grace Kao find that interracial dating is most common among teenagers but becomes infrequent for people approaching 30. They attribute this shift to the increasing importance of family and friends—and their possible disapproval—as we age. When people are ready to be "serious," they tend to fall in love with people who are just like themselves.

## INTERRACIAL COHABITATION AND MARRIAGE

Who pairs up with whom depends partly on the size of the different racial groups in the United States. The larger the group, the more likely members are to find marriageable partners of their own race. The U.S. Census Bureau classifies race into four major categories: whites, African Americans, Asian Americans and American Indians. Hispanics can belong to any of the four racial groups.... Whites form the largest group, about 70 percent of the population, and just 4 percent of married whites aged 20–34 in 2000 had nonwhite spouses. The interracial marriage rates are much higher for American-born racial minorities: 9 percent for African Americans, about 39 percent for Hispanics, 56 percent for American Indians, and 59 percent for Asian Americans (who account for less than 4 percent of the total population). Mathematically, one marriage between an Asian American and a white raises the intermarriage rate for Asian Americans much more than for whites, because whites are so much more numerous. Because of their numbers as well, although just 4 percent of whites are involved in interracial marriages, 92 percent of all interracial marriages include a white partner. About half of the remaining 8 percent are black-Hispanic couples. Racial minorities have more opportunities to meet whites in schools, workplaces, and neighborhoods than to meet members of other minority groups.

Some interracial couples contemplating marriage avoid family complications by just living together. In 2000, 4 percent of married white women had nonwhite husbands, but 9 percent of white women who were cohabiting had non-white partners (see figure 1). Similarly, 13 percent of married black men had nonblack spouses, but 24 percent of cohabiting black men lived with nonblack partners. Hispanics and Asian Americans showed the same tendency; only American Indians showed the opposite pattern. Black-white combinations are particularly notable. Black-white pairings accounted for 26 percent of all cohabiting couples but only 14 percent of all interracial marriages. They are more likely to cohabit than other minority-white couples, but they are also less likely to marry. The long history of the ban on interracial marriage in the United States, especially black-white marriage, apparently still affects black-white relationships today.

Given differences in population size, comparing rates of inter-marriage across groups can be difficult. Nevertheless, statistical models used by social scientists can account for group size, determine whether members of any group are marrying out more or less often than one would expect given their numbers, and then discover what else affects intermarriage. Results show that the lighter the skin color, the higher the rate of intermarriage with white Americans. Hispanics who label themselves as racially "white" are most likely to marry non-Hispanic whites. Asian Americans and American Indians are next in their levels of marriage with whites. Hispanics who do not consider themselves racially white have low rates of intermarriage with whites. African Americans are least likely of all racial minorities to marry whites. Darker skin in America is associated with discrimination, lower educational attainment, lower income, and segregation. Even among African Americans, those of lighter tone tend to experience less discrimination.

## RACE AND EDUCATION

Most married couples have similar levels of education, which typically indicates that they are also somewhat similar in social position, background, and values. Most interracial couples also have relatively equal educational attainments.

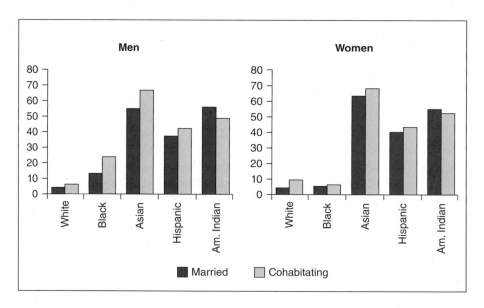

**FIGURE 1** Interracial marriage remains rare among white and black Americans. This figure shows the percentage of Americans in couples married to (dark bars) or cohabiting with (light bars) someone of a different race.

SOURCE: 2000 Census

However, when interracial couples do differ in their education, a hierarchy of color is apparent. The darker the skin color of racial minorities, the more likely they are to have married whites "below" them, that is, with less education than themselves. Six of ten African Americans who marry whites with different levels of education marry whites less educated than themselves. Hispanics also tend to marry whites less educated than themselves, but Asian Americans marry whites at about the same educational level.

Highly educated minority members often attend integrated colleges, and their workplaces and neighborhoods are integrated. Although they often develop a strong sense of their group identity in such environments, they also find substantial opportunities for interracial contact, friendship, romance, and marriage. College-educated men and women are more likely to marry interracially than those with less education. The fact that Asian Americans attend college at unusually high rates helps explain their high level of intermarriage with whites. The major exceptions to the interracial influence of higher education are African Americans.

Although middle-class African Americans increasingly live in integrated neighborhoods, they are still much more segregated than other minorities. Well-educated African Americans are less likely to live next to whites than are well-educated Hispanics and Asian Americans. One reason is that middle-class, black Americans are so numerous that they can form their own middle-class black neighborhoods, while middle-class Hispanic and Asian-American communities are smaller and often fractured by differences in national origin and language. In addition, studies show clearly that whites resist having black neighbors much more than they resist having Hispanic or Asian American neighbors....

Residential and school segregation on top of a long and relentless history of racial discrimination and inequality reduce African Americans' opportunities for interracial contact and marriage. The geographic distance between blacks and whites is in many ways rooted in the historical separation between the two groups. In contrast, the distance of Hispanics and Asian Americans from whites has more to do with their current economic circumstances; as those improve, they come nearer to whites geographically, socially, and matrimonially.

## A MAN AND A WOMAN

Black–white couples show a definite pattern: 74 percent involve a black husband and a white wife. Asian American–white couples lean the other way; 58 percent involve an Asian-American wife. Sex balances are roughly even for couples that include a white and a Hispanic (53 percent involve a Hispanic husband) or a white and an American Indian (49 percent involve Indian husbands).

I mentioned before that most black–white couples have similar educations; nonetheless, white women who marry black men "marry up" more often than those who marry white men. This is especially striking because the pool of highly educated white men greatly outnumbers the pool of highly educated black men. More than half of black husbands of white women have at least some college education, but only two-fifths of black husbands of black women do. In that sense, white wives get more than their "share" of well-educated black husbands. This further reduces the chances that black women, especially highly educated black women, will marry, because they often face shortages of marriageable men. African-American women often resent this. Interviewed by Maria Root, one black man in such a relationship reported being accused of "selling out" and "dis-sing his black sisters."

Half a century ago, Robert Merton proposed a "status exchange" theory to explain the high proportion of marriages between black men and white women. He suggested that men with high economic or professional status who carry the stigma of being black in a racial caste society "trade" their social position for whiteness by marriage. On the other hand, some social scientists argue that racialized sexual images also encourage marriages between white women and black men. Throughout Europe and the West, people have long seen fair skin tone as a desirable feminine characteristic, and African Americans share those perceptions. For example, Mark Hill found that black interviewers participating in a national survey of African Americans rated black women interviewees with lighter skin as more attractive than those with darker skin. But they did not consider male interviewees with light skin any more attractive than darker-skinned men.

Asian Americans show a different pattern; in most of their marriages with whites, the husband is white. Although Asian-American men are typically more educated than white men, in the mixed couples, white husbands usually have more education than their Asian-American wives. As with white wives of black men, the wives have "married up" educationally. Some speculate that Asian-American women tend to marry white men because they perceive Asian-American men to be rigidly traditional on sex roles and white men as more nurturing and expressive. The emphasis in Asian cultures on the male line of descent may pressure Asian-American men to carry on the lineage by marrying "one of their own." But what attracts white men to Asian-American women? Some scholars suggest that it is the widespread image of Asian women as submissive and hyper-feminine (the "Madame Butterfly" icon).

## THE FUTURE OF INTERRACIAL MARRIAGE

Rates of interracial marriage in the future will respond to some conflicting forces: the weakening of barriers between groups; increasing numbers of Hispanics and Asians in the nation; and possible rising ethnic consciousness. The continued progress of racial minorities in residential integration and economic achievement promotes contact between members of different races as equals. The color line, however, probably will not disappear. Marriage between African Americans and whites is likely to remain rare. Stubborn economic differences may be part of the reason for the persistence of this barrier, but cultural experiences also play a role.

In recent years, the middle-class African-American population has grown, yet the persistence of residential segregation reduces the opportunities for contact between blacks and whites. African Americans also maintain a strong racial identity compared to that of other minorities. In the 1990 census, for example, less than 25 percent of children born to a black-white couple were identified by their parents as white—a much lower percentage than for other biracial children. In the 2000 census, blacks identified themselves or their children as multiracial much less often than did other racial minorities. The stronger racial identities of African Americans, forged by persistent inequality, discrimination, and residential isolation, along with continued white resistance, will hold down the increase in marriages across the black-white divide.

Increases in the relatively high marriage rates of Hispanics and Asian Americans with whites may slow as new immigrants keep arriving from their homelands. Immigration expands the marriage pools for the native-born, who are more able to find spouses in their own racial or ethnic groups. These pools are expanded further by the way the wider society categorizes Hispanics and Asian Americans. They distinguish among themselves by national origin (Cuban versus Mexican or Thai versus Chinese), but whites tend to lump them into two large groups. Common experiences of being identified as the same, along with anti-Latino and anti-Asian prejudice and discrimination, help create a sense of pan-ethnic identity. This in turn inhibits marriage with whites, fosters solidarity within the larger group, and increases marriage rates between varieties of Hispanics and Asian Americans. Interethnic marriage is frequent among American-born Asians despite small group sizes and limited opportunities for contact. For example, in 1990, 18 percent of Chinese-Americans and 15 percent of Japanese-Americans aged 20–34 married spouses of other Asian ethnic groups (compared to 39 percent and 47 percent who married whites).

Many people view the increasing number of interracial marriages as a sign that racial taboos are crumbling and that the distances between racial groups in American society are shrinking. However, marriages across racial boundaries remain rarer than those that cross religious, educational, or age lines. The puzzle is whether interracial marriages will develop as marriages between people of different nationalities did among European immigrants and their descendants in the early 20th century. Diverse in many ways when they entered the country, these 20th-century European Americans, such as Italians, Poles, and Greeks, reached the economic level of earlier immigrants within a couple generations. Their success blurred ethnic boundaries and increased the rate of interethnic marriage. Many of their descendants now define themselves simply as white despite their diverse national origins. For most white Americans, ethnic identities have become largely symbolic.

Similar trends for interracial marriages are unlikely in the near future. The experiences of European Americans show the importance of equal economic achievement in dissolving barriers, so what happens economically to recent immigrants and African Americans will be important. Even then, the low levels of interracial marriage for middle-class African Americans suggest that this particular color line will persist as a barrier to marriage. And the continuing influx of Asian and Latino immigrants may reinforce those groups' barriers to intermarriage.

## DISCUSSION QUESTIONS

1.  How has acceptance of interracial dating and marriage changed over time? What obstacles remain?

2.  What social factors are related to people's acceptance of interracial dating and marriage? Have you seen evidence of this in your own experience?

# Student Exercises

1.  Families are central to raising young people and helping them move into adulthood. Consequently, the lack of a family creates major challenges. Children in foster care are wards of the state, with their foster families ending their commitments at age eighteen, when most foster-care children age out. Each year about 25,000 young people age out of the system. Put yourself in the shoes of a young person aging out of the system. You have been with your foster care family for two years. They have been nice, but you are turning eighteen and have to make your own way. Your foster care family can no longer offer economic support or even housing, because they have to begin to care for an aging parent. You need a job and also housing. How do you make this critical transition? As a high school graduate, how are you able to find a job and a place to live, and begin to develop a way to survive and also take some steps towards a more stable future? You can use this opportunity to see what services your community and state offer to young people coming out of foster care. How does your research into the prospect of being in someone else's shoes help you think about the support and resources that your own family provides for you?

2.  Ideas about relationships outside of one's own race and ethnicity are changing. We are also seeing a growing number of people who are multiracial or multi-ethnic. What is happening in your own family? What are your attitudes about being friends with someone from another race and ethnicity group? What about dating or perhaps marrying someone from another race and ethnicity group? How do people in your friendship circle think? What about your parents' expectations and possible acceptance? Families have many different reasons for wanting their children to marry within their racial or ethnic group, while others are more readily supportive of their children having relationships outside of their groups. What are the dynamics in your own family? Can you relate your family's position to historical or contemporary trends, such as patterns of segregation and integration, as well as to your family's history?

# Section X

# How We Live and Learn:
## Segregation, Housing, and Education
## Elizabeth Higginbotham
## and Margaret L. Andersen

In its 1954 *Brown v. Board of Education of Topeka, Kansas* decision, the Supreme Court declared segregation unconstitutional. The *Brown* decision overturned an 1898 Supreme Court decision, *Plessy v. Ferguson,* that had allowed segregation in public facilities based on the principle of "separate but equal." But, of course, public facilities—schools, transportation, hospitals, and so forth—were nowhere near equal. Yet, until *Brown,* segregation was legally sanctioned, that is, **de jure segregation** (meaning segregation by law). Under de jure segregation (known less formally as Jim Crow segregation), the American South, in particular, maintained strictly segregated lives for people based on race. In the North, people were also socially and spatially separated, but rather than the separation being legally enforced, **de facto segregation** (that is, segregation "in fact," if not in law) was the practice. Both systems caused harm through the unequal facilities that people experienced and through the separation in relationships that segregation produced.

Since the *Brown* decision, the nation has slowly recognized that de jure segregation is wrong, but have we desegregated society—its housing, schools, health care systems, and other institutions? Hardly. People of different racial and ethnic backgrounds are still socially and spatially separated from each other to large degrees—and, in some cases, to very high degrees, such as in our nation's schools, as we see in the readings in this section. And although "separate but equal" is unconstitutional, some of the basic facts of how we live and what opportunities are available to us are highly unequal.

Inequality in housing and education is also fundamental to other life opportunities—whether your home accumulates in value over time (thereby affording you the opportunity to make further investments, such as in the education of your children or the improvement of your property—or indeed the acquisition of more property). Especially in a society so dependent on technological innovation and skilled workers, not providing a quality education to a large portion of the citizenry can be a recipe for international disaster. More and more, as we saw in the Section VIII, the economy requires skilled, educated workers to keep the nation economically healthy. Without our own population receiving the education needed, the United States relies more on workers from other nations to fuel the nation's economic health. Meanwhile, millions of our own citizens are left without opportunities for advancement, sometimes leading to the only option they may have: crime (as we see in Section XII).

Segregation in housing and education is thus lethal to the nation's health, and segregation means that people's daily lives are structured very differently. These differences are linked to opportunity structures as well as to a group's or person's abilities to be safe, to provide for their families, to advance educationally, and to enjoy the fruits of society.

Segregation also shapes the relationships that people have—or do not have. Among other things, as we have seen in the section on racial representations, segregation can distort people's ideas about each other and make them more susceptible to accepting racial stereotypes. This has consequences for both White people and people of color. With racial segregation, there is also less sharing of culture, histories, ideas, and caring across racial lines. Even friendships are affected by segregation. A team of sociologists studied cross-race friendships in several high schools around the nation and found that cross-race friendships increase when school populations are more diverse (Quillian and Campbell 2003). And, increasingly, there is a large body of scholarly research indicating that having more diverse working groups—that is, creating work and educational spaces where people are not segregated—means a number of positive outcomes: more learning by everyone (Gurin et al. 2004), more innovation in ideas (Page 2007), even more profit in companies with a more diverse labor force (Herring 2009).

In this section, we explore two areas where segregation remains a huge problem: housing and education. Over forty years ago, in 1968, the U.S. Congress passed fair housing legislation to address discrimination in housing. Yet, **residential segregation** still prevails, isolating people from one another, but also meaning that the value of people's homes varies *because of* patterns of racial segregation. John E. Farley and Gregory D. Squires ("Fences and Neighbors: Segregation in the 21st Century") detail the patterns of residential segregation that characterize our cities and surrounding areas. Moreover, **hypersegregation**—that is, the concentration of people of color in poor, urban areas—clusters poverty in certain communities, thereby shielding those in more advantaged areas from understanding the current realities of race and class.

Melvin L. Oliver and Thomas M. Shapiro ("Sub-Prime as a Black Catastrophe") show the consequences of housing segregation for what the nation experienced during the recent economic recession. Although the recession affected many, if not most, people in society, as Oliver and Shapiro show its effects were disastrous for African Americans. One major basis for the crisis was the sub-prime mortgage system. A sub-prime mortgage is one in which the rate of interest is higher than the "prime" rate; sub-prime rates are charged when lending institutions think there is a greater economic risk—or simply when people are being exploited. Those with sub-prime loans during the housing crisis in 2010 were at far greater risk of housing foreclosure —that is, having their homes repossessed by lending institutions. As shown by Oliver and Shapiro, African Americans—even controlling for their economic status—had far higher rates of sub-prime loans and thus were far more likely to lose homes during the recession. Oliver and Shapiro refer to this as a "segregation tax" because some of the assessment of risk comes from practices of housing segregation. As they further point out, having a less valuable home (or losing one) seriously affects one's ability to accumulate value over time and thus finance other things (such as one's children's education) or to pass financial assets on to the next generation. All in all, not having the benefit of a home that maintains and increases its value over time has repercussions not just for individual families, but in fact for whole communities.

Residential segregation is also played out in the nation's school system. School segregation follows from residential segregation. Moreover, schools are heavily funded through local property taxes. The devaluation of real estate in a given area means fewer resources for schools, thus creating a vicious cycle where children cannot get ahead, families fall behind, and society suffers.

The nation has tried to deal with educational inequality through desegregation of the schools and educational reform. However, now schools are becoming increasingly segregated. The court-ordered desegregation plans that followed the *Brown vs. Board of Education* Supreme Court decision are being abandoned, particularly as the nation's courts have become more conservative and less willing to allow race-specific policies for school desegregation. This trend is well documented in the important work being done at the UCLA Civil Rights Project—a research center that is providing some of the best studies of ongoing school segregation. As part of this work, Gary Orfield and Chungmei Lee's article here ("Historic Reversals, Accelerating Resegregation, and the Need for New Integration Strategies") reviews the legal history of desegregation efforts, but even more particularly, documents how recent court decisions are reversing this trend. Their research shows an extensive retreat from national efforts to desegregate schooling, such that now the vast majority of Black and Latino students in public schools attend overwhelmingly minority schools, while White students attend segregated schools wherein they do not interact much with students of color. Not only do "majority-minority" schools receive fewer resources, but they and the segregated, predominantly White schools foster isolation of students from one another.

Segregation is also supported by the actions of private individuals whose choices about residential and educational options reflect their own racial privileges. Heather Beth Johnson and Thomas M. Shapiro ("Good Neighborhoods, Good Schools: Race and the 'Good Choices' of White Families") show this pattern in their research. They find that White parents' decisions about schooling for their children are founded in racial matters, even though the parents do not think of their decisions this way—at least not overtly. This provides a good example of how **color-blind racism** reproduces unequal relationships of race. As long as schools attended by predominantly racial minorities are so unequal, it is unlikely that the nation will achieve the dream established by the *Brown* decision.

## REFERENCES

Gurin, Patricia, A. Nagda Biren, and Gretchen E. Lopez. 2004. "The Benefits of Diversity in Education for Democratic Citizenship." *Journal of Social Issues* 60(1): 17–34.

Herring, Cedric. 2009. "Does Diversity Pay? Race, Gender, and the Business Case for Diversity." *American Sociological Review* 74(2): 208–224.

Page, Scott. 2007. *The Difference: How the Power of Diversity Creates Better Groups, Firms, Schools, and Societies.* Princeton, NJ: Princeton University Press.

Quillian, Lincoln and Mary E. Campbell. 2003. "Beyond Black and White: The Present and Future of Multiracial Friendship Segregation." *American Sociological Review* 68(4): 540–566.

## FACE THE FACTS: WHO IS SEGREGATED IN SCHOOLS?

**Racial Composition of Schools Attended by the Average Student of Each Race, 2005-2006**

| Percent Race in Each School | White Student | Black Student | Latino Student | Asian Student | American Indian Student |
|---|---|---|---|---|---|
| % White | 77 | 30 | 27 | 44 | 44 |
| % Black | 9 | 52 | 12 | 12 | 7 |
| % Latino | 9 | 14 | 55 | 21 | 12 |
| % Asian | 4 | 3 | 5 | 23 | 3 |
| % American Indian | 1 | 1 | 1 | 1 | 35 |

SOURCE: Orfield, Gary, and Chungmei Lee. 2007. *Historic Reversals, Accelerating Resegregation and the Need for New Integration Strategies.* Los Angeles, CA: UCLA Civil Rights Project.

*Think about it*: School segregation is increasing in the United States, as shown in one of the readings in this section. But look at which groups are most likely to be segregated, as this table details. What does this suggest to you about how different groups are affected by race relations and the segregation of the schools?

# 41

# Fences and Neighbors:

## Segregation in 21st-Century America

JOHN E. FARLEY AND GREGORY D. SQUIRES

*Residential segregation is key to understanding a host of other dimensions of race relations, including access to schooling, as well as the formation of friendships. Farley and Squires review current trends in residential segregation and discuss the fair-housing movement.*

"Do the kids in the neighborhood play hockey or basketball?"
—ANONYMOUS HOME INSURANCE AGENT, 2000

America became less racially segregated during the last three decades of the 20th century, according to the 2000 census. Yet, despite this progress, despite the Fair Housing Act, signed 35 years ago, and despite popular impressions to the contrary, racial minorities still routinely encounter discrimination in their efforts to rent, buy, finance, or insure a home. The U.S. Department of Housing and Urban Development (HUD) estimates that more than 2 million incidents of unlawful discrimination occur each year. Research indicates that blacks and Hispanics encounter discrimination in one out of every five contacts with a real estate or rental agent. African Americans, in particular, continue to live in segregated neighborhoods in exceptionally high numbers.

What is new is that fair-housing and community-development groups are successfully using antidiscrimination laws to mount a movement for fair and equal access to housing. Discrimination is less common than just ten years ago; minorities are moving into the suburbs, and overall levels of segregation have gone down. Yet resistance to fair housing and racial integration persists and occurs today in forms that are more subtle and harder to detect. Still, emerging coalitions using new tools are shattering many traditional barriers to equal opportunity in urban housing markets.

*Segregation* refers to the residential separation of racial and ethnic groups in different neighborhoods within metropolitan areas. When a metropolitan area is highly segregated, people tend to live in neighborhoods with others of their own

SOURCE: From *Contexts* 4(1): 33–39, 2005. Reprinted by permission of University of California Press.

group, away from different groups. The index of dissimilarity (D), a measure of segregation between any two groups, ranges from 0 for perfect integration to 100 for total segregation. For segregation between whites and blacks (imagining, for the sake of the example, that these were the only two groups), a D of 0 indicates that the racial composition of each neighborhood in that metropolitan area is the same as that of the entire area. If the metropolitan area were 70 percent white and 30 percent black, each neighborhood would reflect those percentages. A D of 100 would indicate that every neighborhood in the metropolitan area was either 100 percent white or 100 percent black. In real metropolitan areas, D always falls somewhere between those extremes. For example, the Chicago metropolitan area is 58 percent non-Hispanic white and 19 percent non-Hispanic black. Chicago's D was 80.8 in 2000. This means that 81 percent of the white or black population would have to move to another census tract in order to have a D of 0, or complete integration. On the other hand, in 2000 the Raleigh-Durham, N.C. metropolitan area, which is 67 percent non-Hispanic white and 23 percent non-Hispanic black, had a D of 46.2—a little more than half that of Chicago.

## SEGREGATION: DECLINING BUT NOT DISAPPEARING

Although segregation has declined in recent years, it persists at high levels, and for some minority groups it has actually increased. Social scientists use a variety of measures to indicate how segregated two groups are from each other....

Although African Americans have long been and continue to be the most segregated group, they are notably more likely to live in integrated neighborhoods than they were a generation ago. For the past three decades, the average level of segregation between African Americans and whites has been falling, declining by about ten points on the D scale between 1970 and 1980 and another ten between 1980 and 2000. But these figures overstate the extent to which blacks have been integrated into white or racially mixed neighborhoods. Part of the statistical trend simply has to do with how the census counts "metropolitan areas." Between 1970 and 2000, many small—and typically integrated—areas "graduated" into the metropolitan category, which helped to bring down the national statistics on segregation. More significantly, segregation has declined most rapidly in the southern and western parts of the United States, but cities in these areas, especially the West, also tend to have fewer African Americans. At the same time, in large northern areas with many African-American residents, integration has progressed slowly. For example, metropolitan areas like New York, Chicago, Detroit, Milwaukee, Newark, and Gary all had segregation scores in the 80s as late as 2000. Where African Americans are concentrated most heavily, segregation scores have declined the least.... In places with the highest proportions of black population, segregation decreased least between 1980 and 2000. Desegregation has been slowest precisely in the places African Americans are most likely to live. There, racial isolation can be extreme. For example, in the Chicago, Detroit, and Cleveland metropolitan areas, most African Americans live in census tracts (roughly, neighborhoods) where more than 90 percent of the residents are black and fewer than 6 percent are white.

Other minority groups, notably Hispanics and Asian Americans, generally live in less segregated neighborhoods. Segregation scores for Hispanics have generally been in the low 50s over the past three decades, and for Asian Americans and Pacific Islanders, scores have been in the low 40s. Native Americans who live in urban areas also are not very segregated from whites (scores in the 30s), but two-thirds of the Native Americans who live in rural areas (about 40 percent of their total population) live on segregated reservations. Although no other minority group faces the extreme segregation in housing that African Americans do, other groups face segregation of varying levels and have not seen a significant downward trend.

## CAUSES OF CONTINUING SEGREGATION

Popular explanations for segregation point to income differences and to people's preferences for living among their "own kind." These are, at best, limited explanations. Black-white segregation clearly cannot be explained by differences in income, education, or employment alone. Researchers have found that white and black households at all levels of income, education, and occupational status are nearly as segregated as are whites and blacks overall. However, this is not the case for other minority groups. Hispanics with higher incomes live in more integrated communities than Hispanics with lower incomes. Middle-class Asian Americans are more suburbanized and less segregated than middle-class African Americans. For example, as Chinese Americans became more upwardly mobile, they moved away from the Chinatowns where so many had once lived. But middle-class blacks, who have made similar gains in income and prestige, find it much more difficult to buy homes in integrated neighborhoods. For example, in 2000 in the New York metropolitan area, African Americans with incomes averaging above $60,000 lived in neighborhoods that were about 57 percent black and less than 15 percent non-Hispanic white—a difference of only about 6 percentage points from the average for low-income blacks.

Preferences, especially those of whites, provide some explanation for these patterns. Several surveys have asked whites, African Americans, and in some cases Hispanics and Asian Americans about their preferences concerning the racial mix of their neighborhoods. A common technique is to show survey respondents cards displaying sketches of houses that are colored-in to represent neighborhoods of varying degrees of integration. Interviewers then ask the respondents how willing they would be to live in the different sorts of neighborhoods. These surveys show, quite consistently, that the first choice of most African Americans is a neighborhood with about an equal mix of black and white households. The first choice of whites, on the other hand, is a neighborhood with a large white majority. Among all racial and ethnic groups, African Americans are the most disfavored "other" with regard to preferences for neighborhood racial and ethnic composition. Survey research also shows that whites are more hesitant to move into hypothetical neighborhoods with large African-American populations, even if those communities are described as having good schools, low crime rates, and other amenities. However, they are much less hesitant about moving into areas with significant Latino, Asian, or other minority populations.

Why whites prefer homogeneous neighborhoods is the subject of some debate. According to some research, many whites automatically assume that neighborhoods with many blacks have poor schools, much crime, and few stores; these whites are not necessarily responding to the presence of blacks per se. Black neighborhoods are simply assumed to be "bad neighborhoods" and are avoided as a result. Other research indicates that "poor schools" and "crime" are sometimes code words for racial prejudice and excuses that whites use to avoid African Americans.

These preferences promote segregation. Recent research in several cities, including Atlanta, Detroit, and Los Angeles, shows that whites who prefer predominantly white neighborhoods tend to live in such neighborhoods, clearly implying that if white preferences would change, integration would increase. Such attitudes also imply tolerance, if not encouragement, of discriminatory practices on the part of real estate agents, mortgage lenders, property insurers, and other providers of housing services.

## HOUSING DISCRIMINATION: HOW COMMON IS IT TODAY?

When the insurance agent quoted at the beginning of this article was asked by one of his supervisors whether the kids in the neighborhood played hockey or basketball, he was not denying a home insurance policy to a particular black family because of race. However, he was trying to learn about the racial composition of the neighborhood in order to help market his policies. The mental map he was drawing is just as effective in discriminating as the maps commonly used in the past that literally had red lines marking neighborhoods—typically minority or poor—considered ineligible for home insurance or mortgage loans.

Researchers with HUD, the Urban Institute, and dozens of nonprofit fair housing organizations have long used "paired testing" to measure the pervasiveness of housing discrimination—and more recently in mortgage lending and home insurance. In a paired test, two people visit or contact a real estate, rental, home-finance, or insurance office. Testers provide agents with identical housing preferences and relevant financial data (income, savings, credit history). The only difference between the testers is their race or ethnicity. The testers make identical applications and report back on the responses they get.... Discrimination can take several forms: having to wait longer than whites for a meeting; being told about fewer units or otherwise being given less information; being steered to neighborhoods where residents are disproportionately of the applicant's race or ethnicity; facing higher deposit or down-payment requirements and other costs;

or simply being told that a unit, loan, or policy is not available, when it is available to the white tester.

In 1989 and 2000, HUD and the Urban Institute, a research organization, conducted nationwide paired testing of discrimination in housing. They found generally less discrimination against African Americans and Hispanics in 2000 than in 1989, except for Hispanic renters (see figure 1). Nevertheless, discrimination still occurred during 17 to 26 percent of the occasions when African Americans and Hispanics visited a rental office or real-estate agent. (The researchers found similar levels of discrimination against Asians and Native Americans in 2000; these groups were not studied in 1989.)

In 2000, subtler forms of discrimination, such as invidious comments by real estate agents, remained widespread. Even when whites and nonwhites were shown houses in the same areas, agents often steered white homeseekers to segregated neighborhoods with remarks such as "Black people do live around here, but it has not gotten bad yet;" "That area is full of Hispanics and blacks that don't know how to keep clean;" or "(This area) is very mixed. You probably wouldn't like it because of the income you and your husband make. I don't want to sound prejudiced."

Given the potential sanctions available under current law, including six- and seven-figure compensatory and punitive damage awards for victims, it seems surprising that an agent would choose to make such comments. However, research shows that most Americans are unfamiliar with fair housing rules, and even those

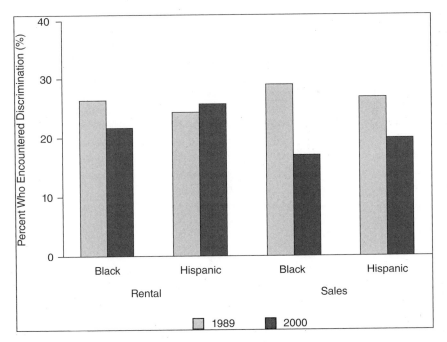

**FIGURE 1**  Percent of auditors who encountered discrimination, 1989 and 2000.

SOURCE: U.S. Department of Housing and Urban Development, 2000 Housing Discrimination Study.

who are familiar and believe they have experienced racial discrimination rarely take legal action because they do not believe anything would come of it. Most real estate professionals do comply with fair housing laws, but those who work in small neighborhoods and rely on word of mouth to get clients often fear losing business if they allow minorities into a neighborhood where local residents would not welcome them. In a 2004 study of a St. Louis suburb, a rental agent pointed out that there were no "dark" people in the neighborhood to a white tester. She said that she had had to lie to a black homeseeker and say that a unit was unavailable because she would have been "run out of" the suburb had she rented to a black family.

Discrimination does not end with the housing search. Case studies of mortgage lending and property insurance practices have also revealed discriminatory treatment against minorities. White borrowers are offered more choice in loan products, higher loan amounts, and more advice than minority borrowers. The Boston Federal Reserve Bank found that even among equally qualified borrowers in its region applications from African Americans were 60 percent more likely to be rejected than those submitted by whites. Other paired-testing studies from around the country conclude that whites are more likely to be offered home insurance policies, offered lower prices and more coverage, and given more assistance than African Americans or Hispanics.

## THE CONTINUING COSTS OF SEGREGATION

Beyond constricting their freedom of choice, segregation deprives minority families of access to quality schools, jobs, health care, public services, and private amenities such as restaurants, theatres, and quality retail stores. Residential segregation also undercuts families' efforts to accumulate wealth through the appreciation of real estate values by restricting their ability both to purchase their own homes and to sell their homes to the largest and wealthiest group in the population, non-Hispanic whites. Just 46 percent of African Americans owned their own homes in 2000, compared to 72 percent of non-Hispanic whites. In addition, recent research found that the average value of single-family homes in predominantly white neighborhoods in the 100 largest metropolitan areas with significant minority populations was $196,000 compared to $184,000 in integrated communities and $104,000 in predominantly minority communities. As a result of the differing home values and appreciation, the typical white homeowner has $58,000 in home equity compared to $18,000 for the typical black homeowner. Segregation has broader effects on the quality of neighborhoods to which minorities can gain access. In 2000, the average white household with an income above $60,000 had neighbors in that same income bracket. But black and Hispanic households with incomes above $60,000 had neighbors with an average income of under $50,000. In effect, they lived in poorer neighborhoods, and this gap has widened since 1990.

Segregation restricts access to jobs and to quality schools by concentrating African Americans and Hispanics in central cities, when job growth and better schools are found in the suburbs. Amy Stuart Wells and Robert Crain found, for example, that black children living in St. Louis who attend schools in the suburbs are more likely to graduate and to go on to college than those attending city schools. Yet only busing makes it possible for these students to attend suburban schools, and America has largely turned away from this remedy to segregation. According to research by Gary Orfield at the Harvard Civil Rights Project, our nation's schools are as segregated today as they were 35 years ago. Most job growth also occurs in suburban areas, and difficulty in finding and commuting to those jobs contributes to high unemployment rates among African Americans and Latinos.

The risks of illness and injury from infectious diseases and environmental hazards are also greater in minority neighborhoods, while resources to deal with them are less available than in mostly white areas. For example, in Bethesda, Md., a wealthy and predominantly white suburb of Washington, D.C., there is one pediatrician for every 400 residents compared to one for every 3,700 residents in Washington's predominantly minority and poor southeast neighborhoods. As John Logan has argued, "The housing market and discrimination sort people into different neighborhoods, which in turn shape residents' lives—and deaths. Bluntly put, some neighborhoods are likely to kill you."...

Finally, segregation helps perpetuate prejudice, stereotypes, and racial tension. Several recent studies show that neighborhood-level contact between whites and African Americans reduces prejudice and increases acceptance of diversity. Yet with today's levels of housing segregation, few whites and blacks get the opportunity for such contact. More diverse communities generally exhibit greater tolerance and a richer lifestyle—culturally and economically—for all residents.

## A GROWING MOVEMENT

In 1968, the U.S. Supreme Court ruled that racial discrimination in housing was illegal, characterizing it as "a relic of slavery." In the same year, Congress passed the Fair Housing Act, providing specific penalties for housing discrimination along with mechanisms for addressing individual complaints of discrimination. These legal developments laid the groundwork for a growing social movement against segregation that has brought limited but real gains. Members of the National Fair Housing Alliance, a consortium of 80 nonprofit fair housing organizations in 30 cities and the District of Columbia, have secured more than $190 million for victims of housing discrimination since 1990 by using the Federal Fair Housing Act and equivalent state and local laws. In addition, they have negotiated legal settlements that have transformed the marketing and underwriting activities of the nation's largest property insurance companies, including State Farm, Allstate, Nationwide, American Family, and Liberty. The key investigative technique members of the alliance have used to secure these victories is paired testing.

Community reinvestment groups have secured more than $1.7 trillion in new mortgage and small business loans for traditionally underserved low- and moderate-income neighborhoods and minority markets since the passage of the Community Reinvestment Act (CRA). The CRA was passed in order to prevent lenders from refusing to make loans, or making loans more difficult to get, in older urban communities, neighborhoods where racial minorities are often concentrated. Under the CRA, third parties (usually community-based organizations) can formally challenge lender applications or requests by lenders to make changes in their business operations. Regulators who are authorized to approve lender applications have, in some cases, required the lender to respond to the concerns raised by the challenging party prior to approving the request. In some cases, just the threat of making such challenges has provided leverage for community organizations in their efforts to negotiate reinvestment agreements with lenders. Community groups have used this process to generate billions of new dollars for lending in low-income and minority markets. Sometimes, in anticipation of such a challenge, lenders negotiate a reinvestment program in advance. For example, shortly after Bank One and JP Morgan Chase announced their intent to merge in 2004, the lenders entered into an agreement with the Chicago Reinvestment Alliance, a coalition of Chicago-area neighborhood organizations. The banks agreed to invest an average of $80 million in community development loans for each of the next six years. Research by the Joint Center for Housing Studies at Harvard University indicates that mortgage loans became far more accessible in low-income and minority neighborhoods during the 1990s and that the CRA directly contributed to this outcome.

Many housing researchers and fair-housing advocates have criticized fair-housing enforcement authorities for relying too heavily on individual complaints and lawsuits to attack what are deeper structural problems. Currently, most testing and enforcement occurs when individuals lodge a complaint against a business rather than as a strategic effort to target large companies that regularly practice discrimination. Reinvestment agreements recently negotiated by community groups and lenders illustrate one more systemic approach. More testing aimed at detecting what are referred to as "patterns and practices" of discrimination by large developers and rental management companies would also be helpful. Such an undertaking, however, would require more resources, which are currently unavailable. Despite the limits of current enforcement efforts, most observers credit these efforts with helping to reduce segregation and discrimination.

Resistance to fair housing and integration efforts persists. For example, lenders and their trade associations continually attempt to weaken the CRA and related fair-housing rules. Yet fair-housing and community-reinvestment groups like the National Fair Housing Alliance and the National Community Reinvestment Coalition have successfully blocked most such efforts in Congress and among bank regulators. As more groups refine their ability to employ legal tools like the CRA and to litigate complex cases under the jurisdiction of the Fair Housing Act, we can expect further progress. The struggle for fair housing is a difficult one, but with the available tools, the progress we have made since 1970 toward becoming a more integrated society should continue.

## DISCUSSION QUESTIONS

1. Look at the neighborhood where you grew up. Would you describe it as racially segregated or racially integrated? What effect did this have on the kinds of relationships you formed with people as you grew up?

2. What if someone calculated the index of dissimilarity on your campus? In a general sense, what figure would result? Were you in charge of campus life, what might you do if you think this needs to change?

# 42

# Sub-Prime as a Black Catastrophe

MELVIN L. OLIVER AND THOMAS M. SHAPIRO

*Home ownership is the primary basis for even modest wealth for anyone, regardless of race. Here, Oliver and Shapiro document the discrimination against African Americans in the housing market and show how this affects racial differences in wealth. This pattern of racial difference in wealth has meant that African Americans were particularly hard hit by the recent economic recession.*

No other recent economic crisis better illustrates the saying "when America catches a cold, African Americans get pneumonia" than the sub-prime mortgage meltdown. African Americans, along with other minorities and low-income populations, have been the targets of the sub-prime mortgage system. Blacks received a disproportionate share of these loans, leading to a "stripping" of their hard-won home-equity gains of the recent past and the near future. To fully understand how this has happened, we need to place this in the context of the continuing racial-wealth gap, the importance of home equity in the wealth portfolios of African Americans, and its intersection with the new financial markets of which sub-prime is but one manifestation.

Family financial assets play a key role in poverty reduction, social mobility, and securing middle-class status. Income helps families get along, but assets help them get and stay ahead. Those without the head start of family assets have a much steeper climb out of poverty. This generation of African Americans is the first one afforded the legal, educational, and job opportunities to accumulate financial assets essential to launching social mobility and sustaining well-being throughout the life course.

Despite legal gains in civil rights, however, asset inequality in America has actually been growing rapidly during the last 20 years. The assets that current generations own are heavily dependent on the legacies of their families of origin. Today's blacks still suffer from the fact that their parents and grandparents grew up in a rigidly segregated America, where opportunities to accumulate human and financial capital were strictly limited. So housing wealth is a disproportionate share of total black wealth.

Despite some income gains, African Americans own only 7 cents for every dollar of net worth that white Americans own; for Hispanics the figure is only slightly higher at 9 cents for every dollar. Even when middle-class accomplishments

SOURCE: *The American Prospect*, September 22, 2008.

like income, job, and education are comparable, the racial-wealth gap is stuck stubbornly at about a quarter on the dollar.

## HOUSING WEALTH: LAST IN, FIRST OUT

Prior to the sub-prime meltdown, the advances of African Americans in accumulating wealth depended heavily upon housing wealth. Home equity is the most important reservoir of wealth for average American families and disproportionately so for African Americans. For black households, home equity accounts for 63 percent of total average net worth. In sharp contrast, home equity represents only 38.5 percent of average white net worth. According to the Economic Policy Institute's *State of Working America 2008/2009*, black homeownership rates dropped a full percentage point between 2005 and 2007—the largest decrease for any racial or ethnic group.

Even though homeownership rates for African Americans are lower, and even though a "segregation tax" is in play because homes appreciate far more slowly in minority or even integrated neighborhoods, housing wealth is still a more prominent engine of wealth for African American families. Given this centrality of homeownership as a source of wealth accumulation for black families, and the racialized dynamics of housing markets, sub-prime has been a special disaster for black upward mobility.

Families use home equity to finance retirement, start small businesses, pay for college educations, and tide themselves over during hard times. Information from the Federal Reserve Flow of Funds on home equity cashed out during refinancing (loans refinanced above 105 percent of the balance on the original loan) tells a dramatic story. Refinancing resulting in cash pullouts skyrocketed from $15 billion in 1995 to $327 billion in 2006. This is housing wealth converted, potentially, into other investments, used to launch social mobility, pay down credit-card and store debt, or finance consumption.

Especially in minority and immigrant communities, home equity often is the collateral for small business start-ups. But, it appears that about half of this enormous recent cash pullout was not used for retirement or to invest in mobility but to pay down past debt. While the conversion of housing wealth into cash and then the infusion of it into the economy provided a boost to the economy, in reality it disguised a less-rosy economic picture. Often it was the only way for families to keep pace with rising essential costs and burgeoning debt.

Between 2003 and 2007, the amount of housing wealth extracted more than doubled from the previous period, as families pulled out $1.19 trillion—an incredible sum that allowed families to adjust to shrinking purchasing power and that significantly boosted gross national product. So, while homeownership reached historic highs, families today actually own a lesser share of their homes than at any previous time, because they have borrowed against their housing wealth.

Families typically spend more as house values increase and they can borrow against their equity. Then, as prices fall and credit is tightened, they spend less. For a time, up until the sub-prime meltdown, equity withdrawals acted as an engine of growth on the economy. The opposite is true now—the sharp drop in housing prices has become a drag on the economy. Real home equity fell 6.5 percent to $9.6 trillion in 2007. The 2008 State of the Nations Housing study reports that the switch from housing appreciation to depreciation, plus the 2007 slowdown in home equity withdrawals, trimmed about one-half of a percentage point from real consumer spending and more than one-third of a percentage point from total economic growth. Worse is still to come.

## RACIAL TARGETING—AGAIN

Changes in the mortgage market, of which the current sub-prime meltdown is the most visible part of a larger pattern, were not racially neutral. Sub-prime loans were targeted at the African American community. With the recognition that average American families were accumulating trillions of dollars in housing wealth, "financial innovation" soon followed. New financial instruments, which relaxed (and sometimes ignored) rules and regulations, became the market's answer to broadening homeownership.

But the industry-promoted picture of sub-prime as an instrument of home-ownership opportunity for moderate income buyers is highly misleading. First, homeownership rates reached their historic highs before the zenith of sub-prime lending; and, second, increased access to credit brought homeownership opportunities within the reach of groups that had historically been denied access to credit. The issue became the terms of credit.

In hindsight, many critics now describe the sub-prime crisis as the consequence of bad loans to unqualified borrowers. In fact, the issue needs to be reframed to focus on the onerous terms of these loans. Data from the longest natural experiment in the field—the Community Advantage Program, a partnership of Self-Help, Fannie Mae, and the Ford Foundation, where tens of thousand of loans were made beginning over a decade ago—show that home loans to apparently riskier populations, like lower-income, minority, and single-headed households, do not default at significantly higher rates than conventional loans to middle-class families do, as long as they are not the handiwork of predators.

The difference is that loans such as ones made through the Community Advantage Program had terms that were closer to conventional mortgages as opposed to the risky terms that have characterized sub-prime mortgages. The latter had high hidden costs, exploding adjustable rates, and prepayment penalties to preclude refinancing. When lower-income families have similar terms of credit as conventional buyers, and they are linked with a community-based social and organizational infrastructure that helps them become ready for home-ownership, they pay similar interest rates and default at similar rates.

## PREDATORY LENDING AND WEALTH STRIPPING

Minority communities received a disproportionate share of sub-prime mortgages. As a result, they are suffering a disproportionate burden of the harm and losses.

According to a Dēmos report, *Beyond the Mortgage Meltdown* (June 2008), in addition to being the target of mortgage companies specializing in sub-prime lending, minorities were steered away from safer, conventional loans by brokers who received incentives for jacking up the interest rate. Worst of all, African Americans who qualified for conventional mortgages were steered to riskier, and more profitable, sub-prime loans.

Households of color were more than three times as likely as white households to end up with riskier loans with features like exploding adjustable rates, deceptive teaser rates, and balloon payments. Good credit scores often made no difference, as profit incentives trumped sound policy. The line from redlining to sub-prime is direct, as is the culpability. Even many upper-income African Americans were steered into sub-prime mortgages.

The Center for Responsible Lending (and other groups) projects that 2.2 million borrowers who bought homes between 1998 and 2006 will lose their houses and up to $164 billion of wealth in the process. African American and Latino homeowners are twice as likely to suffer sub-prime-related home foreclosures as white homeowners are. Foreclosures are projected to affect one in 10 African American borrowers. In contrast, only about one in 25 white mortgage holders will be affected. African Americans and Latinos are not only more likely to have been caught in the sub-prime loan trap; they are also far more dependent, as a rule, on their homes as financial resources.

The Dēmos report finds that home equity, at its current total value of $20 trillion, represents the biggest source of wealth for most Americans, and, as we have noted, it is even more important for African Americans. The comparatively little bit of wealth accumulation in the African American community is concentrated largely in housing wealth.

One recent estimate places the total loss of wealth among African American households at between $72 billion and $93 billion for sub-prime loans taken out during the past eight years.

Forty years after the Fair Housing Act of 1968, housing markets are still segmented by class and race, what realtors politely call location, location, location. Homes appreciate most in value when they are situated in predominantly white communities, and they appreciate least in value when situated in low-income minority or integrated communities, except when those communities undergo gentrification (and often become predominantly white).

This perverse market logic is also reflected in the sub-prime crisis. Sub-prime loans and foreclosures are not randomly distributed but spatially concentrated in low-to-moderate income communities, especially minority communities. Thus, the wealth-stripping phenomenon, of which sub-prime lending schemes are the latest financial innovation to tap new sources of wealth, is even more devastating in African American and minority communities. In turn, foreclosures and the terms of credit in African American neighborhoods bring down home values in the entire community. The community impact adds an institutional level to the personal tragedies and downstream consequences.

This devastating impact is not confined to just those who have suffered foreclosures; there is a spillover effect in addition to the direct hit of 1.27 million foreclosures. An additional 40.6 million neighboring homes will experience devaluation because of sub-prime foreclosures that lake place in their community.

The Center for Responsible Lending estimates that the total decline in house values and the tax base from nearby foreclosures will be $202 billion. The direct hit on housing wealth for homeowners living near foreclosed properties will cause property values to decrease by $5,000 on average.

It is not possible to analyze specifically the full spillover impact of sub-prime foreclosures on African Americans, largely because these data are not available yet. However, communities of color will be especially harmed, since these communities receive a disproportionate share of sub-prime home loans. We estimate that this lost home value translates into a decrease in the tax base, consumer expenditures, investment opportunities, and money circulating in communities of color. United for a Fair Economy estimates that borrowers of color have collectively lost between $164 billion and $213 billion in housing wealth as a result of sub-prime loans taken during the past eight years.

Whatever the exact figures, the bottom line is clear—after centuries of being denied any opportunity to accumulate wealth, after a few decades of having limited opportunities, and after a generation during which African American families accumulated significant wealth, the African American community now faces the greatest loss of financial wealth in its history. Institutional processes and racialized policy are trumping hard-earned educational, job, and income advances.

## DISCUSSION QUESTIONS

1. The recent economic recession affected multiple groups, though in different ways. What special consequences were there for African Americans, according to Oliver and Shapiro? What do you think were the consequences for other groups, and how are they related to racial, ethnic, and class status?

2. Most Americans accumulate wealth by purchasing a home. Why is this path problematic for many African Americans, according to Oliver and Shapiro?

# 43

## Historic Reversals, Accelerating Resegregation, and the Need for New Integration Strategies

GARY ORFIELD AND CHUNGMEI LEE

*Although the principle of "separate but equal" in schools and other public facilities is now unconstitutional, schools are now rapidly re-segregating in the face of court decisions in recent years. White students are thus highly segregated from other groups, and African Americans and Latinos are highly likely to be attending poorly supported "majority-minority" schools.*

American schools, resegregating gradually for almost two decades, are now experiencing accelerating isolation and this will doubtless be intensified by the recent decision of the U.S. Supreme Court. In 2007, the Supreme Court handed down its first major decision on school desegregation in 12 years in the Louisville and Seattle cases.[1] A majority of a divided Court told the nation both that the goal of integrated schools remained of compelling importance but that most of the means now used voluntarily by school districts are unconstitutional. As a result, most voluntary desegregation actions by school districts must now be changed or abandoned. As educational leaders and citizens across the country try to learn what they can do, and decide what they will do, we need to know how the nation's schools are changing, what the underlying trends are in the segregation of American students, and what the options are they might consider.

The Supreme Court struck down two voluntary desegregation plans with a majority of the Justices holding that individual students may not be assigned or denied a school assignment on the basis of race in voluntary plans even if the intent is to achieve integrated schools—and despite the fact that the locally designed plans actually fostered integration. A majority of the Justices, on a Court that divided 4-4-1 on the major issues, also held that there are compelling reasons for school districts to seek integrated schools and that some other limited techniques such as choosing where to build schools are permissible. In the process, the Court reversed nearly four decades of decisions and regulations which had permitted and even required that race be taken into account because of the earlier failure of desegregation plans that did not do that. The decision also called

SOURCE: Civil Rights Project/*Proyecto Derechos Civiles*, UCLA, August 2007.

into question magnet and transfer plans affecting thousands of American schools and many districts. In reaching its conclusion the Court's majority left school districts with the responsibility to develop other plans or abandon their efforts to maintain integrated schools. The Court's decision rejected the conclusions of several major social science briefs submitted by researchers and professional associations which reported that such policies would foster increased segregation in schools that were systematically unequal and undermine educational opportunities for both minority and white students. The Court's basic conclusion, that it was unconstitutional to take race into account in order to end segregation, represented a dramatic reversal of the rulings of the civil rights era which held that race must be taken into account to the extent necessary to end racial separation.

The trends ... are those of increasing isolation and profound inequality. The consequences become larger each year because of the growing number and percentage of nonwhite and impoverished students and the dramatic relationships between educational attainment and economic success in a globalized economy. Almost nine-tenths of American students were counted as white in the early 1960s, but the number of white students fell 20 percent from 1968 to 2005, as the baby boom gave way to the baby bust for white families, while the number of blacks increased 33 percent and the number of Latinos soared 380 percent amid surging immigration of a young population with high birth rates. The country's rapidly growing population of Latino and black students is more segregated than they have been since the 1960s and we are going backward faster in the areas where integration was most far-reaching. Under the new decision, local and state educators have far less freedom to foster integration than they have had for the last four decades. The Supreme Court's 2007 decision has sharply limited local control in this arena, which makes it likely that segregation will further increase.

Compared to the civil rights era we have a far larger population of "minority" children and a major decline in the number of white students. Latino students, who are the least successful in higher education attainment, have become the largest minority population. We are in the last decade of a white majority in American public schools and there are already minorities of white students in our two largest regions, the South and the West. When today's children become adults, we will be a multiracial society with no majority group, where all groups will have to learn to live and work successfully together. School desegregation has been the only major policy directly addressing this need and that effort has now been radically constrained.

The schools are not only becoming less white but also have a rising proportion of poor children. The percentage of school children poor enough to receive subsidized lunches has grown dramatically. This is not because white middle class students have produced a surge in private school enrollment; private schools serve a smaller share of students than a half century ago and are less white. The reality is that the next generation is much less white because of the aging and small family sizes of white families and the trend is deeply affected by immigration from Latin American and Asia. Huge numbers of children are growing up in families with very limited resources, and face an economy with deepening inequality of income distribution, where only those with higher education are securely in the middle class. It is a simple statement of fact to say that the country's future depends on finding ways to prepare groups of students who have traditionally fared badly in American schools to perform at much higher levels and to prepare all young Americans to live and work in a society vastly more diverse than ever in our past. Some of our largest states will face a decline in average educational levels in the near future as the racial transformation proceeds if the educational success of nonwhite students does not improve substantially.

From the "excellence" reforms of the Reagan era and the Goals 2000 project of the Clinton Administration to the No Child Left Behind Act of 2001, we have been trying to focus pressure and resources on making the achievement of minority children in segregated schools equal. The record to date justifies deep skepticism. On average, segregated minority schools are inferior in terms of the quality of their teachers, the character of the curriculum, the level of competition, average test scores, and graduation rates. This does not mean that desegregation solves all problems or that it always works, or that segregated schools do not perform well in rare circumstances, but it does mean that desegregation normally connects minority students with schools which have many potential advantages over segregated ghetto and barrio schools especially if the children are not segregated at the classroom level.

Desegregation is often treated as if it were something that occurred after the *Brown* decision in the 1950s. In fact, serious desegregation of the black South only came after Congress and the Johnson Administration acted powerfully under the 1964 Civil Rights Act; serious desegregation of the cities only occurred in the 1970s and was limited outside the South. Though the Supreme Court recognized the rights of Latinos to desegregation remedies in 1973, there was little enforcement as the Latino numbers multiplied rapidly and their segregation intensified.

Resegregation, which took hold in the early 1990s after three Supreme Court decisions from 1991 to 1995 limiting desegregation orders, is continuing to grow in all parts of the country for both African Americans and Latinos and is accelerating the most rapidly in the only region that had been highly desegregated—the South. The children in United States schools are much poorer than they were decades ago and more separated in highly unequal schools. Black and Latino segregation is usually double segregation, both from whites and from middle class students. For blacks, more than a third of a century of progress in racial integration has been lost—though the seventeen states which had segregation laws are still far less segregated than in the 1950s when state laws enforced apartheid in the schools and the massive resistance of Southern political leaders delayed the impact of *Brown* for a decade. For Latinos, whose segregation in many areas is now far more severe than when it was first measured nearly four decades ago, there never was progress outside of a few areas and things have been getting steadily worse since the 1960s on a national scale. Too often Latino students face triple segregation by race, class, and language. Many of these segregated black and Latino schools have now been sanctioned for not meeting the requirements of No Child Left Behind and segregated high poverty schools account for most of the "dropout factories" at the center of the nation's dropout crisis....

One would assume that a nation which now has more than 43 percent non-white students, but where judicial decisions are dissolving desegregation orders and fostering increasing racial and economic isolation must have discovered some way to make segregated schools equal since the future of the country will depend on the education of its surging nonwhite enrollment which already accounts for more than two students of every five. You would suppose that it must have identified some way to prepare students in segregated schools to live and work effectively in multiracial neighborhoods and workplaces since experience in many racially and ethnically divided societies show that deep social cleavages, especially subordination of the new majority, could threaten society and its basic institutions. Those assumptions would be wrong. The basic judicial policies are to terminate existing court orders, to forbid most race-conscious desegregation efforts without court orders, and to reject the claim that there is a right to equal resources for the segregated schools. Not only do the federal courts not require either integration or equalization of segregated schools but this means that they forbid state and local officials to implement most policies that have proven effective in desegregating schools. State and local politics will determine what, if anything, happens in terms of equalizing resources between segregated schools and privileged schools....

While the courts are terminating desegregation plans, statistics show steadily increasing separation. After three decades of preparing reports on trends in segregation in American schools, the most disturbing element of this year's report is the finding that the great success of the desegregation battle—turning Southern education, which was still 98 percent segregated in 1964, into the most desegregated part of the nation—is being rapidly lost. These new data show that the South has lost the leadership it held as the most desegregated region for a third of a century, even as the region becomes majority nonwhite and faces a dramatic Latino immigration. It took decades of struggle to achieve desegregated schools in the South, our most populous region, and no one would have predicted during the civil rights era that leaders in some Southern communities once forced to desegregate with great difficulty would, in the early 21$^{st}$ century, wish to remain desegregated but be forbidden to maintain their plans by federal courts. Yet that is exactly what happened in a number of districts.

The basic educational policy model in the post-civil rights generation assumes that we can equalize schools without dealing with segregation through testing and accountability. It is nearly a quarter century since the country responded to the Reagan Administration's 1983 report, "A Nation at Risk," warning of dangerous shortcomings in American schools and demanding that "excellence" policies replace the "equity" policies of the 1960s. Since then almost every state has adopted the recommendations for the more demanding tests and accountability and more required science and math classes the report recommended. Congress and the last three Presidents have established national goals for upgrading and equalizing education. The best evidence indicates that these efforts have failed, both the Goals 2000 promise of equalizing education for nonwhite students by 2000 and the NCLB promise of closing the achievement gap with mandated minimum yearly gains so that everyone would be proficient by 2013. In fact, the previous progress in narrowing racial achievement gaps from the 1960s well into the 1980s has ended and most studies find that there has been no impact from NCLB on the racial achievement gap. These reforms have been dramatically less effective in that respect than the reforms of the 1960s and '70s, including desegregation and anti-poverty programs. On some measures the racial achievement gaps reached their low point around the same time as the peak of black–white desegregation in the late 1980s.

Although the U.S. has some of the best public schools in the world, it also has too many far weaker than those found in other advanced countries. Most of these are segregated schools which cannot get and hold highly qualified teachers and administrators, do not offer good preparation for college, and often fail to graduate even half of their students. Although we have tried many reforms, often in confusing succession, public debate has largely ignored the fact that racial and ethnic separation continues to be strikingly related to these inequalities. As the U.S. enters its last years in which it will have a majority of white students, it is betting its future on segregation. The data coming out of the No Child Left Behind tests and the state accountability systems show clear relationships between segregation and educational outcomes but this fact is rarely mentioned by policy makers.

The fact of resegregation does not mean that desegregation failed and was rejected by Americans who experienced it. Of course the demographic changes made full desegregation with whites more difficult, but the major factor, particularly in the South, was that we stopped trying. Five of the last seven Presidents actively opposed urban desegregation and the last significant federal aid for desegregation was repealed 26 years ago in 1981. The last Supreme Court decision expanding desegregation rights was handed down in 1973, more than a third of a century ago, one year before a decision rejecting city-suburban desegregation. This second decision in 1974 meant that desegregation was impossible in much of the North since the large majority of white students in many areas were already in the suburbs and stable desegregation was impossible within city boundaries, as Justice Thurgood Marshall accurately predicted in his dissent in the 1974 *Milliken v. Bradley* decision.

The *Milliken* decision could be seen as the return of the doctrine of "separate but equal" for urban school children in a society where four of five Americans live in metropolitan areas. The problem is that the Supreme Court held in the 1973 *Rodriguez* case that there is no federal right to an equal education, so "separate but equal" could not be enforced either. With the 2007 rejection of most of the techniques that have preserved a modicum of desegregation by voluntary local action, the doctrine is basically one of separation and local political control, except if local governments want to pursue voluntary integration strategies, which are now largely prohibited....

One of the deepest ironies of this period is that never before has there been more evidence about the inequalities inherent in segregated education, the potential benefits for both nonwhite and white students, and the ways in which those benefits could be maximized. The evidence submitted to the Supreme Court regarding the Louisville and Seattle cases was many times more compelling than that the Court relied on in striking down the segregation system of the South in 1954. This evidence does not claim that desegregation will eliminate inequalities, since those are based on social and economic issues that reach far beyond the schools but it does show that the policies provide important benefits in both educational attainment and life chances—and that there are no harms and some large benefits for white as well nonwhite students, and for society and its institutions. Yet we are dismantling plans that actually work in favor of an alternative, double and triple segregation, that has never worked on any substantial scale....

Nearly 40 years after the assassination of Dr. Martin Luther King, Jr., we have now lost almost all the progress made in the decades after his death in desegregating our schools. It was very hard won progress that produced many successes and enabled millions of children, particularly in the South to grow up in more integrated schools. Though it was often imperfectly implemented and sometimes poorly designed, school integration was, on average, a successful policy, linked to a period of social mobility and declining gaps in achievement and school completion and improved attitudes and understanding among the races. The experience under No Child Left Behind and similar high stakes testing and accountability policies that ignore segregation has been deeply disappointing and the evidence from those tests show the continuing inequality of segregated schools even after many years of fierce pressure and sanctions on those schools and students.

It is time to think very seriously about the central proposition of the *Brown* decision, that segregated education is "inherently unequal" and think about how we can begin to regain the ground that has been lost. The pioneers whose decades of investigations and communication about the conditions of racial inequality helped make the civil rights revolution possible a half century ago should not be honored merely by naming schools and streets or even holidays after them but should be remembered as a model of the work that must be done, as many times as necessary, for as long as it takes, to return to the promise of truly equal justice under law in our schools, to insist that we have the kind of schools that can build and sustain a successful profoundly multiracial society....

## NATIONAL SEGREGATION TRENDS

Across the country, segregation is high for all racial groups except Asians. While white students are attending schools with slightly more minority students than in the past, they remain the most isolated of all racial groups: the average white student attends schools where 77 percent of the student enrollment is white. Black and Latino students attend schools where more than half of their peers are black and Latino (52% and 55% respectively), a much higher representation than one would expect given the racial composition of the nation's public schools and substantially less than a third of their classmates are white. Whites had been even more segregated back in 1990, when they constituted a significantly larger share of the total enrollment.

## WHITES: STILL THE MOST SEGREGATED

Though white students in 2005–6 were in schools with more minority students than in the past, they were still the most segregated population, being in schools that were 77 percent white, on average, in a country with 57 percent white students. Almost no attention has been given in the discussion of desegregation strategies and neighborhood schools about the consequences of ending city- and county-wide desegregation plans for white students living in city and inner suburban areas. In the absence of desegregation plans, much of the racial contact that exists is accounted for either by the small but significant number of whites in heavily minority schools or reflects the temporary diversity produced by residential racial transition as blacks and Latinos move very rapidly into some sectors of suburbia. A transitional neighborhood is a highly unstable process of a sort all too familiar during the decades when residential resegregation converted thousands of white city neighborhoods to minority communities. Under neighborhood schools or magnet school plans without desegregation guidelines more of these urban white students are going to end up isolated in high poverty, very high minority schools, a process that could well undermine some stably integrated residential areas and further limit the options of poor whites if choice plans are not operating. Unrestricted choice plans in the past have often accelerated residential resegregation when white students from integrated neighborhoods transferred out to whiter schools, helping tip the neighborhood school toward resegregation and making the neighborhood less attractive for white home seekers. When the courts and federal civil rights officials prohibited choice plans without desegregation standards in the 1960s they were very conscious of these problems and often found unrestricted choice strategies to be contributors to segregation. Now, as a result of the recent Court decision, we will have more such plans.

## ASIANS: THE MOST INTEGRATED STUDENTS

When considering issues of immigration, the success of Asian students is often compared to the academic challenges facing Latino students. One of the significant differences is the level of segregation. Asian students are in schools where, on average, less than a fourth of fellow students are Asian and, since Asians speak many languages, they are far less likely to be in a school where their language is a major factor. Asians typically attend schools that are 48 percent white, compared to 32 percent for Latinos (pp. 312 ). However, despite the fact that Asians represent only five percent of the total student enrollment, the average Asian attends a school that is 24 percent Asian.

Likely due to their high residential integration and relatively small numbers outside the West, Asian students, on average, are the most integrated group and the group which attends school where their own ethnicity is least represented. Asians are also the most integrated racial group in residential patterns. U.S. immigration policies have tended to produce a very highly educated immigration from Asia. When educated middle class migrations have taken place from Latin America, such as the first wave of Cuban migration, their experience has been similar to the average Asian experience, but most Latino immigration is of people with far fewer resources and lower levels of education.

The Asian experience, however, is a complex one. Although on average Asians are more educated and have higher family incomes than whites, some Asian groups, particularly refugee Indochinese populations who entered after the Vietnam War experience very different patterns of education and mobility, much more similar to those of typical disadvantaged Latino immigrants. Particularly in the West where Asians already outnumber African Americans and are a very visible presence in the schools it will be increasingly important to understand these differences.

## DESEGREGATION TRENDS FOR BLACK STUDENTS

As previously mentioned, national statistics for black students show very slow progress the first decade after *Brown*, then a substantial decline in black segregation from white from the mid-60s through the early 1970s. There was gradual improvement through most of the 1980s, but then a reversal and a steady gradual rise in segregation since the early 1990s, a rise which is accelerating in the South. In terms of enrollment in majority white schools, most of the progress from urban desegregation has now been lost. The level of extreme segregation of black students in schools with 0–10% whites, however, remains far lower than it was before the civil rights era, though it also is rising. Table 1 shows a sharp rise in the percentage of black students in majority nonwhite schools since the 1980s and by far the largest increase takes place in the South.

**T A B L E   1**    **Percentage of Black Students in Predominantly (>50%) Minority Schools by Region, 1968–2007**

| Region | 1968 | 1980 | 1988 | 1991 | 2005 |
|---|---|---|---|---|---|
| South | 81 | 57 | 57 | 60 | 72 |
| Border | 72 | 59 | 60 | 59 | 70 |
| Northeast | 67 | 80 | 77 | 75 | 78 |
| Midwest | 77 | 70 | 70 | 70 | 72 |
| West | 72 | 67 | 67 | 69 | 77 |
| US Total | 77 | 63 | 63 | 66 | 73 |

SOURCE: U.S. Department of Education office for Civil Rights data in Orfield, Public School Desegregation in the United States, 1980-1; 1988-9, 1991-2; 2005-6 NCES Common Core of Data

## LATINO SEGREGATION

On a national level, the segregation of Latino students has grown the most since the civil rights era. Since the early 1970s, the period in which the Supreme Court recognized Latinos' right to desegregation there has been an uninterrupted national trend toward increased isolation. Latino students have become, by some measures, the most segregated group by both race and poverty and there are increasing patterns of triple segregation—ethnicity, poverty and linguistic isolation. No national administration has made a serious effort to desegregate Latinos and there have been few court orders addressing this problem, the most important of which have now been terminated—those in Denver and Las Vegas. In comparative terms, by 2005 Latinos were most likely to be in schools with less than half whites (78%) and in intensely segregated schools (39%).

## NOTE

1.  *Parents Involved In Community Schools V. Seattle School District No. 1 Et Al.* June 28, 2007.

## DISCUSSION QUESTIONS

1.  What changes in law since *Brown v. Board of Education* have affected the re-segregation of schools?

2.  How would you describe the pattern of integration and segregation in your high school? How was the racial composition related to the neighborhood where you lived? Did the pattern persist into college?

# 44

## Good Neighborhoods, Good Schools:
### Race and the "Good Choices" of White Families

HEATHER BETH JOHNSON AND THOMAS M. SHAPIRO

*"School choice" to most means being able to select a high-quality school for one's child. But Johnson and Shapiro point to a pernicious side of the choices White parents make for good schooling—that is, the perpetuation of racial inequality in education.*

The community and school decisions that parents make are some of the most monumental decisions in the lives of their families. These decisions, made at critical points along the life course, not only affect the everyday lives of parents and their children but contribute significantly to a family's life chances, their future prospects, and their identities. In a system of persistently segregated neighborhoods and vastly unequal and increasingly resegregated schooling, the stakes are high: accessing "good" schools and "good" neighborhoods is not always easy to do, and ultimately not everyone gets the chance. While these are indeed individual choices, the community and school decisions made by parents are highly structured in a context of systematic socioeconomic and educational inequality.

... We discuss what white parents say about why they live in the neighborhoods that they do and send their children to the schools that they do, focusing on the role that race plays in their decision-making about two of the most important decisions they will ever make. Here we are interested in understanding what seems to be "going on" in the minds of whites in contemporary U.S. society when it comes to race, schools, and neighborhood choice, and how this mind-set informs behavior and action. This topic is important for understanding the processes through which race and class inequality are reproduced and how structures of segregation and inequality are recreated in everyday life.

SOURCE: Doane, Ashley W. and Eduardo Bonilla-Silva, eds. 2003. *Whiteout: The Continuing Significance of Racism.* New York: Routledge.

Our research findings suggest that the process of school and community choice can be seen as a key mechanism that reproduces race and class stratification. Our data show that white parents' decisions are often based on race and made possible by their own race and class privilege. The choices are made within an arena that is rigidly stratified and socially structured to reward those in advantageous positions for making decisions that will further their advantage and similarly situate their children. In this context of inequality, individual family's "choices" serve to perpetuate existing inequalities by passing along advantage (or disadvantage, as the case may be), to the next generation and thus contribute to the reproduction of social stratification.

The interview data presented here reveal depths of racialized actions previously hidden by survey research and public opinion polls. Our data show that in-depth interviewing can encourage candid discussions about race to occur, especially around concrete events in people's lives....

## INTERVIEWING WHITE FAMILIES ABOUT RACE: RESEARCH APPROACH

The data for this paper come out of a larger project on race and wealth in America for which we interviewed approximately two hundred families in Boston, St. Louis, and Los Angeles. We completed interviews between January 1998 and June 1999, identifying participants through a structured snowball sampling method. We conducted interviews with black and white families representing a broad socioeconomic spectrum ranging from poor to working class to middle and upper middle class. Each participating family had at least one school-aged (or pre-school-aged) child living in the home, and we conducted interviews with parents—both couples and single heads of households. Interviews took place in participating families' homes or another place of their choosing and were semi-structured, in-depth conversations that lasted between one and three hours each. We asked parents in depth about assets, income, family background, how they came to live where they do and send their children to the schools that they do, and the roles that wealth and race played in their decision-making. The interview sessions were recorded, were transcribed to over seven thousand pages of text, and were coded using NUD*IST qualitative data analysis software. The resulting qualitative data set is extensive, covering a broad range of topics related to the project's focus, and is the first of its kind.

This paper is based on interviews with the seventy-five white families who participated in the study. This particular set of data is uniquely well suited to provide insight into the role of race in the ways that white Americans make decisions about the communities and schools they choose and raise their children in. Our interviews provide clear evidence that race is paramount in the minds of white parents when they make school and community choice decisions for their families. We argue that the role their "choices" play in the social reproduction of racial stratification looms large in contemporary U.S. society.

## SCHOOLS, NEIGHBORHOODS, AND RACE:
## RESEARCH FINDINGS

Laurie:  Now it's not necessarily fair and it's not necessarily right, but I think certain neighborhoods are better, certain schools are better, and your children will have a better childhood and better educational background because of where they go. But it's not right. I don't think it's necessarily right, but I think everyone should have the same opportunities my children do, but they don't.... O.k., I'll rephrase that. I don't think it's right that my children get to go to a private school and get to wear Adidas and, and, there are other children living in the city who aren't even fed breakfast, who wear raggy, holey clothes, who have teachers who don't want to be there, and they get no educational benefits whatsoever.

Q:  Do you feel like race has played a role in any of the decisions that you all have made?

Laurie:  I have to be honest and ... I'm probably wrong for even saying it, but truthfully, it's in the back of my mind, yes.... But I do want to clarify one thing. If there was a nice black family who my husband worked with at General Electric and they bought the house next door to us and had the same values and the same desires and goals that we had, I would have nothing, and I wouldn't be afraid to have my children carpool or sit by them. I guess I am racist deep down inside, and I feel guilty for admitting that, but those poor inner-city kids whose parents are on crack and who don't care about them and don't feed them and have drugs and guns lying around for them to bring to school, I'm afraid of them. I am afraid of them. And maybe I want to shelter my kids until they're older and they can handle it better. When they're young, I don't want them to be exposed to that type of situation.... Well, I shouldn't say that they have to work at General Electric, they don't have to work at General Electric, but they have to work and have to save and have to strive and try to better themselves. But if they're out selling drugs on the corner, I don't want my kids to be around that. And I don't want my kids being shipped into a school like that.... I feel guilty because I'm not doing anything to make their life better or try to help them. I'm hiding out here in my little nice neighborhood and my little private school and I'm like sticking my head in the sand and pretending like these problems don't exist. So I do have a sense of guilt over it.

Laurie, a parent interviewed for our study, lives in a white suburb of St. Louis with her husband and three children. Her explanation of her family's decision to live where they do and send their children to the schools that they do is strikingly candid and reveals a depth of racialized beliefs and action that was not uncommon in our interviews with white parents. For these parents, schools, neighborhoods, and race are intricately tied in complex ways that ubiquitously surround and concretely impact their decisions about community and education.

When asked how they chose their neighborhood, most white parents explained right away that it was for the school district. This is the case even for parents who send their children to private schools and do not use the public school system. For the vast majority of families, looking for a "good" neighborhood means looking for one that is in a "good" school district. And because property taxes fund school districts, residential location and school quality tend to go hand in hand. "Good" schools are in "good" neighborhoods and "good" neighborhoods have "good" schools.

Barbara, one of the parents interviewed, says that the main reason that she and her husband chose the community they did was "schools for the kids." Paul says: "We looked at the city, and the bottom line was we weren't happy with the schools. We wanted to be in a public school in the country." Another parent explained: "we specifically did not look in certain areas because the school districts weren't that good."

Q:  So, did you ask your broker to only look in certain areas?

Steve:  Yes.

Q:  And the area was south county?

Steve:  Yes. Uh, well, specifically, uh, the Lindbergh school district.

These families are trying to make "good choices" for their families, but what do these phrases—"good neighborhoods" and "good schools"—mean to them? Inevitably, in explaining their thinking and decision-making, race became an integral part of parents' explanations. Interviews reveal that almost always, in the case of neighborhoods and schools, "good" is woven with ideas about race, and race is clearly a primary dimension of whites' choices. The specific kind of environment white parents want for their children (or do not want) and attempting to control that environment were integral dimensions of their school and community choices.

Q:  How are you going to decide where to live?

Jean:  I guess race would affect it a little bit. I don't know.... There's some parts of the city and just from the location that they are, and the incomes there, I really wouldn't want her to go to those types of schools. I don't know what kinds of education they have in those schools. I know most of them are probably public, and that would be something I would have to check out. I guess it's kind of sad to say that I would want her in an area maybe, I wouldn't want to say more *white*, but I guess it would really depend on the area, because I mean, it's like I said, there's some places ... that I would be scared for her to go to school. I'm scared walking down by the school and things, and so I guess it would play a part, just in the fact ... of where we buy our home.

Our interviews suggest that in contemporary American society, the school and community choices of white families are thoroughly permeated by race.

## "WE TRY TO STAY WITH OUR OWN TYPE"

Many of our interview participants displayed open and overt racism in discussing their decisions about where to live and send their children to school. Our data reveal deep levels of explicit racism by whites against blacks in particular. This racism is often translated directly into action: often, in our interviews with whites, race alone was a major explanation for school and community choice, and overt racism itself propelled much of their decision-making.

Such is the case with Merilyn and Duane Fisher. The Fishers have a nineteen-month-old daughter; Merilyn does clerical/secretarial work, and Duane is a butcher. They spoke openly of having considered moving into only a community that was predominantly or exclusively white.

Duane:  My first impression of somebody is probably going to be the color of the skin....Then I go on how they act, what kind of character they have...like I said, I wouldn't live in East St. Louis. The crime rate's too high there. There's also the race, you know the one, one race. I hate to say it, but that's, that's the issue....I mean, I don't believe I am a bigot but I mean, in a lot of ways I gotta be because the first thing I think of is, you know, um, type.

Q:  What do you mean?

Duane:  Type cast, I guess....It's hard not to look at someplace like East St. Louis and say, "well, you know, it's a black community" and people know what you mean when you say that. ... It means that, um, maybe it's a little run down and the crime rate's a little higher and it's not a very desirable place to even drive through and you know?

Q:  So as far as choosing a community, was that a factor for you? Were you looking for more of a white community?

Duane:  Yeah.

In terms of school choice, the Fishers say that they *might* consider putting their daughter Sarah into a public school that had *some* black students. They say that as long as the black students "came from a neighborhood that has nice neat lawns and stuff, then it's OK."

Marilyn Masterson is a divorced mother of two children ages twelve and seven who works as a child care center administrator. The fact that the area Marilyn has recently moved her family into is much less racially diverse than where they lived previously is no coincidence:

Marilyn:  I've known blacks, I've dated blacks.... But, but there is a category, you don't want to call them niggers anymore, but I, it's, it's lower than blue collar, it's a certain mental attitude toward, toward everybody, toward the world.... And I see it more in black people than in white. ... The families aren't together, extended or otherwise, or, or the extended women half of the family are together, raising the kids, you know, the unmarried daughter with her baby and the mom and the

grandmom kind of thing. But there's just not a lot of real sticking together protecting their own in, in this particular black community …
the ones I don't want to call niggers.

Q: Is that part of the reason why you wanted to move?

Marilyn: Because it is a part of the black population in "U City" [a section of
the St. Louis metropolitan area]Yes.

Marilyn's openly racist rhetoric may seem shocking, but in fact was not
uncommon in our interviews. She, like many of the whites we interviewed,
expresses her overt racism with some apologies but at the same time seems to
think that her views are completely justified. In explaining their move to a white
suburb from a city neighborhood one father said: "We just thought, the *mix* of
the group and then all of these people going to the same school, it didn't fit with
what we wanted for our family." To avoid exposing their children to people that
they find "undesirable" (as one participant explained it), many white parents simply moved. Many others decide to send their children to predominantly or
exclusively white private schools.

Other white families find themselves living in what used to be exclusively
white neighborhoods but which are now diversifying as more blacks and other
minorities move in. In these cases the neighborhood and school discussions were
steeped with worry and concern about what this meant for the future of the
neighborhood and schools. In such cases the whites we interviewed expressed
serious fear and often anger about what they perceive happening before their
very eyes: as they see black families moving in, they see their white neighbors
moving out. And often even more upsetting to these families is the neighborhood "deterioration" they believe to go hand in hand with the color change of
their neighborhoods….

For some, a "good" community and school truly means, at least in part,
having a certain amount of distance from people who are nonwhite. Others
have shown that perceived risk of crime is directly related to perceived racial
and ethnic composition of neighborhoods, and that for whites criminal threat is
associated with perceived proximity of racial others (Chiricos, McEntire, and
Gertz 2001). Our interviews support this "social threat" perspective by providing
insight that moves beyond survey research about attitudes to the real-life mind-
sets of a sample of white families.

… Many others we interviewed spoke of their concerns about their neighborhoods deteriorating due to blacks moving in and told stories of their successful "escapes" from such situations. When financially possible, these families
tended to flee to white suburbs or to other predominantly white neighborhoods
with what they consider "decent," middle-class values to raise their children in.
Here, they often find the "safe" schools they were looking for, away from the
"bad elements" and "negative influences," which in their minds tend to be representative of more racially diverse—but particularly black—communities. Their
choices for neighborhoods and schools—whether acted upon or simply dreamt
about—were often motivated by a clear and explicit racist belief system. Within
a structure of unequal schools and greatly varying neighborhoods, acting on this

belief system is generally rewarded with often genuinely superior school systems and more ample community resources.

## "I WOULDN'T CARE, BUT ... IT JUST HAPPENS
## TO WORK OUT THAT WAY"

Alice and Philip Hutcheson live in St. Louis with their two children, ages three years and eight months. Alice describes herself as a stay-at-home mom, and Philip is a medical resident. Alice and Philip moved to St. Louis from Utah to complete Philip's residency. They chose their neighborhood because they had friends from medical school who lived there, their real estate agent recommended the neighborhood, and it was close to a church they planned to join. In all of these ways, because their social network is white and middle-class, the decision to choose a white, middle-class neighborhood was structurally *encouraged* before the Hutchesons's decision-making (racially motivated or not) ever even entered the arena. But they were also looking for a certain type of neighborhood and were purposely avoiding others; Alice explains her desire to not be in a black neighborhood as rooted in the fact that "the nicer schools are in the white neighborhoods." ...

Alice further explained that she "wouldn't care a bit having some black neighbors" but she would never put her children in "inner-city" schools because she has seen reports in the media about school violence and would not expose her children to that. She acts surprised by the degree of racial segregation: on one hand she says that she would not mind having black neighbors and would be proud to "branch out," but on the other hand she believes that the nicer schools are in white neighborhoods so she chooses white neighborhoods only. Alice is telling us that she chose her neighborhood for the "nicer" school and the fact that it is all-white is by default. But the fact that it is all-white is Alice's way of knowing that it is "nicer," and that the schools are "good" ones. Thus race, fundamentally, is a primary dimension of how she and her husband are making community and school choices. The structure of the unequal and racially segregated school system is such that indeed, Alice will be rewarded for choosing the white neighborhood: she will be able to access higher-quality education for her children.

Within this highly structured context, this kind of sense-making and logic in terms of racially defining "good" neighborhoods and schools dominated many of our interviews. In such cases, white families we interviewed stood in contrast to the examples of overt racism above because they were more conflicted in their racialized explanations for school and community choice. While their choices were still race-based, in these cases our white participants explain that is not racial minorities *per se* that they are concerned about, it is simply the fact that the white neighborhoods and schools are the "good" ones....

Color is a defining factor in determining a "good" neighborhood and a "good" school. In the minds of most of the whites we interviewed, the whiter, the better. But even more particularly, the fewer blacks, the better. A little diversity doesn't hurt (in fact, it might be nice), but in general our white parti-cipants saw the "nicer" neighborhoods and schools as being the least black. They claim that they would not care if racial minorities lived in their neighborhoods or schools but that in the best interest of their children's education it just happens to work out that the best schools and communities are the white ones. This is not necessarily overt racism, rather it is more of a perception-based logic. Others have shown similar findings on whites' claims of the "naturalization of racial matters" (Bonilla-Silva 2001:149) and whites' claims that "segregation just hap-pened" or is viewed as "natural" and therefore somehow "acceptable" (Doane 1996:44). But again, regardless of their explanations, this logic is encouraged by a social structure that rewards white families for perpetuating segregation through their racialized decisions. As one interview participant said when asked about the diversity of the schools with the better reputations, test scores, and facilities: it just so happens that they are "A lot whiter!" ...

## WHITE FAMILIES, "GOOD CHOICES," AND RACE: RESEARCH CONCLUSIONS

Our interviews with white families tell us that race is a major factor, if not *the* major factor, in determining community and school choice. This concept is not new, but the data presented here move beyond survey research to give insight into white racialized belief systems, white racism, and ongoing systems of neigh-borhood and school segregation. Race is a key dimension of how whites define "good" neighborhoods and "good" schools, and the two are so intricately inter-twined that they cannot be seen as separate choices. Interview participants were clear about their views, and it is obvious that race looms large in their minds when they make school and community choice decisions.

Some have argued that white racism is on the decline and that race relations have been improving in the United States in recent years (Thernstrom and Thernstrom 1997). But most of the recent literature on racial attitudes and beliefs has disagreed, arguing that while *explicit* racism is on the decline, we have moved into a new era of white racism and racialized belief systems defined as "symbolic racism" or "color-blind racism" (for example Bonilla-Silva 2001; Schuman et al. 1997; Sears, Sidanius, and Bobo 2000; Sleeper 1997). Based on this solid body of quantitative and qualitative research, one may be surprised by the depth and cen-trality of race to whites' thinking about community and schools revealed in these candid discussions....

An important dimension of our findings is the fact that *because* the interviews highlighted here are with white families, for the most part they *do* have relatively much more opportunity to act on their perceptions. Due to the structurally advantaged/privileged position of whites as a social group, and in particular be-cause of the financial capabilities that many of them have in terms of assets, they are able to make choices and act on their choices in very real ways (H. B. Johnson 2001; Shapiro 2003; Shapiro and Johnson 2003). Assets enable more options. This stands in contrast to other social groups, such as black Americans, who due to structural disadvantages, in particular lack of assets, are not able to make or act on their "choices" in the same ways. Although everyone, to some degree, has constrained choices, whites have relatively more choices and are relatively more able to act on them for themselves and their children. Within these structural circumstances, "choice-making" by whites can have very real consequences for the society as a whole.

Neighborhoods and schools are not equal: they are segregated and stratified. Property values differ dramatically, community services and quality of life differ drastically, and everyone knows that schools are segregated and vastly unequal. When white parents choose the "good" neighborhoods and the "good" schools, they choose to separate themselves from their black peers. While they may be colleagues at work, they are not neighbors, and their children are not classmates. Sometimes this might be unintended: some white Americans would like to participate in diversity but simply are not willing to sacrifice (as they see it) the best chances possible for themselves and their children. But sometimes this re-creation of segregation is not a coincidence. As we have seen, white Americans often conscientiously decide to isolate themselves in white environments precisely to avoid diversity.

Race is on the minds of white Americans, and they are thinking about it when it comes to neighborhood and school choice. By making the choices they do, white Americans reproduce segregation in neighborhoods and re-create segregated, unequal schooling for their children. The opportunities for white families to choose and their individual decisions to choose the "good" neighborhoods and the "good" schools can hence be seen as a mechanism through which contemporary racial stratification is perpetuated in the United States. The structural context rewards whites for acting in a racist manner and a manner that perpetuates race and class inequality. Most whites do not challenge structural inequality or aspects of it such as systematic school districting and funding that maintains inequalities. Rather they participate, and are rewarded for doing so—often through the unearned benefits of simply being white. Ultimately, if the cycle continues, it is the children of those we interviewed who will go on to perpetuate further the systems of advantage and disadvantage that they are currently reaping the benefits of. One day they too will face the same decisions that their parents explain here.

## REFERENCES

Bonilla-Silva, Eduardo. 2001. *White Supremacy and Racism in the Post-Civil Rights Era.* Boulder, CO: Lynne Rienner.

Chiricos, Ted, Ranee McEntire, and Marc Gertz. 2001. "Perceived Racial and Ethnic Composition of Neighborhood and Perceived Risk of Crime." *Social Problems* 48: 322–340.

Doane, Ashley W., Jr. 1996. "Contested Terrain: Negotiating Racial Understanding in Public Discourse." *Humanity and Society* 20(4): 32–51.

Johnson, Heather Beth. 2001. "The Ideology of Meritocracy and the Power of Wealth: School Selection and the Reproduction of Race and Class Inequality." Ph.D. dissertation, Department of Sociology, Northeastern University, Boston, MA.

Schuman, Howard, Charlotte Steeh, Lawrence Bobo, and Maria Krysan. 1997. *Racial Attitudes in America: Trends and Interpretations.* Cambridge, MA: Harvard University Press.

Sears, David O., Jim Sidanius, and Lawrence Bobo, eds. 2000. *Racialized Politics: The Debate About Racism in America.* Chicago: University of Chicago Press.

Shapiro, Thomas M. 2003. *Racial Legacies: The Reproduction of Inequality.* New York: Oxford University Press.

Shapiro, Thomas M., and Heather Beth Johnson. 2003. "Family Assets and School Access: Race and Class in the Structuring of Educational Opportunity." In *Inclusion in Asset Building: Research and Policy*, ed. Michael Sherraden. New York: Oxford University Press.

Sleeper, Jim. 1997. *Liberal Racism.* New York: Viking.

Thernstrom, Stephen, and Abigail Thernstrom. 1997. *America in Black and White: One Nation, Indivisible.* New York: Simon and Schuster.

## DISCUSSION QUESTIONS

1. If White parents make decisions to send their child to a "good" school, is this racism according to Johnson and Shapiro? How or how not?

2. What are the implications of Johnson and Shapiro's argument for social policies about education?

# Student Exercises

1.  How would you describe the neighborhood where you live? Is it racially integrated or segregated? If it is segregated, what are the factors that influence this residential pattern? If it is integrated, has it always been so or have there been specific changes that have resulted in this? Is it likely to continue?

2.  Similar to Exercise 1, was the high school you attended racially integrated or racially segregated? How and why?

# Section XI

# Do We Care? Race, Health Care, and the Environment

Elizabeth Higginbotham
and Margaret L. Andersen

What does a nation's health care system tell us about the cultural values and beliefs of the society? Quite a lot, say many health care experts. You can look at how care is delivered, who is served, and how much people pay and understand quite a lot about what the society values. In some nations, people would find it unimaginable to have to pay for health care; in others, not only do people pay, but they pay a lot. In some societies, health care is provided through a national health care program; in others, health care is privatized and managed through large, profitable corporations. In some societies, sick people have access to care regardless of their ability to pay; in others, the quality of care depends on one's economic and social resources. What cultural values and norms structure the U.S. system of health care?

Answering this question requires first knowing some basic facts about health and health care availability. What you will see in this section is that social factors associated with racial inequality, as well as class and gender inequality, strongly influence the health care system in the United States. Consider this:

- 255 million people in the United States (prior to health care reform passed in 2010) had no health insurance; this is 15 percent of the total U.S. population (DeNavas–Walt et al. 2009).

- Whether or not you have health insurance is strongly influenced by your race. Twenty percent of African Americans, 30 percent of Hispanics, and 18 percent of Asian Americans have no health insurance (DeNavas–Walt et al. 2009).

- The likelihood of having health care insurance is also influenced by age. People between age 18 and 34 are the age group least likely to have health care insurance. Poor children are twice as likely to be uninsured as other children (DeNavas–Walt et al. 2009).

- Social class is also a strong determinant of health care coverage: 25 percent of those earning under $25,000 a year have no health insurance, compared to 8 percent of those earning $75,000 or more (DeNavas-Walt et al. 2009).

But it is not just health insurance that reveals such disparities. The very likelihood of getting sick or injured is a result of social factors. On measures of infant mortality, thought by experts to be a good indicator of a nation's health, the United States ranks quite low relative to other nations—that is, the number of infant deaths is higher in the United States than in most other industrialized nations. Furthermore, our status in this regard has dropped significantly since 1960, when we were twelfth among a group of other nations (number one having the lowest infant mortality); now we are twenty-eighth. Worse is that among African Americans in the United States, infant mortality is comparable to that in some Third World, poor nations (National Center for Health Statistics 2009; CIA World Fact Book 2005).

Racial inequality, along with class and gender inequality, is a huge part of the inequality that we see in the health of American people. How long you live, the nutrition in your diet, what kind of care you receive, even how you die are all things that are related to your race. Thus, on average, White men live six years longer than African American men, and White women four years longer than African American women. Among Hispanic adults, twice as many have no usual source of health care as compared to both White and African American adults. Also, White and Asian Americans are much more likely to be covered by private health insurance than are African Americans, Hispanics, and Native Americans (National Center for Health Statistics 2009).

As the authors in this section show, health disparities have been a persistent fact of life (and death!) in American society. The reasons are many, as detailed by H. Jack Geiger in "Health Disparities: What Do We Know? What Do We Need to Know? What Should We Do?". Geiger's article shows that racial bias is present in how providers view their clients and is one source of health care disparities. But there is also widespread inequality in the quality of care provided in different communities and regions of the country. Thus, factors that we have earlier examined here, such as racial segregation, also influence access to quality of health care. Geiger argues that we need more attention paid to interventions that will bring a higher standard of care to all groups. He argues for more education in "cultural competency" so that health care providers will be more sensitive to the needs of diverse populations.

Related to Geiger's call for cultural competency is Shirley A. Hill's essay ("Cultural Images and the Health of African American Women"). Hill explains how cultural stereotypes of African American women influence their health. She details the major stereotypes that have framed dominant perceptions of Black women, showing the cost of such stereotypes for the well-being of Black women. Stereotypes of Black women as strong and maternal, Hill argues, produce stress that compromises Black women's health. Even the beauty standards (what she calls "the beauty mandate") imposed by dominant cultural beliefs produce risks for African American women, such as obesity, hypertension, and even body satisfaction.

Beyond the health of people in the society, what about the society itself? How healthy are the environments in which different people live? Ultimately, the health of a society is tied to the health of the planet. As many now know, the health of the planet is threatened. This is a massive problem that affects all people, but as the authors here show, there are particular effects for people of color. David Naguib Pellow and Robert J. Brulle ("Poisoning the Planet: The Struggle for Environmental Justice") document that people of color are more likely than other groups to live near hazardous waste sites. An important point shown by Pellow and Brulle is that change is possible when people mobilize to address problems in their local community. Their article describes how people have organized to eliminate some of the toxic waste dumping that plagues their neighborhoods. The **environmental justice movement** is the term used to refer to the broad coalition of local groups who have organized to take action against the vulnerability of their communities to toxic waste.

The vulnerability of poor communities of color to environmental disasters was perhaps nowhere more clear than during and after Hurricane Katrina. Although multiple groups were severely impacted by the devastation of Katrina, the preexisting poverty of African American people in New Orleans made them particularly vulnerable. Furthermore, the failure of the government to respond quickly and effectively added to the death and misery of many poor African American people trapped by the rising floodwaters when the levees broke. Now, in the aftermath of Katrina, Robert D. Bullard and Beverly Wright ("Race, Place, and the Environment in Post-Katrina New Orleans") show how the recovery from Katrina has continued to disadvantage people of color in the Gulf region. Hurricane Katrina showed that even so-called natural disasters reflect social realities. As other environmental risks unfold, such as the BP oil leak in the Gulf of Mexico, we need to ask how different racial-ethnic and class communities are affected and what national responses can help protect the health and environment of us all.

# REFERENCES

Central Intelligence Agency. 2005. *The World Factbook*. Washington, DC: Central Intelligence Agency. www.cia.gov

DeNavas-Walt, Carmen, Bernadette D. Proctor, and Jessica C. Smith. 2009. *Income, Poverty, and Health Insurance Coverage in the United States: 2008.* Washington, DC: U.S. Department of Commerce.

National Center for Health Statistics. 2009. *Health United States 2009.* Atlanta, GA: Centers for Disease Control and Prevention.

## FACE THE FACTS: PROJECTED LIFE EXPECTANCY BY RACE

### Projected Life Expectancy at Birth, 2025 (in years)

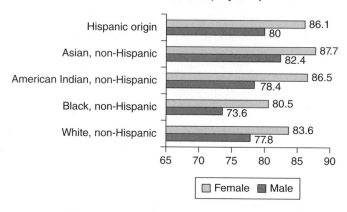

SOURCE: U.S. Census Bureau. 2009. *Statistical Abstract of the United States.* Washington, DC: U.S. Department of Commerce. www.census.gov

***Think about it:*** What specific factors influence the likelihood of a longer or shorter life? How does race connect to such factors?

# 45

# Health Disparities:

## What Do We Know? What Do We Need to Know? What Should We Do?

### H. JACK GEIGER

*One way that racial inequality is manifested is in the health disparities between groups. Here, Geiger reviews the major research findings that document health differences between racial-ethnic groups in the United States.*

One of the most persistent facts of life—and death—in the United States is that African Americans, Native Americans, Hispanics, and members of many Asian subgroups live sicker, die younger, and more often experience inferior health care, in comparison with the white majority. These are the phenomena frequently described as "health disparities," a term that has multiple meanings and is used too often without precise definition. Its major categories are disparities in health status and disparities in health care, but it is sometimes employed to refer to disparities in access to care, and frequently it is used nonspecifically to describe any or all three of these categories. It is worth observing, furthermore, that *disparities* carries at best a faint connotation of moral concern—stronger than the studiously neutral *differences* but weaker than *inequality* or *inequity*. These stronger terms place such phenomena more clearly in the broader American social context of differentiation—by race, ethnicity, social class, and gender—in power, privilege, and resources, and in the different life experiences and life opportunities that result.

... I briefly review the evidence of disparities in health status—the burdens of excess morbidity, mortality, and health-related quality of life that characterize poor and minority population groups in the United States. Next, I review disparities in health care—the differences in the quality, intensity, and comprehensiveness of the medical care afforded to patients of these same poor and minority populations. I point out that these two kinds of disparities are only indirectly related, since the medical care of any population makes only a modest contribution to its mortality experience or life expectancy. Both, however, are deeply rooted in, and reflect, a larger context: the pervasive and continuing power of race, social class, and gender as divisive and differentiating forces in defining the social determinants of health and the social, biological, and physical environments in which different American

SOURCE: From *Gender, Race, Class & Health: Intersectional Approaches,* edited by Amy J. Schulz and Leith Millings. San Francisco: Jossey-Bass, 2005.

groups live. I examine the existing evidence of causation for each type of disparity and identify important gaps in our knowledge and critical needs for further research. Finally, I propose policy changes, health system actions, and research programs that should be undertaken now to begin the process of eliminating inequity in both health status and health care.

## DISPARITIES IN HEALTH STATUS

At no time in the history of the United States has the health status of African Americans, Native Americans, Hispanics, and members of many Asian subgroups equaled or even approximated that of white Americans (Geiger, 2003; Byrd and Clayton, 2000; National Center for Health Statistics, 2003). While the health of all American racial, ethnic, and social class groups has improved dramatically over that long time span, and particularly during the past six decades, people of color and the less affluent continue to experience excess morbidity and mortality (Arias and Smith, 2003; Hajat, 2000). These facts of illness, disability, and premature death continue to be particularly severe for African Americans, whose mean life expectancy is almost six years shorter than that of non-Hispanic whites, and on some indicators, such as infant mortality rates, the black-white gap has recently been increasing (National Center for Health Statistics, 2003; U.S. Department of Health and Human Services, 2000; National Center for Education in Maternal and Child Health, 2000).

These gaps have been stubbornly resistant to change and have profound individual and social costs. If African Americans instead had experienced the same age-adjusted mortality rates as whites during the decade 1991–2000 alone, according to one recent estimate, as many as 866,000 black deaths would have been averted (Woolf, 2004). This is the cumulative toll of excess African American deaths (as compared with whites) from cardiovascular disease, cerebrovascular disease, cancer, diabetes, HIV, pneumonia and influenza, and complications of pregnancy (National Center for Health Statistics, 2001). Comparable cumulative estimates for preventable premature death could be made for Hispanic Americans, who have higher death rates than non-Hispanic whites for diabetes, liver disease, and HIV, and for many categories of excess mortality for Asian/Pacific Islanders and Native Americans/Alaska Natives.

These data and countless other findings of health status disparities by race, ethnicity, social class, gender, migration, acculturation status, and other variables are described and analyzed in detail in thousands of peer-reviewed journal articles, numerous government reports, and other ... research sources. Chapters in this book. What is of further significance is the evolution of explanations for these differences over time.

## WHAT CAUSES DISPARITIES IN HEALTH STATUS?

Throughout the nineteenth century and well into the twentieth, the conventional public wisdom, buttressed by elaborate, pseudo-scientific articles in the medical literature, was that there were biologically and genetically distinct human races—and that African Americans (along with southern and eastern Europeans, Chinese, and any darker-skinned immigrants of other nationalities) were biologically and intellectually inferior, more susceptible to disease (Cartwright, 1851), and, implicitly, a threat to the health of the body politic. The American eugenics movement of the 1920s, incorporating these beliefs, presaged the later full-blown claims of Nazi theories of racial superiority. Alleged biological inferiority and attendant racist stereotyping in turn were used to justify "the settled fact" of rigid segregation and second-class status in citizenship, education, employment, housing, and health (Finnegan, 2003).

Minority scholars throughout this period offered vigorous arguments rejecting biological determinism, but for the most part, this work was ignored by mainstream science and most of the public. In 1944, however, Gunnar Myrdal and his colleagues published *An American Dilemma*, a comprehensive examination of the conflict between the nation's formal commitments to equality and the reality of its racial practices. It offered a powerful argument for the social and structural causes—in particular, racism and discrimination—of the massive disparities it documented in education, employment, economic achievement, and other major sectors. It included a presentation of the then-current data on African American health status and explained that "area for area, class for class, Negroes cannot get the same advantage in the way of prevention and cure of disease that whites can.... Discrimination increases Negro sickness and death both directly and indirectly, and manifests itself both consciously and unconsciously" (pp. 171–172). "Discrimination" referred to both de jure and de facto practices in the larger society and to the structure of a segregated health care system. In the light of what we now know, these were astonishingly prescient statements.

Other, more variegated explanations of disparities in health status have dominated health policy and social science research in more recent decades. Health status disparities have been associated with socioeconomic status (SES) as measured by income, education, and employment; childhood poverty and social position; geographical variation; lack of access to health care; exposure to environmental and occupational hazards; racial/ethnic residential segregation and its effects on income, wealth, social isolation, and access to resources and opportunities; institutional and individual discrimination; lifestyle choices and behavioral risks; inferior housing and poor nutrition; differing beliefs about health and illness; and the consequences of income inequality (Williams, 2004; Berkman and Kawachi, 2000; Hofrichter, 2003; Krieger, 2004; Kawachi, Kennedy, and Wilkinson, 1999). Collectively, these give substance and detail to what has become classic public health doctrine: that the major determinants of population health status, and the primary explanations of disparities among population groups, lie in the social, physical, biological, economic, and political environments, which are determined by the larger society's norms, values, social stratification systems, and political economy (King, 1996; Menefee, 1996)....

In sum, the evidence of differential treatment for African Americans, Hispanics, Native Americans, and Asian/Pacific Islanders is credible and robust, in studies that are appropriately controlled for such major confounding variables as insurance status, income and education, age, gender, stage and severity of disease, and presence or absence of comorbid illness. Furthermore, these studies examine the care of patients who are already in the health care system, minimizing access to care as a factor. This pattern has been found not only for such high-technology interventions as angioplasty and coronary artery bypass grafting, advanced cancer chemotherapy, and renal transplantation but also for such routine processes as general medical and surgical procedures and the treatment of asthma, diabetes, congestive heart failure, and pneumonia. Disparities have been noted even in such basic elements of the clinical care of minority and poor patients as the adequacy of clinical history taking, physical examinations, and routine laboratory procedures (Kahn and others, 1994). A growing literature is documenting failure to provide adequate pain medication for minority patients—for example, for Hispanics with bone fractures (Todd, Samarov, and Hoffman, 1993) and African Americans with colorectal cancer (Cleeland and others, 1997)....

## GENDER, RACE, AND CLASS:
## INTERSECTORAL CLUES

Analyses of variations in health status by the major categories of race and class are important and enormously informative, but they are relatively crude approaches to understanding their causes, the wider social contexts in which they are embedded, or their impact on (or origins in) the everyday life experiences of women and men in these groups. Many categories are absurdly broad. "Hispanics" lumps together women of Puerto Rican, Dominican, Cuban, Central American, and Mexican origin (along with Brazilians, Chileans, Argentinians, and some actual Spaniards)— groups that vary in both social circumstance and various aspects of health status. "Asians" may refer to Chinese, Japanese, Korean, Vietnamese, Cambodian, Laotian, or Hmong groups, as well as to women of Indian, Pakistani, Bangladeshi, or other origins. The life experiences and health status of African Americans vary along an urban–rural continuum and differ from those of immigrants born in Africa or the Caribbean. In sum, every individual has multiple social identities—by "race," social class, age, primary language, cultural belief system, geographical region and neighborhood residence, and other variables—that affect health status in ways that may be cumulative, differentiated, or contradictory....

In every population group, women live longer than men but report more chronic illness during their lifetimes, but those figures vary by race and ethnicity. A recent Kaiser Foundation study (2004) found that nearly 20 percent of African American women and 29 percent of Latinas described their health status as only "fair" or "poor," compared with 13 percent of white women. Nearly 60 percent of African American women aged forty-five to sixty reported a diagnosis of hypertension, compared with 28 percent of white women in the same age group, and 40 percent of African American women of all ages said they had arthritis, as did 33 percent of Latinas and 32 percent of white women. Almost twice as many

African Americans and Latinas as white women had diabetes (16 percent, 17 percent, and 9 percent, respectively), and women of color were far likelier to suffer limb amputations, blindness, and end-stage renal disease. In contrast, roughly 25 percent of all women have been diagnosed with depression, but there is little variation by race/ethnicity (Kaiser Family Foundation, 2004). While low-income women of all race/ethnicity groups were twice as likely as the more affluent to be in fair or poor health, they also reported experiencing more barriers to health care because of lower health insurance coverage, problems with transportation, and the demands of work and child care. While I have previously noted that medical care makes only a small contribution to group health status, it has powerful effects on disability, pain and suffering, and quality of life....

Age, gender, race, and social circumstance intersect dramatically in the HIV/AIDS epidemic in the United States. AIDS has become the leading cause of death for African American women aged twenty-five to forty-four (Bernstein, 2003). In government studies of twenty-nine states, an African American woman was twenty-three times as likely to be infected with AIDS as a white woman, and black women accounted for roughly 72 percent of all new HIV cases in women from 1999 to 2002. Most of the studies exploring the causes of these findings have focused on individual-level risk factors, with relatively little consideration of social context. A recent study, in contrast, argues for the primacy of structural violence and community-level factors to explain racial and ethnic differences in infection rates. These include the disproportionate incarceration rates of African men, creating an African American sex ratio in which women outnumber men in many local communities; residential segregation and constraints on access to sexually transmitted disease services; and social norms stigmatizing homo sexuality, among other factors (Lane and others, 2004)....

## WHAT DO WE NEED TO KNOW?

Our knowledge of the existence of these inequities in care outstrips our understanding of their causes, though research has provided strong clues. Most of the earlier studies documenting unequal treatment included rather cursory and speculative discussions of possible causes but offered no hard data on what really influenced these specific and inequitable clinical decisions. Research in recent years has made it clear that there is no single smoking gun; causation, in the academic phrase, is multifactorial, and many different forces are in play (Smedley, Stith, and Nelson, 2003; Physicians for Human Rights, 2003; van Ryn and Fu, 2003). There is general agreement that these can be grouped into five broad categories: patient factors, provider factors, institutional factors, community factors, and geographical factors (Fiscella, 2004).

Patient factors frequently cited to explain disparities in care include patient choice or preference, usually implying a patient's refusal to accept a clinician's recommendation (Katz, 2001; Ayanian, Cleary, Weissman, and Epstein, 1999); cultural beliefs about health and medical care thought to be held by members of different minority groups; minority mistrust of the medical care system, based at least in part on high reported rates of perceived past discrimination (Corbie-Smith, Thomas, and St. George, 2002); language barriers (Fiscella, Franks, Doescher, and Saver, 2002); difficulties in cross-racial/ethnic, cross-gender, and cross-class physician–patient communication (Horowitz, Davis, Palermo, and Vladeck, 2000); inadequate health literacy (American Medical Association, 1999); and alleged biological differences in clinical presentation....

Provider factors include lack of cultural competence—that is, a physician's familiarity with systems of belief about health and illness among both majority and minority patients of different racial or ethnic groups, socioeconomic position, primary language, gender roles, sexual orientation, and age. It also includes some understanding of the physical and social environments in which different patients live, the wide varieties of family structure, and the constraints of time and money that affect the working poor, as well as an understanding that physicians and other providers bring to the consulting room their own cultural backgrounds and beliefs (Carillo, Green, and Betancourt, 1999). Provider factors also involve physicians' participatory decision-making practice styles (Kaplan and others, 1996); clinical uncertainty about the findings in the medical history or symptom presentation of minority patients, especially for conditions in which there are no clear treatment guidelines (van Ryn, 2002; Balsa and McGuire, 2001; Bloche, 2001); overt racism (Finucane and Carrese, 1990); and both conscious and unconscious racial/ethnic bias and negative racial stereotyping influencing clinical decisions (van Ryn and Fu, 2003).

The issue of bias, racial/ethnic stereotyping, and other forms of stereotyping by gender, social class, sexual orientation, or any combination of these, is central to understanding inequities in care and intersectoral concerns.... Bias and stereotyping are vehemently denied by many physicians, who are committed to the belief that they, their peers, and their institutions treat everyone fairly. And they are methodologically difficult to establish, since direct solicitation of providers' racial/ethnic views is likely to elicit socially acceptable answers and since a rich literature in cognitive psychology affirms that such stereotyping can occur in the absence of conscious prejudice and, indeed, below the level of conscious awareness. These studies also show that stereotyping is facilitated by time pressure and cognitive overload—having to think about many things at once—an apt description of clinical work....

Institutional factors include workforce diversity; availability of interpreters; geographical location of hospitals, physicians' offices, and other facilities; attitudes of nonphysician support staff; culturally appropriate signs, written materials, and health promotion programs; policies with regard to copayments, deductibles, and other out-of-pocket costs; and the presence or absence of intensive quality monitoring programs (Fiscella, 2004). The constraints on time for each patient visit, overall patient load, paperwork requirements, and other administrative burdens in managed care settings may also affect racially disparate outcomes in primary care, reducing the content and quality of physician–patient communication and increasing stereotyping (Barr, 1995).

Community factors include the proximity and availability of providers, the availability of regular sources of care (other than emergency rooms or walk-in clinics), the opportunities for continuity of care, the presence or absence of safety net providers, community levels of health knowledge and trust in the health care system, and community rates of insurance. A recent finding speaks to the power of residential segregation—and a previously unrecognized de facto segregation of medical care itself. In a national cross-sectional study of black and white Medicare patients, most primary care visits by black patients were with a small group of physicians (80 percent of their visits were accounted for by 22 percent of physicians), who provided only a small percentage of their care to white patients (Bach and others, 2004). In effect, to a large extent, these insured black and white patients were treated by different physicians. Furthermore, the physicians treating black patients were much more likely than those visited by white patients to report great difficulty in obtaining access for their patients to high-quality subspecialists, high-quality diagnostic imaging, and admission to the hospital, thus limiting the overall ability to provide high-quality care....

A recent cluster of geographical studies revealed yet another dimension of variation in the quality of care. In the United States, one study found, the overall quality of care (for all patients) varies significantly from one geographical area or metropolitan center to another, but variations in racial/ethnic disparities in the quality of care varied independent of the overall quality level. For black patients living in some predominantly black geographical communities, the quality of their care might be low even if black/white disparities were minimal. Pooled national data mask these regional differences. Here too it seems clear that residential racial segregation plays a role in the care that black patients receive (Baicker, Chancra, Skinner, and Wennberg, 2004).

What is insufficiently recognized is that all of these factors interact and may influence one another. Patient behavior, often rooted in community customs, cultural beliefs, and past experience, influences physician responses. Physician practice styles, often similarly influenced by community customs, cultural beliefs, and the cultures of undergraduate and graduate medical training, affect patient preferences. Lack of physician-patient concordance by race, gender, or social class affects communication, clinical decision making, and patient satisfaction. Stereotyping can operate in both directions: physicians' rote ascription of characteristics to minority patients and patients' beliefs that they are being treated with disrespect on the basis of race rather than as part of the frictions and insensitivities that afflict everyone in the health care system. The whole clinical encounter occurs in the context of a society that has multiple systems of social stratification, not least of which is the power imbalance between physicians and patients....

## WHAT SHOULD WE DO?

There are two broad approaches to intervention to reduce disparities in health care. The first is neutral on race/ethnicity, class, and gender: it focuses on the rigorous monitoring of quality of care and implementation of standardized expert guidelines and evidence-based clinical practices for all patients in all types of health care institutions. It draws on mounting evidence of appalling deficiencies in the quality of medical care, including a finding that barely half of the preventive and curative treatments provided to adults in the United States met recommended standards (Kerr and others, 2004; McGlynn and others, 2003; Institute of Medicine, 2001). The second approach, focused on the specific disparities documented by *Unequal Treatment* and so many other sources, proposes both research and policy actions to deal with the issues of race/ethnicity, gender, and class. They are not mutually exclusive, and both types of efforts are already under way.

A minimal but essential requirement is that all clinical records in every health facility be coded by the patient's self-reported race/ethnicity and, possibly, by primary language and indicators of social class. Without such information, it will not be possible to monitor disparities effectively; identify problem providers, institutions, or areas of care; search more precisely for causes; and design interventions. Resistance to such identification is manifested in "color-blindness" ballot initiatives to eliminate all governmental collection of these data, it raises minority suspicions of racial profiling, and it will inevitably contain some imprecision, but it is necessary—in medical care as in multiple other areas of American life—to count race until race no longer counts. Multiple organizational efforts are already under way by hospital associations, managed care organizations, health insurance companies, and others, but such efforts are still in their infancy. Federal and state government leadership—and mandates—will be essential, especially for Medicare and Medicaid records, but that will require political will (and, in all probability, a different national administration...).

Other efforts include widespread programs to teach cultural competency to physicians and other health providers practicing in a nation that is increasingly diverse and will soon reach "minority majority" status. We do not yet know which types of such education are most effective and at which stages of training, and there is no evidence yet of the effect of such provider education on the clinical care they provide, but such research also is in early stages of development and implementation. We know far too little of the values and attitudes students bring to their health professional training, how they may change in the course of that training, the relevant characteristics of the institutions in which they train, and the extent to which they vary by the social and personal characteristics of providers themselves.

# REFERENCES

American Medical Association, Council on Scientific Affairs, Ad Hoc Committee on Health Literacy for the Council on Scientific Affairs. "Health Literacy: Report of the Council on Scientific Affairs." *JAMA,* 1999, *281,* 552–557.

Arias, E., and Smith, B. L. "Deaths: Preliminary Data from 2001." *National Vital Statistics Report,* 2003, *51,* 1–44.

Ayanian, J. Z., Cleary, P. D., Weissman, S., and Epstein, A. M. "The Effect of Patients' Preferences on Racial Differences in Access to Renal Transplantation." *New England Journal of Medicine,* 1999, *341,* 1661–1669.

Bach, P. B., and others. "Primary Care Physicians Who Treat Blacks and Whites." *New England Journal of Medicine,* 2004, *351,* 575–584.

Baicker, J., Chancra, A., Skinner, J., and Wennberg, J. E. "Who You Are and Where You Live: How Race and Geography Affect the Treatment of Medicare Beneficiaries." *Health Affairs Web Exclusive.* 2004. http://www.healthaffairs.org.

Balsa, A. I., and McGuire, T. G. "Statistical Discrimination in Health Care." *Journal of Health Economics,* 2001, *20,* 881–907.

Barr, D. A. "The Effects of Organizational Structure on Primary Care Outcomes Under Managed Care." *Annals of Internal Medicine,* 1995, *122,* 353–359.

Berkman, L. A., and Kawachi, I. *Social Epidemiology.* New York: Oxford University Press, 2000.

Bernstein, L. A. "Achieving Equity in Women's and Perinatal Health." Dec. 12, 2003. http://www.Medscape.com/womenshealthhome.

Bloche, G. "Race and Discretion in Medical Care." *Yale Journal of Health Policy, Law and Ethics,* 2001, *1,* 95–131.

Byrd, W. M., and Clayton, L. A. *An American Health Dilemma,* New York: Routledge, 2000.

Carrillo, J. E., Green, A. R., and Betancourt, J. R. "Cross-Cultural Primary Care: A Patient-Based Approach." *Annals of Internal Medicine,* 1999, *130,* 829–834.

Cartwright, S. A. "Report on the Diseases and Physical Peculiarities of the Negro Race." *New Orleans Medical and Surgical Journal,* 1851, *8,* 28–37.

Cleeland, C. S., and others. "Pain and Treatment of Pain in Minority Patients with Cancer." *Annals of Internal Medicine,* 1997, *127,* 813–816.

Corbie-Smith, G., Thomas, S. B., and St. George, D. M. "Distrust, Race and Research." *Archives of Internal Medicine,* 2002, *162,* 2458–2463.

Finnegan, W. "The Fire Last Time." *New York Times Book Review,* Mar. 23, 2003, p. 12.

Finucane, T. E., and Carrese, J. A. "Racial Bias in Presentation of Cases." *Journal of General Internal Medicine,* 1990, *5,* 120–121.

Fiscella, K. "Within Our Reach: Equality in Health Care Quality." Paper presented at the Roundtable on Racial and Ethnic Disparities in Health Care Treatment, Harvard Law School Civil Rights Project, Boston, May 18, 2004.

Fiscella, K., Franks, P., Doescher, M. P., and Saver, B. "Disparities in Health Care by Race, Ethnicity and Language Among the Insured: Findings from a National Sample." *Medical Care,* 2002, *40,* 52–59.

Geiger, H. J. "Racial and Ethnic Disparities in Diagnosis and Treatment: A Review of the Literature and a Consideration of Causes." In B. D. Smedley, A. Y. Stith, and A. R. Nelson (eds.), *Unequal Treatment: Confronting Racial and Ethnic Disparities in Health* Care. Washington. D.C. National Academy Press, 2003.

Hajat, A. "Health Outcomes Among Hispanic Subgroups: Data from the National Health Interview Survey, 1992–1995." *Vital Health Statistics,* 2000, *310,* 111–114.

Hofrichter, R. (ed.). *Health and Social Justice: Politics, Ideology, and Inequity in the Distribution of Disease.* San Francisco: Jossey-Bass, 2003.

Horowitz, C.R., Davis, M. M., Palermo, A. S., and Vladeck, B. C. "Approaches to Eliminating Sociocultural Disparities in Health." *Health Care Financing Review,* 2000, *21,* 57–73.

Institute of Medicine. *Crossing the Quality Chasm: A New Health System for the Twenty-First Century.* Washington, D.C.: National Academy Press, 2001.

Kahn, K. L., and others. "Health Care for Black and Poor Hospitalized Medicare Patients." *Journal of the American Medical Association,* 1994, *271,* 1169–1174.

Kaiser Family Foundation. "Kaiser Women's Health Survey." Mar. 2004. http://www.kff.org/womenshealth.

Kaplan, S. H., and others. "Characteristics of Physicians with Participatory Decision-Making Styles." *Annals of Internal Medicine,* 1996, *124,* 497–504.

Katz, J. N. "Patient Preference and Healthcare Disparities." *Journal of the American Medical Association,* 2001, *286,* 1506–1509.

Kawachi, I., Kennedy, B. P., and Wilkinson, R. C. (eds.). *Income Inequality and Health.* New York: New Press, 1999.

Kerr, E. A., and others. "Profiling the Quality of Care in Twelve Communities: Results from the CQI Study." *Health Affairs,* 2004, *23,* 247–256.

King, G. "Institutional Racism and the Medical/Health Complex: A Conceptual Analysis." *Ethnicity and Disease,* 1996, *6,* 30–46.

Krieger, N. "Self-Reported Experiences of Racial Discrimination and Black/White Differences in Preterm and Low Birth Weight Deliveries: The CARDIA Study." *Journal of Epidemiology and Community Health,* 2004, *58,* 938–943.

Lane, S. D., and others. "Structural Violence and Racial Disparity in HIV Transmission." *Journal of Health Care for the Poor and Underserved,* 2004, *15,* 319–335.

McGlynn, E. A., and others. "The Quality of Health Care Delivered to Adults in the United States." *New England Journal of Medicine,* 2003, *348,* 2635–2645.

Menefee, L. T. "Are Black Americans Entitled to Equal Health Care? A New Research Paradigm." *Ethnicity and Disease,* 1996, *6,* 56–68.

Myrdal, G. *An American Dilemma.* New York: HarperCollins, 1944.

National Center for Education in Maternal and Child Health. *Knowledge Path: Infant Mortality.* Mar. 2000. http://www.necmh.org/pubs.

National Center for Health Statistics. *Health, United States, 2003.* Hyattsville, Md.: National Center for Health Statistics, 2003.

Physicians for Human Rights. "The Right to Equal Treatment: An Action Plan to End Racial and Ethnic Disparities in Clinical Diagnosis and Treatment in the United States." 2003. http://www.phrusa.org/righttoequaltreatment.pdf.

Smedley, B. D., Stith, A. Y., and Nelson, A. R. (eds.). *Unequal Treatment: Confronting Racial and Ethnic Disparities in Health Care.* Washington, D.C.: National Academy Press, 2003.

Todd, K. H., Samarov, N., and Hoffman, J. R. "Ethnicity as a Risk Factor for Inadequate Emergency Department Analgesia." *Journal of the American Medical Association,* 1993, *269,* 1537–1539.

U.S. Department of Health and Human Services. *Healthy People 2010.* Washington, D.C.: Government Printing Office, 2010.

van Ryn, M. "Research on the Provider Contribution to Race-Ethnicity Disparities in Medical Care." *Medical Care,* 2002, *40,* 140–151.

van Ryn, M., and Burke, J. "The Effect of Patient Race and Socioeconomic Status on Physicians' Perception of Patient." *Social Science and Medicine,* 2000, *50,* 813–828.

van Ryn, M., and Fu, S. S. "Paved with Good Intentions: Do Public Health and Human Service Providers Contribute to Racial/Ethnic Disparities?" *American Journal of Public Health,* 2003, *93,* 248–255.

Williams, D. R. "Racial Disparities in Health." Paper presented at the Conference on Racial/Ethnic and Socioeconomic Disparities in Health: Implications for Action, Washington, D.C., Apr. 29, 2004.

Woolf, S. "Society's Choice: The Tradeoff Between Efficacy and Equity and the Lives at Stake." *American Journal of Preventive Medicine,* 2004, 49–56.

## DISCUSSION QUESTIONS

1.  What different causes of health care disparities does Geiger identify? Which do you think are the most compelling?

2.  What benefits does the current national policy on health care provide for different groups?

# 46

# Cultural Images and the Health of African American Women

SHIRLEY A. HILL

*Health, for African American women, is conditioned by some of the cultural constructions of Black womanhood. Hill reviews the health status of Black women and links it to dominant beliefs about Black women and strength, the "motherhood mandate," and Black women's beauty images.*

… Sociologists have long observed that patterns of sickness, health, and mortality are primarily influenced by social inequalities based on social class, race, and gender. This research supports the merits of an intersectionality perspective on health, which contends that multiple social inequalities, especially those based on race, gender, and social class, intersect and mutually interact to create the specific social location and the life chances of groups and individuals (Baca Zinn and Dill 1996; P. H. Collins 1990). These systems of social inequality and domination are rooted in a complex set of economic, political, and ideological forces, and these structural forces influence the behaviors, cultural practices, and identities of individuals. For example, poverty is one of the strongest predictors of health and mortality: Poor people everywhere are sicker and have shorter life spans (Cockerham 2007; Payne 2006; Weitz 2010). Social factors such as malnutrition, poor housing, inadequate education, hazardous neighborhoods and working conditions, and high levels of stress explain why poor people are sicker. But poverty also shapes individual and cultural behaviors that lead to poor health, for example, violence, substance abuse, and the lack of exercise and preventive health care. This helps to explain the high death rate of African Americans, who have a poverty rate (24.3 percent) that is more than twice as high as that of whites (10.3 percent; CDC 2009, 175). The death rate for African Americans is higher than whites' for eight of the 10 leading causes of death (Spalter-Roth, Lowenthal, and Rubio 2005). But race and racism also negatively affect the health of African Americans. Middle-class African Americans have more serious illnesses and poorer self-related health than do their white counterparts (Farmer and Ferraro 2005) and are less likely to receive quality health care (Spalter-Roth, Lowenthal, and Rubio 2005).

SOURCE: *Gender & Society*, Vol. 23 No. 6, December 2009: 733–746.

Although they live longer than men, women have higher rates of sickness. This gender gap in health status begins at puberty, when girls start to report more anxiety, physical sickness, and depression than do boys and to evaluate their health as poorer (Lorber and Moore 2002; Sweeting 1995). It extends over the life course, even when reproductive and contraceptive issues are excluded. The poorer health of women is especially evident in their self-reported health status and the prevalence of mental health problems (Lorber and Moore 2002; Payne 2006). Feminist explanations of the gender gap in health have focused on male-dominated medicine, the medicalization of women's health issues, the secondary social and economic roles of women, and cultural images that depict them as naturally weaker and sicker than men (Zimmerman and Hill 2000). Feminists have also noted that cultural ideals about female attractiveness shape sickness and health care usage among women. Illnesses ranging from anorexia nervosa to depression are related to gender norms, and optional procedures such as lipo-suction, breast augmentation, and hormone therapy increase the use of medical services by women and can have adverse effects on women's health. The lives of women have changed in recent decades, yet physicians still medicalize women's health complaints and link their illnesses to the violation of gender traditions. For example, physicians have described endometriosis as the disease of career women who are "intelligent, living with stress [and] determined to succeed at a role other than 'mother' in early life" (Roberts 1997, 255).

Racial, class, and gender inequalities intersect in creating poor health profiles for African American women, who experience high rates of serious illnesses such as hypertension, heart disease, diabetes, and sexually transmitted diseases (C. F. Collins 2006; Weitz 2010). One study summed up the health status of African American women by noting that they live "truncated lives with a significant number of years spent in poor health" (Hayward and Heron 1999, 88). Researchers have attributed the health status of African Americans to multiple social inequalities (Lovejoy 2001) and have focused on social-structural inequalities and a lack of health-promoting behaviors (C. F. Collins 1996). Fewer, however, have examined the role of cultural images in shaping the health behaviors of African American women. The cultural construction of Black womanhood has been distinctively different from that of white womanhood. There is, for example, scant evidence of societal efforts to exclude African American women from productive labor or idealize them as housewives. And far from being seen as naturally fragile and prone to sicknesses, African American women have often been characterized as having unusual stamina and strength, a natural immunity to some diseases, and less sensitivity to pain (Herndl 1995; Payne 2006). Such images were used to justify the exploitation of the reproductive and productive labor of African American women, but once that became more difficult, they were defined with demeaning images such as jezebels, matriarchs, mammies, and welfare queens (P. H. Collins 1990). Such images have since been thoroughly contested by research exploring the historic and contemporary experiences of African American women. Including the voices of women of color has enriched feminist research and added credence to its key premise—that gender is socially constructed. But such research has also reinforced the search for racial and cultural differences among rather than similarities among women and propagated images of African American women that have important health consequences.

## THE SOCIAL CONSTRUCTION OF AFRICAN AMERICAN WOMANHOOD

Since the 1970s, the array of negative images constructed to characterize Black women gradually has been inverted, valorized, and repackaged as reflecting a distinct set of African-based cultural values. This more recent cultural construction of African American womanhood rejected racist stereotypes of Black women and distanced them from the dominant notions of white womanhood and femininity. For example, nothing in the historic experiences of African American women resonated with the notion that women were innately passive and dependent, and there was much to support the idea that they had been exempt from such notions of femininity and depicted as unusually strong. Similarly, when white feminists implicated motherhood and child rearing as contributing to the oppression and secondary status of women, some Black scholars argued that motherhood was a cultural value rooted in their African heritage. African American women, it was argued, often found motherhood a source of pride and power and shared the work of child rearing in female-centered families (P. H. Collins 1990). African American women also challenged notions of physical attractiveness that were based on white standards by embracing African-inspired attire and hairstyles and accepting their full body figures. Indeed, big bodies are associated with the notion of African American women as strong (Beauboeuf-LaFontant 2003).

These culturally constructed images of African American womanhood have been challenged in recent years but still inform scholarship on poor Black families and shape the identities of many African American women. Yet while strength, motherhood, and big bodies have been recast as elements of the cultural heritage of African Americans, this cultural construction of Black womanhood merely inverts derogatory stereotypes and has important health consequences. For example, African American women are no longer matriarchs but rather are strong; they are no longer mammies but value motherhood. Ironically, these cultural images of African American womanhood are typically built on the experiences of the poorest and least powerful women and are more prescriptive than descriptive. Thus, they stand to promote the behaviors that jeopardize health. Here, I examine notions of strength, motherhood, and full-figured bodies in terms of their potential health consequences.

### The Strength Mandate

The dominant and most pernicious dimension of the cultural construction of African American womanhood has been the notion of strength. Some have traced the strength mandate to the value placed on the productive labor of Black women in precolonial Africa and during slavery (Terborg-Penn 1986), but the notion is also imbued with connotations of surviving hardship and exploitation. Slavery treated women as harshly as men, so there was much hardship to endure: Most enslaved women were field workers who were denied the right to legal marriages and apt to see their children sold as property. In some cases, African American women were forced to prioritize the care of white children and families and were easy targets of sexual abuse. Even after slavery ended, racial and economic exploitation reduced the benefits and stability of marriage for African Americans, and single-mother families became common. It was within this context that African American sociologist E. Franklin Frazier (1939/1957) coined "the Black matriarch" concept, which not only labeled poor Black families as pathological but also impugned the women who headed them. As a matriarch, the Black woman was seen as free from conventional gender norms and as a "strong and independent person who placed little value on marriage, engaged without conscience in free sexual activity, and had no notion of male supremacy" (Dill 1979, 545).

The Black matriarch concept became the most politically contentious issue in the study of African American families during the civil rights era, and it was gradually replaced with a cultural strength perspective on Black families. As the heads of many of these families, Black women were also redefined as strong and capable. Beyond their family roles, scholars reinforced the strength image by highlighting the social activism of African American women, their ability to combine labor market and family work (Jones 1985), and their egalitarian marital relationships (Scanzoni 1977). Thus, the negative matriarch image was inverted into a closely related but more acceptable image of African American women as "strong" (Beauboeuf-LaFontant 2003). Scholars have defined their strength as having a sense of self-reliance and resourcefulness (P. H. Collins 1990) and independence and power (King 1988), and one study found that African American women continue to be depicted in the popular media as strong, independent, tough, streetwise, and powerful (Emerson 2002).

The strength mandate perpetuates other myths about African American women, such as the assumption that they can "go it alone" without others, a notion that fosters silence and social isolation among those who feel they are "less than a woman" if they show signs of weakness and vulnerability. Yet research shows that African American women experience high levels of chronic stress and are more likely than other women to feel powerless and depressed (Gibbs and Fuery 1994). The strength mandate denies that African American women need companionship, social support, or emotional intimacy and is often a barrier to relationships (Franklin 2000). It may also imply that there is gender equality between African American men and women, a notion that finds little support in empirical research. African American men are more likely than white men to have traditional gender ideologies (Blee and Tickamyer 1995), and African American women do more housework and child care than do their male partners (John and Shelton 1997). In fact, Black women, perhaps because of their dual family and labor market roles, have lower levels of marital happiness and well-being than do white women (Blum and Deussen 1996; Edin 2000;

Goodwin 2003). African American women are more likely to be victimized by severe domestic violence and less likely to trust authorities to help them (Hill 2005). One woman I interviewed several years ago, explaining why she did not leave an abusive relationship, said, "I thought if I was a strong enough person, I could make this relationship work" (Hill 2005, 192).

The notion of strength rarely recognizes that the cultural power of poor Black women has declined, as have their traditional support networks. As Deborah White (1999, 119) once pointed out, "If [Black women] seemed exceptionally strong, it was partly because they often functioned in groups and derived strength from numbers." In recent decades, many factors have challenged the traditional cultural power of African American women, such as the demise of Black communities, social-class polarization, changing social policies, and the male-dominated Black nationalist movement. K. Y. Scott (1991) described the crisis of Black womanhood as their declining ability to shape Black culture and keep people alive. Their strength, borne out of anger, bitterness, and disappointment, has endowed them with a "habit of surviving"—but the consequences are often dire.

### The Motherhood Mandate

Motherhood is a central value for women: A significant majority have or plan to have children. But motherhood has especially been seen as a strong value for African American women, often rooted in the power linked to female fertility in precolonial African societies (Christain 1994). Slavery reinforced and exploited the fertility of Black women and usually made the mother–child relationship the basis of African American families (Hill 2008). It also created the mammy image by vaunting the ability of African American women to rear and relate to white children—a notion repeatedly reproduced in the popular media. Once slavery ended, the pattern of nonmarital childbearing it had fostered became a "social problem," and African American women were blamed for creating dysfunctional families (Frazier 1939/1957) and subjected to coercive birth control and sterilization practices (Roberts 1997). Both of these issues proved contentious during the civil rights era, and scholars challenged pathological images of single-mother families and defended the reproductive rights of all women, especially women of color. This scholarship asserted the cultural value of motherhood for African American women and situated it within the context of African societies, where motherhood and childbearing were inherently valued and the concept of illegitimacy was unknown. Thus, it is not surprising that the white feminist critique of motherhood met with resistance from feminists of color.

But as Dorothy Roberts (1997) has explained, birth control was not entirely something that was thrust on African American women against their will; rather, Black people, especially those who were leaders and members of the middle class, often saw fertility control as vital to racial progress. Rates of childbirth among African American women declined after slavery ended and have continued to do so, with the white–Black gap in fertility rates reaching a historic low in 2005: 66.3 for white women and 69.0 for Black women (CDC 2009, 177). The current racial gap mostly reflects the higher fertility rate of African American teenagers and is related to poverty. Young, poor women of all races have babies for the same reason—the lack of other attractive life options (Edin and Kefalas 2005). But for poor African American women, the notion of motherhood as a cultural value may mask the fact that, from the moment of conception through the child-rearing years, motherhood is an especially perilous journey. Overall, they have more coercive sexual encounters, more unwanted pregnancies and abortions, higher rates of infant and maternal mortality, and less support for raising their children.

Many studies have shown that teenage pregnancy among African Americans is most often the result of the casual use of contraceptives, the desire to nurture and be loved, an effort to gain adult status, and few expectations of a better life (Edin and Kefalas 2005; Jacobs 1994; Plotnick 1992; J. W. Scott 1993). But it is also the case that African American adolescent girls often have little control over their own sexuality: They are more likely than white adolescents to be living apart from or inadequately supervised by parents and involved with older men and are twice as likely to be sexually abused or to engage in nonvoluntary sex (Abma, Driscoll, and Moore 1998). In consensual sexual relationships, they are ineffective in negotiating the use of condoms with their partners (Hill 1994), which helps explain their high rates of sexually transmitted diseases (Lorber and Moore 2002). Fully half of all African American girls between the ages of 14 and 19 (compared to 20 percent of white female teenagers) have a STD, and such diseases are linked to infertility and other health problems (Weitz 2010).

Unplanned and unwanted pregnancies entail other health risks, such as late or no prenatal care: The CDC (2009, 183) reported that about 29 percent of African American women (and 11 percent of white women) did not get prenatal care during the first trimester of their pregnancies. Infant mortality rates are twice as high for African American women, as is the likelihood of having a low-birth-weight infant. Moreover, having children before the age of 20 has a long-term impact on health: It increases the hazards of early death and, among those who reach midlife, correlates with higher rates of cancer, heart disease, and lung disease (Henretta 2007). Those who have more children and those who have poor prenatal and infant nutrition are more likely to have diabetes in adulthood (Henretta 2007). Thus, the notion of motherhood as one of the cultural values of African American women, especially in the lives of those who are young and poor, has important health consequences.

## The Beauty Mandate

Feminists have noted how norms of physical attractiveness disproportionately affect the behaviors and bodies of women, often leading to eating disorders, excessive dieting, and optional surgeries. A strong cultural consensus exists when it comes to defining female beauty, and it is based on white standards of physical attractiveness, for example, straight hair, thin bodies, and white skin. Although many African Americans once tried to comply with those standards, the civil rights era of the 1960s marked a growing pride in Blackness and efforts to redefine or broaden white beauty standards. The "Black is beautiful" initiative affirmed pride in kinky hair, dark skin, and bigger female bodies, but this pride has neither been fully embraced nor erased the social, economic, and health penalties associated with African features. African hairstyles still evoke negative social judgments, and colorism is pervasive among Black people and in the broader society. For African Americans, having light skin correlates with higher levels of self-esteem among teens and adolescents (Robinson and Ward 1995), higher occupational aspirations among college students (Hall 1995), and higher self-efficacy and self-esteem among women (Thompson and Keith 2001). Indeed, women are more likely than men to be penalized by colorism, and having dark skin is related to discrimination and adverse health consequences (Krieger, Sidney, and Coakley 1998).

Obesity is the most serious physical threat to the health of African American women and is linked to the major health problems they experience, such as diabetes, heart disease, and hypertension. More than 78 percent of African American women are classified as overweight, and slightly more than half are obese (C. F. Collins 2006). Disparities in the body weights of Black and white females start in childhood, mostly because of differences in health behaviors, but having big bodies is also tied to the social construction of African American womanhood. Being "heavy" is associated with being resilient, nourished, and able to survive anything (Beauboeuf-LaFontant 2003). It can also be seen as a rejection of dominant norms of physical attractiveness; indeed, Black women are more satisfied with their bodies than are white women and less likely to diet (Lovejoy 2001), and many note that African American men prefer big women. Eating habits conducive to obesity are ensconced in African American cultural norms; for example, having a hearty appetite is seen as more authentically Black than being a picky eater, and dieting is often seen as a "white thing." Sharing a meal of traditional African American foods, which are often fried and fatty, has been depicted in movies such as *Soul Food* as a way of maintaining and solidifying family ties.

This acceptance of obesity has often been theorized as evidence that African Americans have different standards of physical attractiveness than whites and has led many to assume an absence of eating disorders among African American women. But compulsive overeating is an eating disorder and may be rooted in the social stigma and stress experienced by African American women (Lovejoy 2001). Although there is evidence that young, upwardly mobile African American women are becoming more weight and health conscious, they still lag behind when it comes to engaging in health-promoting behaviors (Eyler et al. 1997).

## CONCLUSION

... The soaring cost of health care is a drain on the economy, and the lack of health care insurance for millions of American is unconscionable. But as medical sociologists have contended for decades, social-structural and cultural factors are the fundamental causes of sickness and diseases, and even universal access to care will not completely eliminate health inequities. The gender gap in health exemplifies that fact: Although women are more likely than men to have health care coverage and more likely to use health care resources, they are still sicker than men. In recent years, the gender gap in health and mortality has narrowed (Payne 2006), which may reflect the fact that the social and economic roles of women have changed. Women are more likely to be employed and increasingly to occupy high-level political positions. But gender equality has remained elusive, and the gender disparity in health status has not disappeared (Bird and Rieker 2008). The gender factor, when theorized in the context of racial and class inequities, is amplified for African American women. While social-structural inequalities account for health disparities, I have argued that the cultural images they produce also adversely affect health, especially for African American women. While universal access to health care is a laudable goal, the broader goal of a more equitable society will undoubtedly do more to improve the health outcomes of the poor, women, and racial minorities.

## REFERENCES

Abma, J. C., A. Driscoll, and K. A. Moore. 1998. Young women's degree of control over first intercourse: An exploratory analysis. *Family Planning Perspectives* 30(1): 12–18.

Baca Zinn, Maxine, and Bonnie Thornton Dill. 1996. Theorizing difference from multiracial feminism. *Feminist Studies* 22(2):321–31.

Beauboeuf-LaFontant, Tamara. 2003. Strong and large Black women? Exploring relationships between deviant womanhood and weight. *Gender & Society* 17:111–21.

Blee, K. M., and A. R. Tickamyer. 1995. Racial differences in men's attitudes about women's gender roles. *Journal of Marriage and Family* 57:21–30.

Blum, Linda M., and Theresa Deussen. 1996. Negotiating independent motherhood: Working-class African American women talk about marriage and motherhood. *Gender & Society* 10:199–211.

Centers for Disease Control and Prevention. 2009. *Health, United States, 2008.* US Department of Health and Human Services, National Center for Health Statistics (March 2009), Publication No. 2009-1232.

Christain, Barbara. 1994. An angle of seeing: Motherhood in Buchi Emecheta's *Joys of Motherhood* and Alice Walker's *Meridian.* In *Mothering: Ideology, experience, and agency,* edited by E. N. Glenn, G. Chang, and L. R. Forcey. New York: Routledge.

Cockerham, William C. 2007. *Social causes of health and disease.* Malden, MA: Polity.

Collins, Catherine Fisher. 1996. Commentary on the health and social status of African-American women. In *African-American women's health and social issues,* edited by Catherine Fisher Collins. Westport, CT: Auburn House.

Collins, Catherine Fisher. 2006. Introduction. In *African American women's health and social issues,* 2nd ed., edited by Catherine Fisher Collins. Westport, CT: Praeger.

Collins, Patricia Hill. 1990. *Black feminist thought: Knowledge, consciousness, and the politics of empowerment.* Boston: Unwin Hyman.

Dill, Bonnie Thornton. 1979. The dialectic of Black womanhood. *Signs: Journal of Women in Culture and Society* 4(3):545–55.

Edin, Kathryn. 2000. What do low-income single mothers say about marriage? *Social Problems* 47(1):112–13.

Edin, Kathryn, and Maria Kefalas. 2005. *Promises I can keep: Why poor women put motherhood before marriage.* Berkeley: University of California Press.

Emerson, Rana A. 2002. "Where my girls at?" Negotiating Black womanhood in music videos. *Gender & Society* 16:115–35.

Eyler, A. A., R. C. Brownson, A. C. King, D. Brown, R. J. Donatelle, and G. Heath. 1997. Physical activity and women in the United States: An overview of health benefits, prevalence, and intervention opportunities. *Women & Health* 26:27–39.

Farmer, Melissa M., and Kenneth F. Ferraro. 2005. Are racial disparities in health conditional on socioeconomic status? *Social Science & Medicine* 60:191–204.

Franklin, Donna L. 2000. *What's love got to do with it? Understanding and healing the rift between Black men and women.* New York: Touchstone.

Frazier, E. Franklin. 1939/1957. *The Negro in the United States.* New York: Macmillan.

Gibbs, Jewelle Taylor, and Diana Fuery. 1994. Mental health and well-being of Black women: Toward strategies for empowerment. *American Journal of Community Psychology* 22(4):559–82.

Goodwin, Paula Y. 2003. African American and European American women's marital well-being. *Journal of Marriage and Family* 65:550–60.

Hall, Ronald. 1995. The bleaching syndrome: African Americans' response to cultural domination vis-à-vis skin color. *Journal of Black Studies* 26:172–84.

Hayward, M. D., and M. Heron. 1999. Racial inequality in active life among adult Americans. *Demography* 36:77–91.

Henretta, John C. 2007. Early childbearing, marital status, and women's health and mortality. *Journal of Health and Social Behavior* 48(3): 254–66.

Herndl, D. P. 1995. The invisible (invalid) woman: African-American women, illness, and nineteenth century narrative. *Womens, Studies* 24:553–73.

Hill, Shirley A. 1994. Motherhood and the obfuscation of medical knowledge: The case of sickle cell disease. *Gender & Society* 8:29–47.

Hill, Shirley A. 2005. *Black intimacies: A gender perspective on families and relationships.* Walnut Creek, CA: AltaMira.

Hill, Shirley A. 2008. African American mothers: Victimized, vilified, and valorized. In *Feminist mothering,* edited by A. O. O'Reilly. New York: State University of New York Press.

Jacobs, Janet L. 1994. Gender, race, class, and the trend toward early motherhood. *Journal of Contemporary Ethnography* 22(4):442–62.

John, Daphne, and Beth Anne Shelton. 1997. The production of gender among Black and white women and men: The case of household labor. *Sex Roles* 36(3-4): 171–93.

Jones, Jacqueline. 1985. *Labor of love, labor of sorrow: Black women, work, and the family from slavery to the present.* New York: Basic Books.

King, Deborah. 1988. Multiple jeopardy, multiple consciousness: The context of Black feminist ideology. *Signs: Journal of Women in Culture and Society* 14(1):42–72.

Krieger, N., S. Sidney, and E. Coakley. 1998. Racial discrimination and skin color in CARDIA: Implications for public health research. *American Journal of Public Health* 88:1308–13.

Lorber, Judith, and Lisa Jean Moore. 2002. *Gender and the social construction of illness.* Walnut Creek, CA: AltaMira.

Lovejoy, Meg. 2001. Disturbances in the social body: Differences in body image and eating problems among African American and white women. *Gender & Society* 15:239–61.

Payne, Sarah. 2006. *The health of men & women.* Malden, MA: Polity.

Plotnick, Robert D. 1992. The effects of attitudes on teenage premarital pregnancy and its resolution. *American Sociological Review* 57(6):800–811.

Roberts, Dorothy. 1997. *Killing the Black body: Race, reproduction, and the meaning of liberty.* New York: Pantheon.

Robinson, T. L., and J. V. Ward. 1995. African American adolescents and skin color. *Journal of Black Psychology* 21:256–74.

Scanzoni, John. 1977. *The Black family in modern society: Patterns of stability and security.* Chicago: University of Chicago Press.

Scott, J. W. 1993. African American daughter-mother relations and teenage pregnancy? Two faces of premarital teenage pregnancy. *Western Journal of Black Studies* 17(2): 73–81.

Scott, Kesho Yvonne. 1991. *The habit of surviving: Black women's strategies for life.* New Brunswick, NJ: Rutgers University Press.

Spalter-Roth, Roberta, Terri Ann Lowenthal, and Mercedes Rubio. 2005. Race, ethnicity, and the health of Americans. *ASA Series on How Race and Ethnicity Matter* (July):1–16.

Sweeting, H. 1995. Reversals of fortune? Sex differences in health in childhood and adolescence. *Social Science & Medicine* 40:77–90.

Terborg-Penn, Rosalyn. 1986. Women and slavery in the African diaspora: A cross-cultural approach to historical analysis. *Sage: A Scholarly Journal on Black Women* 3:11–15.

Thompson, Maxine S., and Verna M. Keith. 2001. The blacker the berry: Gender, skin tone, self-esteem, and self-efficacy. *Gender & Society* 15:336–57.

Weitz, Rose. 2010. *The sociology of health, illness, and health care: A critical approach.* Boston: Wadsworth.

White, D. G. 1999. *Too heavy a load: Black women in defense of themselves, 1894–1994.* New York: Norton.

Zimmerman, Mary K., and Shirley A. Hill. 2000. Re-forming gendered health care: An assessment of change. *International Journal of Health Services* 30(4):769–93.

## DISCUSSION QUESTIONS

1. What are the cultural constructions of Black women that Hill identifies, and how are they related to Black women's health?

2. Some say that "history repeats itself." How does Hill see Black women's history as affecting their current health status?

# 47

# Poisoning the Planet:
## The Struggle for Environmental Justice

DAVID NAGUIB PELLOW AND ROBERT J. BRULLE

*Pellow and Brulle identify how the neighborhoods of people of color are often the dumping grounds for toxic waste and other pollutants. These are hazards to people's health and well-being. The authors talk about how people are fighting back as part of an environmental justice movement. They place the activism of people of color within a context of national and international racial politics.*

One morning in 1987 several African-American activists on Chicago's southeast side gathered to oppose a waste incinerator in their community and, in just a few hours, stopped 57 trucks from entering the area. Eventually arrested, they made a public statement about the problem of pollution in poor communities of color in the United States—a problem known as environmental racism. Hazel Johnson, executive director of the environmental justice group People for Community Recovery (PCR), told this story on several occasions, proud that she and her organization had led the demonstration. Indeed, this was a remarkable mobilization and an impressive act of resistance from a small, economically depressed, and chemically inundated community. This community of 10,000 people, mostly African-American, is surrounded by more than 50 polluting facilities, including landfills, oil refineries, waste lagoons, a sewage treatment plant, cement plants, steel mills, and waste incinerators. Hazel's daughter, Cheryl, who has worked with the organization since its founding, often says, "We call this area the 'Toxic Doughnut' because everywhere you look, 360 degrees around us, we're completely surrounded by toxics on all sides."

## THE ENVIRONMENTAL JUSTICE MOVEMENT

People for Community Recovery was at the vanguard of a number of local citizens' groups that formed the movement for environmental justice (EJ). This movement, rooted in community-based politics, has emerged as a significant player at the local, state, national, and, increasingly, global levels. The movement's origins lie in local activism during the late 1970s and early 1980s aimed

SOURCE: From *Contexts* 6(1): 37–41: 2007.

at combating environmental racism and environmental inequality—the unequal distribution of pollution across the social landscape that unfairly burdens poor neighborhoods and communities of color.

The original aim of the EJ movement was to challenge the disproportionate location of toxic facilities (such as landfills, incinerators, polluting factories, and mines) in or near the borders of economically or politically marginalized communities. Groups like PCR have expanded the movement and, in the process, extended its goals beyond removing existing hazards to include preventing new environmental risks and promoting safe, sustainable, and equitable forms of development. In most cases, these groups contest governmental or industrial practices that threaten human health. The EJ movement has developed a vision for social change centered around the following points:

- All people have the right to protection from environmental harm.
- Environmental threats should be eliminated before there are adverse human health consequences.
- The burden of proof should be shifted from communities, which now need to prove adverse impacts, to corporations, which should prove that a given industrial procedure is safe to humans and the environment.
- Grassroots organizations should challenge environmental inequality through political action.

The movement, which now includes African-American, European-American, Latino, Asian-American/Pacific-Islander, and Native-American communities, is more culturally diverse than both the civil rights and the traditional environmental movements, and combines insights from both causes.

## ENVIRONMENTAL INEQUALITIES

Researchers have documented environmental inequalities in the United States since the 1970s, originally emphasizing the connection between income and air pollution. Research in the 1980s extended these early findings, revealing that communities of color were especially likely to be near hazardous waste sites. In 1987, the United Church of Christ Commission on Racial Justice released a groundbreaking national study entitled *Toxic Waste and Race in the United States*, which revealed the intensely unequal distribution of toxic waste sites across the United States. The study boldly concluded that race was the strongest predictor of where such sites were found.

In 1990, sociologist Robert Bullard published *Dumping in Dixie*, the first major study of environmental racism that linked the siting of hazardous facilities to the decades-old practices of spatial segregation in the South. Bullard found that African-American communities were being deliberately selected as sites for the disposal of municipal and hazardous chemical wastes. This was also one of the first studies to examine the social and psychological impacts of environmental pollution in a community of color. For example: across five communities in

Alabama, Louisiana, Texas, and West Virginia, Bullard found that the majority of people felt that their community had been singled out for the location of a toxic facility (55 percent); experienced anger at hosting this facility in their community (74 percent); and yet accepted the idea that the facility would remain in the community (77 percent).

Since 1990, social scientists have documented that exposure to environmental risks is strongly associated with race and socioeconomic status. Like Bullard's *Dumping in Dixie*, many studies have concluded that the link between polluting facilities and communities of color results from the deliberate placement of such facilities in these communities rather than from population-migration patterns. Such communities are systematically targeted for the location of polluting industries and other locally unwanted land uses (LULUs), but residents are fighting back to secure a safe, healthy, and sustainable quality of life. What have they accomplished?

## LOCAL STRUGGLES

The EJ movement began in 1982, when hundreds of activists and residents came together to oppose the expansion of a chemical landfill in Warren County, North Carolina. Even though that action failed, it spawned a movement that effectively mobilized people in neighborhoods and small towns facing other LULUs. The EJ movement has had its most profound impact at the local level. Its successes include shutting down large waste incinerators and landfills in Los Angeles and Chicago; preventing polluting operations from being built or expanded, like the chemical plant proposed by the Shintech Corporation near a poor African-American community in Louisiana; securing relocations and home buyouts for residents in polluted communities like Love Canal, New York; Times Beach, Missouri; and Norco, Louisiana; and successfully demanding environmental cleanups of LULUs such as the North River Sewage Treatment plant in Harlem.

The EJ movement helped stop plans to construct more than 300 garbage incinerators in the United States between 1985 and 1998. The steady expansion of municipal waste incinerators was abruptly reversed after 1990. While the cost of building and maintaining incinerators was certainly on the rise, the political price of incineration was the main factor that reversed this tide. The decline of medical-waste incinerators is even more dramatic.

Sociologist Andrew Szasz has documented the influence of the EJ movement in several hundred communities throughout the United States, showing that organizations such as Hazel Johnson's People for Community Recovery were instrumental in highlighting the dangers associated with chemical waste incinerators in their neighborhoods. EJ organizations, working in local coalitions, have had a number of successes, including shutting down an incinerator that was once the largest municipal waste burner in the Western Hemisphere. The movement has made it extremely difficult for firms to locate incinerators, landfills, and related LULUs anywhere in the nation, and almost any effort to expand existing polluting facilities now faces controversy.

# BUILDING INSTITUTIONS

The EJ movement has built up local organizations and regional networks and forged partnerships with existing institutions such as churches, schools, and neighborhood groups. Given the close association between many EJ activists and environmental sociologists, it is not surprising that the movement has notably influenced the university. Research and training centers run by sociologists at several universities and colleges focus on EJ studies, and numerous institutions of higher education offer EJ courses. Bunyan Bryant and Elaine Hockman, searching the World Wide Web in 2002, got 281,000 hits for the phrase "environmental justice course," and they found such courses at more than 60 of the nation's colleges and universities.

EJ activists have built lasting partnerships with university scholars, especially sociologists. For example, Hazel Johnson's organization has worked with scholars at Northwestern University, the University of Wisconsin, and Clark Atlanta University to conduct health surveys of local residents, study local environmental conditions, serve on policy task forces, and testify at public hearings. Working with activists has provided valuable experience and training to future social and physical scientists.

The EJ movement's greatest challenge is to balance its expertise at mobilizing to oppose hazardous technologies and unsustainable development with a coherent vision and policy program that will move communities toward sustainability and better health. Several EJ groups have taken steps in this direction. Some now own and manage housing units, agricultural firms, job-training facilities, farmers' markets, urban gardens, and restaurants. On Chicago's southeast side, PCR partnered with a local university to win a federal grant, with which they taught lead-abatement techniques to community residents who then found employment in environmental industries. These successes should be acknowledged and praised, although they are limited in their socio-ecological impacts and longevity. Even so, EJ activists, scholars, and practitioners would do well to document these projects' trajectories and seek to replicate and adapt their best practices in other locales.

# LEGAL GAINS AND LOSSES

The movement has a mixed record in litigation. Early on, EJ activists and attorneys decided to apply civil rights law (Title VI of the 1964 Civil Rights Act) to the environmental arena. Title VI prohibits all government and industry programs and activities that receive federal funds from discriminating against persons based on race, color, or national origin. Unfortunately, the courts have uniformly refused to prohibit government actions on the basis of Title VI without direct evidence of discriminatory intent. The Environmental Protection Agency (EPA) has been of little assistance. Since 1994, when the EPA began accepting Title VI claims, more than 135 have been filed, but none has been formally resolved. Only one federal agency has cited environmental justice concerns to protect a community in a significant legal case: In May 2001, the Nuclear Regulatory Commission denied a permit for a uranium enrichment plant in Louisiana because environmental justice concerns had not been taken into account.

With regard to legal strategies, EJ activist Hazel Johnson learned early on that, while she could trust committed EJ attorneys like Keith Harley of the Chicago Legal Clinic, the courts were often hostile and unforgiving places to make the case for environmental justice. Like other EJ activists disappointed by the legal system, Johnson and PCR have diversified their tactics. For example, they worked with a coalition of activists, scholars, and scientists to present evidence of toxicity in their community to elected officials and policy makers, while also engaging in disruptive protest that targeted government agencies and corporations.

## NATIONAL ENVIRONMENTAL POLICY

The EJ movement has been more successful at lobbying high-level elected officials. Most prominently, in February 1994, President Clinton signed Executive Order 12898 requiring all federal agencies to ensure environmental justice in their practices. Appropriately, Hazel Johnson was at Clinton's side as he signed the order. And the Congressional Black Caucus, among its other accomplishments, has maintained one of the strongest environmental voting records of any group in the U.S. Congress.

But under President Bush, the EPA and the White House did not demonstrate a commitment to environmental justice. Even Clinton's much-vaunted Executive Order on Environmental Justice has had a limited effect. In March 2004 and September 2006, the inspector general of the EPA concluded that the agency was not doing an effective job of enforcing environmental justice policy. Specifically, he noted that the agency had no plans, benchmarks, or instruments to evaluate progress toward achieving the goals of Clinton's Order. While President Clinton deserves some of the blame for this, it should be no surprise that things have not improved under the Bush administration. In response, many activists, including those at PCR, have shifted their focus from the national level back to the neighborhood, where their work has a more tangible influence and where polluters are more easily monitored. But in an era of increasing economic and political globalization, this strategy may be limited.

## GLOBALIZATION

As economic globalization—defined as the reduction of economic borders to allow the free passage of goods and money anywhere in the world—proceeds largely unchecked by governments, as the United States and other industrialized nations produce larger volumes of hazardous waste, and as the degree of global social inequality also rises, the frequency and intensity of EJ conflicts can only increase. Nations of the global north continue to export toxic waste to both do-mestic and global "pollution havens" where the price of doing business is much lower, where environmental laws are comparatively lax, and where citizens hold little formal political power.

Movement leaders are well aware of the effects of economic globalization and the international movement of pollution and wastes along the path of least resistance (namely, southward). Collaboration, resource exchange, networking, and joint action have already emerged between EJ groups in the global north and south. In the last decade EJ activists and delegates have traveled to meet and build alliances with colleagues in places like Beijing, Budapest, Cairo, Durban, The Hague, Istanbul, Johannesburg, Mumbai, and Rio de Janeiro. Activist colleagues outside the United States are often doing battle with the same transnational corporations that U.S. activists may be fighting at home. However, it is unclear if these efforts are well financed or if they are leading to enduring action programs across borders. What is certain is that if the EJ movement fails inside the United States, it is likely to fail against transnational firms on foreign territory in the global south.

Although EJ movements exist in other nations, the U.S. movement has been slow to link up with them. If the U.S. EJ movement is to survive, it must go global. The origins and drivers of environmental inequality are global in their reach and effects. Residents and activists in the global north feel a moral obligation to the nations and peoples of the south, as consumers, firms, state agencies, and military actions within northern nations produce social and ecological havoc in Latin America, the Caribbean, Africa, Central and Eastern Europe, and Asia. Going global does not necessarily require activists to leave the United States and travel abroad, because many of the major sources of global economic decisionmaking power are located in the north (corporate headquarters, the International Monetary Fund, the World Bank, and the White House). The movement must focus on these critical (and nearby) institutions. And while the movement has much more to do in order to build coalitions across various social and geographic boundaries, there are tactics, strategies, and campaigns that have succeeded in doing just that for many years. From transnational activist campaigns to solidarity networks and letter-writing, the profile of environmental justice is becoming more global each year.

After Hazel Johnson's visit to the Earth Summit in Rio de Janeiro in 1992, PCR became part of a global network of activists and scholars researching and combating environmental inequality in North America, South America, Africa, Europe, and Asia. Today, PCR confronts a daunting task. The area of Chicago in which the organization works still suffers from the highest density of landfills per square mile of any place in the nation, and from the industrial chemicals believed to be partly responsible for the elevated rates of asthma and other respi-ratory ailments in the surrounding neighborhoods. PCR has managed to train local residents in lead-abatement techniques; it has begun negotiations with one of the Big Three auto makers to make its nearby manufacturing plant more eco-logically sustainable and amenable to hiring locals, and it is setting up an envi-ronmental science laboratory and education facility in the community through a partnership with a major research university.

What can we conclude about the state of the movement for environmental justice? Our diagnosis gives us both hope and concern. While the movement has accomplished a great deal, the political and social realities facing activists (and all of us, for that matter) are brutal. Industrial production of hazardous wastes continues to increase exponentially; the rate of cancers, reproductive illnesses, and respiratory disorders is increasing in communities of color and poor communities; environmental inequalities in urban and rural areas in the United States have remained steady or increased during the 1990s and 2000s; the income gap between the upper classes and the working classes is greater than it has been in decades; the traditional, middle-class, and mainly white environmental movement has grown weaker; and the union-led labor movement is embroiled in internecine battles as it loses membership and influence over politics, making it likely that ordinary citizens will be more concerned about declining wages than environmental protection. How well EJ leaders analyze and respond to these adverse trends will determine the future health of this movement. Indeed, as denizens of this fragile planet, we all need to be concerned with how the EJ movement fares against the institutions that routinely poison the earth and its people.

## DISCUSSION QUESTIONS

1. What is the vision of the environmental justice movement and how is an understanding of racism central to this vision?

2. What do you learn from the environmental justice movement about the activism of communities of color?

3. Why do environmental justice activists in the United States have to connect with activists in other nations?

# 48

# Race, Place, and the Environment in Post-Katrina New Orleans

ROBERT D. BULLARD AND BEVERLY WRIGHT

*The poverty of African Americans in New Orleans prior to Hurricane Katrina made them uniquely vulnerable to the devastating consequences of this storm. Even though Katrina hurt many groups, racial inequalities still abound in the recovery and rebuilding efforts in the post-Katrina New Orleans.*

The year 2005 saw the worst Atlantic hurricane season since record keeping began in 1851 (Cuevas 2005). An average season produces ten named storms, of which about six become hurricanes and two or three become major hurricanes. But 2005 saw the most named storms ever, 27, topping the previous record of 21 in 1933—and 13 hurricanes—breaking the old record of 12 in 1969 (Tanneeru 2005). And on August 29, 2005, of course, Hurricane Katrina laid waste to New Orleans. Katrina's death toll of 1,836 and counting made it the third most deadly hurricane in U.S. history, after the 1928 Okeechobee hurricane in Florida, which killed 2,500, and the 1900 Galveston hurricane, which killed 8,000 (Ho 2005, A1). The disaster in New Orleans after Katrina was unnatural and man-made. Flooding in the New Orleans metropolitan area largely resulted from breached levees and flood walls (Gabe, Falk, McCarthy, and Mason 2005). A May 2006 report from the Russell Sage Foundation, *In the Wake of the Storm: Environment, Disaster, and Race After Katrina*, found these same groups often experience a "second disaster" after the initial storm (Pastor et al. 2006).

Hurricane Katrina demonstrated that negative effects of climate change fall heaviest on the poor and people of color (Dyson 2006; Pastor et al. 2006). Eighty percent of New Orleans was flooded. Low-income and people-of-color neighborhoods were hardest hit. Pre-storm vulnerabilities limit participation of thousands of Gulf Coast low-income communities of color in the after-storm reconstruction, rebuilding, and recovery. In these communities, days of hurt and loss are likely to become years of grief, dislocation, and displacement....

New Orleans, like most major urban centers, was a city in peril long before Hurricane Katrina's floodwaters devastated the city (Pastor et al. 2006; Dyson 2006). New Orleans (Orleans Parish) had a population of 484,674 in 2000. Of

SOURCE: Robert D. Bullard and Beverly Wright, eds. 2009. *Race, Place, and Environmental Justice After Hurricane Katrina.* Boulder, CO: Westview Press.

this total, 325,947 (68 percent) were African Americans, 135,956 (28 percent) were non-Hispanic whites, and 22,871 (4 percent) were of other ethnic groups. Like many great cities, New Orleans also had its share of problems. The economic structure of the city made it difficult to provide jobs with wages high enough to support a family. New Orleans' economy was built around low-wage service jobs in the tourism sector....

## CLEANING UP AFTER KATRINA

Before Katrina, over 50 percent (some studies place this figure at around 70 percent) of children living in the inner-city neighborhoods of New Orleans had blood lead levels above the current guideline of 10 micrograms per deciliter (mcg/dl) (Mielke 1999). Childhood lead poisoning in some New Orleans black neighborhoods was as high as 67 percent (Rabiro, White, and Shorter 2004)....

Katrina has been called one of the worst environmental disasters in U.S. history. A September 2005 *Business Week* commentary described the handling of the untold tons of "lethal goop" as the "mother of all toxic cleanups" (2005). However, the billion-dollar question facing New Orleans is which neighborhoods will get cleaned up, which ones will be left contaminated, and which ones will be targeted as new sites to dump storm debris and waste from flooded homes....

## DESTRUCTION OF LOW-INCOME AND
## WORKING-CLASS HOUSING

All eyes are watching New Orleans' rebuilding efforts, especially how it addresses the repopulation of its historically African-American neighborhoods and its strategically sited public housing. The Housing Authority of New Orleans was dismantling traditional public housing for nearly a decade before Katrina through Hope VI, a Clinton-era program that favors vouchers and mixed-income developments. Dramatic population shifts occurred in New Orleans as a result of the Hope VI project, which displaced thousands of public housing residents. Gentrification of historically black areas was becoming a problem for many citizens.

The St. Thomas redevelopment in New Orleans in the late 1990s became the prototype for elite visions of the city's future. Strategically sited public housing projects like the St. Thomas homes were demolished to make way for neo-traditionalist townhouses and stores (in the St. Thomas case, a Wal-Mart) in the New Urbanist spirit. These "mixed-use, mixed-income" developments were typically advertised as little utopias of diversity, but—as in St. Thomas in New Orleans, Olympic Village (formerly Techwood Homes) in Atlanta, and similar places around the country—the real dynamic is exclusionary rather than inclusionary, with only a few project residents being rehoused on the development site.

After Katrina, HUD announced it would invest $154 million in rebuilding public housing in New Orleans and assist the city to bring displaced residents home, but critics fear that government officials and business leaders are quietly planning to demolish the old projects and privatize public housing. Ten months after Katrina, 80 percent of public housing in New Orleans remained closed. Six of ten of the largest public housing developments in the city were boarded up, with the other four in various states of repair....

New Orleans' homeless population has skyrocketed since Katrina—reaching an unprecedented 4 percent of the total population in 2008—12,000 homeless people, nearly double the pre-Katrina count. New Orleans' homeless rate is more than four times that of most U.S. cities. The cities with homeless rates closest to that of New Orleans are Atlanta (1.4 percent) and Washington (0.95 percent), both majority-black cities (Jervis 2008a)....

Powerful forces have been trying to demolish public housing in New Orleans for decades. When Katrina emptied New Orleans of public housing residents, the *Wall Street Journal* reported U.S. Congressman Richard Baker, a ten-term Republican from Baton Rouge, telling lobbyists: "We finally cleaned up public housing in New Orleans. We couldn't do it, but God did" (Babington 2005, A04). The demolition of four sprawling public housing projects—the St. Bernard, C. J. Peete, B. W. Cooper, and Lafitte housing developments—represents more than half of all of the conventional public housing in the city, where only 1,097 units were occupied ten months after the storm....

Although Katrina did not discriminate, a May 2008 progress report from the Louisiana Family Recovery Corps found a wide disparity in adaptation and recovery between black and white storm victims: "There is great disparity in the progress towards recovery, disruption from the storms and levels of progress between black and white households, even for those with similar incomes. On nearly every indicator, the storm impact and recovery experience for black households is significantly different than for whites, even after examining these issues by income levels" (Alfred 2008, 12)....

## A "SAFE" ROAD HOME

Katrina and the failures of the federal levee system displaced more than 378,000 people from New Orleans, creating "one of the largest disaster diasporas in U.S. history" (Jervis 2008c, 1A). Three years after Katrina, population estimates vary on how many people have actually made it back. Some demographers place the total population of the city between 315,000 and 320,000 residents, estimated by utility and water hookups, mail delivery, and other public service accounts. In August 2008, the Brookings Institution estimated that New Orleans had reached 72 percent of its 453,726 pre-Katrina level (Liu and Plyer 2008). The storm cut deeper for African-American households than for white households as 47 percent of African-American households live someplace different, compared to only 19 percent of white households (Alfred 2008, 16).

Since Katrina, the New Orleans African-American population has plummeted by 57 percent, while the white population has fallen less, by only 36 percent. African Americans now make up 58 percent of New Orleans compared to 67 percent before the storm. New Orleans has been a predominately black city for three decades, but now some well-known African-American communities are a fraction of what they were, and others see their very existence threatened. For example, the Lower Ninth Ward has seen only 9.9 percent of its population return. A traditionally mixed-race neighborhood within the Lower Ninth, Holy Cross, has fared better with a 37 percent return, benefiting from the work of preservationists who seek to restore the federally declared historic district....

Katrina hit New Orleans' mostly African-American blue-collar workers, individuals who never lived in public housing and who often made ends meet by working two jobs, especially hard. With limited plans to replace rental units lost in the storm, the city is at risk of losing an entire tier of workers. It is no surprise that such a large share of the African-American working-class population is still stranded three years after the storm. This trend can be observed in job vacancy rates in the cleaning and maintenance sector that are up from 4.1 percent before Katrina to 13.1 percent now, in the restaurant sector from 3.6 percent to 13.4 percent, and in other service jobs from 6.3 percent to 16.7 percent (Gonzales 2008b).

The government has been slow to invest in bricks-and-mortar housing for working-class families. By March 2008, FEMA had paid to Louisiana 93 percent of the $6.6 billion infrastructure allocation, but only 47 percent had actually reached localities. Overall, Katrina relief and rebuilding funds have only trickled down to local governments and residents. Given the enormity and urgency of the need, one would think much more would have been done after three years.

FEMA even withheld disaster relief supplies from Katrina victims. In June 2008, nearly three years after the storm, the first truckload of $85 million in federal relief supplies, lost in a bureaucratic hole, arrived in Louisiana and were distributed to those still displaced by Katrina and Rita. The supplies had been stored in Fort Worth for two years, and FEMA finally deemed them surplus goods early in 2008 after the building's owner decided to demolish the structure.

The road home for many Katrina survivors has been bumpy, largely due to slow government actions to distribute the $116 billion in federal aid to residents to rebuild. Only about $35 billion has been appropriated for long-term rebuilding. Most of the Katrina money coming from Washington hasn't gotten to those most in need—and the funding squeeze is stopping much of the Gulf Coast from coming back (Kromm and Sturgis 2007)....

## DYING FOR A HOME—TOXIC FEMA TRAILERS

Right after Katrina, FEMA purchased about 102,000 travel trailers for $2.6 billion, or roughly $15,000 each (Spake 2007). Soon there were reports of residents becoming ill in these trailers due to the release of potentially dangerous levels of formaldehyde, a known carcinogen (Hampton 2006). In fact, formaldehyde was omnipresent in the glues, plastics, building materials, composite wood, plywood panels, and particle board used to manufacture the trailers.

In Mississippi, FEMA received 46 complaints by individuals who had symptoms of formaldehyde exposure, including eye, nose, and throat irritation, nausea, skin rashes, sinus infections, depression, inflamed mucus membranes, asthma attacks, headaches, insomnia, intestinal problems, memory impairment, and breathing difficulties (Schwartz 2007; Spake 2007; Hampton 2006; Johnson 2007). The Sierra Club conducted tests of 31 trailers and found that 29 had unsafe levels of formaldehyde (Hampton 2006; Damon 2007; Brunker 2006). According to the Sierra Club, 83 percent of the trailers tested in Alabama, Louisiana, and Mississippi had formaldehyde levels above the EPA limit of 0.10 parts per million (Schwartz 2007; Brunker 2006).

Even though FEMA received numerous complaints about toxic trailers, the agency only tested one occupied trailer to determine the levels of formaldehyde in it (Committee on Oversight and Government Reform 2007). The test confirmed that the levels of formaldehyde were extraordinarily high and presented an immediate health risk to the occupants (Committee on Oversight and Government Reform 2007). Unfortunately, FEMA did not test any more occupied trailers and released a public statement discounting any risk associated with formaldehyde exposure.

According to findings from a congressional committee hearing, FEMA deliberately neglected to investigate any reports of high levels of formaldehyde in trailers so as to bolster FEMA's litigation position in case individuals affected by their negligence decided to sue them (Damon 2007; Babington 2007). In fact, more than 500 hurricane survivors and evacuees in Louisiana are pursuing legal action against the trailer manufacturers for formaldehyde exposure. Two years after Katrina, more than 65,000 Gulf Coast families, an estimated 195,000 people, were living in FEMA trailers. The vast majority of the trailers, about 45,000, were in Louisiana (Alberts 2007; Damon 2007; Babington 2007).

In July 2007, FEMA stopped buying and selling disaster relief trailers because of the formaldehyde contamination (Johnson 2007). FEMA administrator R. David Paulison admitted that the trailers used by displaced Katrina residents were toxic and concluded that the agency should have moved faster in addressing the health concerns of residents (Cruz 2007). In August 2007, FEMA began moving families out of the toxic trailers and finding them new rental housing. Testing of FEMA travel trailers for formaldehyde and other hazards began in September 2007 (Treadway 2007). The Centers for Disease Control and Prevention was tasked with developing parameters for testing the travel trailers.

In February 2008, more than two and a half years after residents of FEMA trailers began complaining of breathing difficulties, nosebleeds, and persistent headaches, CDC officials announced that long-awaited government tests had found potentially hazardous levels of toxic formaldehyde gas in travel trailers and mobile homes provided by FEMA. CDC tests found that levels of formaldehyde gas in 519 trailers and mobile homes tested in Louisiana and Mississippi were—on average—about five times what people are exposed to in most modern homes (Maugh and Jervis 2008). More than 38,000 families, or roughly

114,000 individuals, were living in FEMA provided travel trailers or mobile homes along the Gulf Coast at the time of the CDC tests—down from a high of about 144,000 families....

## LET THEM FIND FOOD

Before Katrina, predominantly African-American communities in New Orleans were struggling with the mass closings of shopping centers and grocery stores. Many watched in horror at the explosion of chain-store fast-food restaurants, liquor stores, dollar stores, pawn shops, and check-cashing shops in their neighborhoods. Having to travel great distances for the ordinary amenities of life made life more and more difficult. After Katrina, middle- and upper-middle-class black neighborhoods have fallen victim to the same fate. All must drive long distances to white neighborhoods for supermarkets, shopping centers, and quality restaurants.

In a 2007 survey of low-income Orleans Parish residents, nearly 60 percent were more than three miles from a supermarket while only 50 percent owned cars. Additionally, of those surveyed, 70 percent reported that they "would buy" or "might buy" fresh produce items if they were available in their neighborhoods. Moreover, the study showed that low-income people "like" to eat fruit and vegetables as much as or more than unhealthy foods (The New Orleans Food Policy Advisory Committee 2007).

Access to fresh, nutritious food was inadequate in New Orleans even before Katrina. At that time, there were about 12,000 residents per supermarket while the nation's average was 8,800 residents (New Orleans Food Policy Advisory Committee 2007). Now, nearly three years after Katrina, the availability of these types of foods has only gotten worse. Today, there are nearly 18,000 residents per supermarket. There are presently only 18 supermarkets open in New Orleans. Adding to this woeful lack of stores is the fact that the smaller stores that have reopened are not meeting the demand for fresh produce....

Access to fresh healthy foods, like fruits and vegetables, high in nutrients and low in salt, fat, and calories, is vital to the good health of the people in our communities. Research in New Orleans' Central City neighborhood revealed that greater access to fresh vegetables has led to increased consumption of these foods by residents of the neighborhood (New Orleans Food Policy Advisory Committee 2008). Improving access to healthy foods would lead to better dietary practices and the resultant better health of individuals and families in underserved communities. In the rebuilding of New Orleans, we must reverse this trend of poor access to healthy foods leading to poor dietary health.

## UNEQUAL LEVEE PROTECTION

The Army Corps of Engineers is working to fix or replace 220 miles of levees and floodwalls, build new flood gates and pump stations at the mouths of three outfall canals, and strengthen existing walls and levees at important points. By May 2008, the Corps had spent $4 billion of the $14 billion set aside by Congress to repair and upgrade the metropolitan area's hundreds of miles of levees by 2011. Some outside experts say that there are leaks in the new levees, that some of the work already completed may need to be redone, and that billions more will be needed (Burdeau 2008).

The latest report including flood maps produced by the Army Corps of Engineers shows no increase in levee protection to New Orleans East residents since Katrina (Army Corps of Engineers Interagency Performance Evaluation Task Force 2007).

A disproportionately large swath of black New Orleans once again is left vulnerable to future flooding. After nearly two years and billions spent on levee repairs, the Army Corps of Engineers has estimated that there is a 1-in-100 annual chance that about a third of the city will be flooded with as much as six feet of water (Schwartz 2007).

Mostly African-American parts of New Orleans are still likely to be flooded in a major storm. Increased levee protection maps closely correspond with race of neighborhoods, black neighborhoods such as the Ninth Ward, Gentilly, and New Orleans East receiving little, if any, increased flood protection. These disparities could lead insurers and investors to redline and think twice about supporting the rebuilding efforts in vulnerable black areas.

The Lakeview-area resident can expect 5.5 feet of increased levee protection. This translates into 5.5 feet less water than what they received from Katrina. Lakeview is mostly white and affluent, New Orleans East is mostly black and middle class. This same scenario holds true for the mostly black Lower Ninth Ward, Upper Ninth Ward, and Gentilly neighborhoods. There is a racial component to the post-Katrina levee protection. Whether you are rich, poor, or middle class, if you are a black resident of New Orleans, you are less protected and you have received less increased flood protection from the federal government than the more white and affluent community of Lakeview.

Racism has taken an unmeasured toll on the lives of minorities and the poor. We say unmeasured because institutionalized racism has influenced policy that discriminates in ways that better serve the white and more affluent populations and communities. Katrina and its impacts, in a very powerful and revealing way, showed the world how race and class are intrinsically tied to policy. Moreover, it pointedly displayed how government policy can actually be harmful to the health and well-being of vulnerable populations (racial minorities, the poor, the sick and elderly, and children).

The scenes of stranded New Orleanians trapped on the roof of the crumbling Superdome and people dying on the street outside the Superdome and the New Orleans Convention Center are visions tragically etched in our collective memory. What was obvious to all was that policies for responding to disasters were woefully inadequate and needed to change.

What the New Orleans recovery process is also showing is that policies intended to be race-neutral can accelerate rather than alleviate the destructiveness of a disaster for the most vulnerable populations if the policies are not also race-sensitive....

# CONCLUSIONS

… We can only speculate on what progress could have been made toward rebuilding New Orleans and returning most of its citizens if the environmental clean-up that we deserved had been done. What if the same priority for clean-up and safety given to the French Quarter, the Central Business District, and the racetrack had been given to the Lower Ninth Ward, New Orleans East, and other hard-hit sections of the city?

Just after the storm, an article in the *Dallas Morning News* quoted the Army Corps of Engineers as saying that it would take the Corps three months to scrape the city clean of all contaminated soil and sediment (Loftis 2005). This, of course, did not happen. What did occur was politics as usual, and the losers were the citizens of New Orleans, with African Americans taking the biggest hit.

Residents of devastated New Orleans neighborhoods do not need government agencies debating the "chicken or egg" contamination argument ("Which came first, the contamination or Katrina?"). They need the government to clean up the mess. All levels of government have a golden opportunity to get it right this time. Clean-up and reconstruction efforts in New Orleans have been shamefully sluggish and patchy, and environmental injustice may be compounded by rebuilding on poisoned ground.

The opportunities are only fading as Katrina slowly slips off the political radar. It is no accident that not one word about Katrina and the Gulf Coast reconstruction was mentioned in President Bush's State of the Union address in January 2007—seventeen months after the devastating storm. Displaced residents need a "road home" program that is not only fair but also safe. It is immoral—and should be illegal—to unnecessarily subject Katrina survivors to contamination—whether the pollution was there before or after the storm.

Clearly, prevention and precaution should be the driving force behind the environmental clean-up in post-Katrina New Orleans. Either we all pay now or we all pay later. It will cost more in terms of dollars and ill health if we wait. The nation cannot allow another immoral, unethical, and illegal "human experiment" to occur in New Orleans and the Gulf Coast. The solution is prevention. In July 2008, FEMA sought immunity from lawsuits over potentially dangerous fumes in government-issued trailers that have housed tens of thousands of Gulf Coast hurricane victims (Kunzelman 2008). Lawyers for the trailer home plaintiffs want the cases certified as a class action on behalf of tens of thousands of current and former trailer occupants in Alabama, Louisiana, Mississippi, and Texas. Such cases and legal wrangling often take years to resolve.

# REFERENCES

Alfred, D. 2008. *Progress for Some, Hope and Hardship for Many.* New Orleans: Louisiana Family Recovery Corps. May.

Army Corps of Engineers Interagency Performance Evaluation Task Force. 2007. "Risk and Reliability Report." June 20. Available at http://nolarisk.usace.atmy.mil (accessed July 1, 2008).

Babington, C. 2005. "Some COP Legislators Hit Jarring Notes in Addressing Katrina." *Washington Post*. September 5.

Brunker, M. 2006. "FEMA Trailers 'Toxic Tin Cans'?" July 23. Available at http://risingfromruin.msnbc.com/2006/07/are_fema_traile.html (accessed August 1, 2006).

*Business Week*. 2005. "The Mother of All Toxic Cleanups." September 26. Available at http://www.businessweek.com/magazine/content/05_39/b3952055.htm (accessed December 21, 2005).

Committee on Oversight and Government Reform. 2007. "Committee Probes FEMA's Response to Reports of Toxic Trailers." July 19. Available at http://oversight.house.gov/story.asp?ID=1413 (accessed July 17, 2008).

Cuevas, F. 2005. "Fla. Eyes on Strengthening Wilma." *Atlanta Journal-Constitution*. October 18. A6.

Damon, A. 2007. "FEMA Covered Up Toxic Danger in Trailers Given to Katrina Victims." July 21. Available at http://www.wsws.org/articles/2007/jul2007/fema-j21.shtml (accessed July 17, 2008).

Dyson, M. E. 2006. *Come Hell or High Water: Hurricane Katrina and the Color of Disaster*. New York: Basic Books.

Gabe, T., G. Falk, M. McCarthy, and V. W. Mason. 2005. *Hurricane Katrina: Social-demographic Characteristics of Impacted Areas*. Washington, DC: Congressional Research Service Report RL33141. November.

Gonzales, J. M. 2008b. "New Orleans Working Class Hit by Cost Squeeze." *Boston Globe*. January 27.

Hampton, M. 2006. "Formaldehyde in FEMA Travel Trailers Making People Sick." August 8. Available at http://www.homelandstupidity.us/2006/08/08/formalde-hyde-in-fema-travel-trailers-making-people-sick (accessed July 17, 2008).

Ho, D. 2005. "The Worst Hurricane Season Ever." *Atlanta Journal-Constitution*. November 30.

Jervis, R. 2008a. "New Orleans' Homeless Rate Swells to 1 in 25." *USA Today*. March 17.

Johnson, A. 2007. "FEMA Suspends Use of 'Toxic' Trailers." Available at http://www.msnbc.msn.com/id/20165754 (accessed July 17, 2008).

Kromm, C., and S. Sturgis. 2007. *Blueprint for Gulf Renewal: The Katrina Crisis and a Community Agenda for Action*. Durham, NC: Institute for Southern Studies.

Liu, A., and A. Plyer. 2008. *The New Orleans Index, Tracking Recovery of the New Orleans Metro Area: Anniversary Edition Three Years After Katrina*. The Brookings Institution and Greater New Orleans Community Data Center.

Maugh, T. H., and J. Jervis. 2008. "FEMA Trailers Toxic, Tests Show." *Los Angeles Times*. February 15.

Mielke, H. 1999. "Lead in the Inner Cities: Policies to Reduce Children's Exposure to Lead May Be Overlooking a Major Source of Lead in the Environment." *American Scientist* 87, no. 1 (January/February).

New Orleans Food Policy Advisory Committee. 2008. *Building Healthy Communities: Expanding Access to Fresh Food Retail*. New Orleans: A Report by the New Orleans Food Policy Advisory Committee.

Pastor, M., R. D. Bullard, J. K. Boyce, A. Fothergill, R. Morello-Frosch, and B. Wright. 2006. *In the Wake of the Storm: Environment, Disaster and Race After Katrina*. NewYork: Russell Sage Foundation.

Rabito, F. A., L. E. White, and C. Shorter. 2004. "From Research to Policy: Targeting the Primary Prevention of Childhood Lead Poisoning." *Public Health Reports* 119 (May/June).

Schwartz, S. M. 2007. "Deja Vu, Indeed: The Evolving Story of FEMA's Toxic Trailers." July 16. Available at http://www.toxictrailerscase.com (accessed July 17, 2008).

Spake, A. 2007. "Dying for a Home: Toxic Trailers Are Making Katrina Refugees Ill." *The Nation.* February 15. Available at http://www.alternet.org/katrina/48004 (accessed July 17, 2008).

Tanneeru, M. 2005. "It's Official: 2005 Hurricanes Blew Records Away." CNN.com. December 30. Available at http://www.cnn.com/2005/WEATHER/12/19/hurricane.season.ender (accessed June 22, 2008).

## DISCUSSION QUESTIONS

1.  Were a disaster to strike your community, which groups would be most vulnerable? How might this be affected by race and class?

2.  Find a current news report on the recovery status of New Orleans and the Gulf Coast. How has race affected the recovery effort up until now?

# Student Exercises

1. Are there specific racial stereotypes about women and men in your racial-ethnic group? What are they and how might they affect the health of each?

2. Using a community where you grew up as an example, describe any threats to this community that come from hazardous waste and/or other forms of pollution. How does this compare to other communities in your state or locale?

# Section XII

# Criminal Injustice? Courts, Crime, and the Law

## Elizabeth Higginbotham and Margaret L. Andersen

In a democratic society, all individuals should be treated fairly by the law, law enforcement officials, and the courts. Beginning with the Fourteenth Amendment to the U.S. Constitution (adopted in 1868), U.S. citizens are guaranteed "equal protection under the law." Furthermore, the nation has laws in place, such as the Civil Rights Bill of 1964, that prohibit discrimination based on race, creed (that is, religion), national origin, and sex. Yet we know that, even after decades of people mobilizing for equal rights, race still plays a role in the treatment of individuals in all institutions—perhaps nowhere more starkly than in the courts and the criminal justice system. There is racial profiling on the highways, on city streets, in airports, and in other public places. Police and other law enforcement officials can be a hostile presence in minority communities. African Americans, Native Americans, and Latinos are more likely than White people to be brought into the criminal justice system and, once there, are generally treated differently than others.

In this section, we ask, how just is the justice system? Is it mostly operating to protect people's rights or is it mostly operating as a system of social control? Reams of research document that African Americans have a unique relationship with law enforcement. Whereas it may be invisible to many White Americans, surveillance of minority communities and differential treatment before the law is a fact for people of color. Now—at a time when the nation is fearful of "foreigners" and sees itself as under threat—minority men in particular have been harshly subjected to stereotypes as potentially violent criminals. Thus, the society is granting more power to the criminal justice system, even though this is a poor substitute for other social reforms such as education. National immigration policy focuses more on border control and the criminalization of immigrants than on social services that would help immigrants gain citizenship. Middle Eastern citizens are subjected to perceptions that they are potential terrorists. Low-income, inner-city neighborhoods are routinely portrayed in evening newscasts as violent, dangerous places—not places where people are trying to make a living and keep their families well and safe. And our nation's prisons are burgeoning in their population—at a time when we spend far more on prisons than on education. Altogether, these realities mean that for many, criminal justice means injustice and, for others, the law has become a source of intimidation and harassment, not a rational system of dispute resolution or protection from harm.

Bruce Western ("Punishment and Inequality") explores this trend in his analysis of the large growth in imprisonment rates of African American men. He argues, in part, that this growth is partially a reaction to the political mobilization within the Black community in the 1960s and 1970s. At the same time, he shows how disruptive imprisonment is for providing economic opportunities to Black men. But beyond individual life chances, this prison boom also disrupts whole communities, robbing people of the chance to get ahead.

Rubén Rumbaut, Robert Gonzales, Golnaz Komaie, and Charlie V. Moran ("Debunking the Myth of Immigrant Criminality: Imprisonment among First and Second Generation Young Men") explore the social myth that immigrants are prone to criminal behavior. Based on a large national study, they actually find that the second generation is far more likely to engage in criminal behavior, which they link to a process of "Americanization." Their research is a strong counterpoint to many of the stereotypes that abound about immigrants, at the same time that it points to the structural conditions that produce crime.

Christina Swarns ("The Uneven Scales of Capital Justice") discusses how race and social class play a role in criminal cases, particular where the death penalty is involved. The U.S. Supreme Court declared the death penalty unconstitutional in 1972, but it has been reinstated in many states across the nation since 1976. Swarm examines the biases that still exist at various levels of the criminal justice system. She explores why unchecked power influences who is likely to be prosecuted, found guilty, and sentenced to death. In addition to identifying the bias in the system, we can see how patterns of segregation originating in some areas reverberate in the courts and prisons.

Race is demonstrated to be paramount in the criminal justice system in an important study by Devah Pager ("The Mark of a Criminal Record"). Pager had four testors, two White men and two Black men, look for work presenting themselves as either having a criminal record or not. She finds that employers are more likely to hire a White man with a criminal record than a Black man without one.

Perhaps the challenge for this new century is to dismantle both the current racial segregation and the ideologies that target people of color as problems. Such injustices keep us from working together to make this nation a land of equality and social justice for all.

## FACE THE FACTS: IMPRISONMENT IN THE U.S. POPULATION

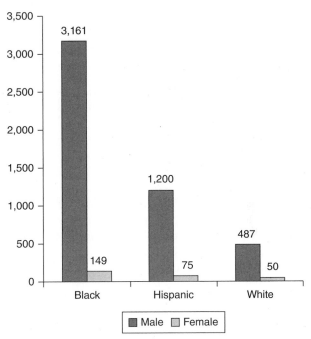

**Imprisonment Rate per 100,000 Persons in the U.S. Population 2008**

SOURCE: Bureau of Justice Statistics. 2009. *Prisoners in 2008*. Washington, DC: U.S. Department of Justice. www.bjs.gov

***Think about it:*** What impact do rates of imprisonment in different racial-ethnic groups have on other life experiences—such as family stability, political participation, and community health?

# 49

# Punishment and Inequality in America

BRUCE WESTERN

*Western argues that the rise in imprisonment of African American men forestalls much chance of economic progress and leaves our nation vulnerable to division and stagnating racial inequality.*

In 1831, Alexis de Tocqueville and Gustave de Beaumont were dispatched to America to study the penitentiary, a novel institution generating great discussion among the social reformers of Europe. At that time, two institutions—Auburn State Prison in New York and the Eastern Penitentiary in Philadelphia—offered leading examples of a new approach to the public management of criminals. The institutions were devised for moral correction. Rigorous programs of work and isolation would remedy the moral defects of criminal offenders so they might safely return to society. The penitentiary was billed as a triumph of progressive thinking that provided a humane and rational alternative to the disorderly prisons and houses of correction in Europe. Tocqueville and Beaumont were just two of many official visitors from Europe who toured the prisons in the 1830s, eager to view the leading edge of social reform.

Grand projects in crime control often spring from deep fissures in the social order. Tocqueville and Beaumont saw this clearly, despairing of "a state of disquiet" in French society. Writing in 1833, they traced the need for prison reform to a restless energy in the minds of men "that consumes society for want of other prey." This moral decline was compounded by the material deprivation of the French working class, "whose corruption, beginning in misery, is completed in prison." Instead of deflecting vice and poverty, the French prisons made things worse—aggravating immiseration and immorality. America offered a fresh alternative.

Although the prisons that provided the pretext for Tocqueville's American tour did not figure in his observations on American democracy, democratic aspirations were faintly inscribed on the Auburn and Pennsylvania penitentiaries. The project of rehabilitation assumed an innate moral equality among men that could be restored to criminals through penal discipline. Rehabilitative institutions comprised part of a primitive social democracy that conferred not just the vote and freedom of association but also a minimal equality of life chances. Despite curtailing freedom (and applying corporal punishment), the prison posed no basic

SOURCE: Bruce Western. 2009. *Punishment and Inequality in America*. New York: Russell Sage Foundation.

threat to democracy because the official ideology of rehabilitation promised to reestablish the social membership of those who had fallen into poverty and crime. In practice, of course, the rehabilitative ideal was regularly compromised and in the South it barely took hold at all. In conception at least, and sometimes in practice, the prison sat comfortably alongside an array of welfare institutions that included not only reformatories and asylums but also public schools, hospitals, and rudimentary schemes for social insurance. Like other welfare institutions, the prison was conceived to rescue the citizenship of the unfortunate, the poor, and the deviant.

... The latest revolution in criminal punishment followed some of the logic of its nineteenth-century predecessor. Shifts in the structure of society and politics forced changes in criminal justice, with large consequences for the quality of American democracy. Through the last decades of the twentieth century, the patchwork system of American criminal justice turned away from the rehabilitative project first attempted in New York and Pennsylvania. By the 1970s, policy experts were skeptical that prisons could prevent crime by reforming their inmates. Incarceration would be used less for rehabilitation than for incapacitation, deterrence, and punishment. Politicians vowed to get tough on crime. State lawmakers abandoned the rehabilitative ideals etched in the law of criminal sentencing and opted for mandatory prison terms, the abolition of parole, and long sentences for felons on their second and third convictions. Tough new sentences were attached to narcotics offenses as the federal government waged first a war on crime, then a war on drugs. Locked facilities proliferated around the country to cope with the burgeoning penal population. Prison construction became an instrument for regional development as small towns lobbied for correctional facilities and resisted prison closure.

Prisons themselves changed as a result of the punitive turn in criminal justice. Budgets tightened for education and work programs. But some social service function remained as the penal system assumed new responsibilities for public health, delivering treatment on a large scale for mental illness, tuberculosis, HIV/AIDS, and hepatitis C. High-risk inmates were gathered in supermax facilities that placed entire prison populations in solitary confinement. In a thousand ways, large and small, the democratic aspirations of rehabilitative corrections were erased and the coercive power of the state penetrated more deeply into the lives of the poor.

Most striking was the increase in the size of the correctional population. Between 1970 and 2003, state and federal prisons grew sevenfold to house 1.4 million convicted felons serving at least one year behind bars, and typically much longer. Offenders held in county jails, awaiting trial or serving short sentences, added another seven hundred thousand by 2003. In addition to the incarcerated populations, another 4.7 million people were under probation and parole supervision. The entire correctional population of the United States totaled nearly seven million in 2003, around 6 percent of the adult male population.

Growth in the penal population signaled more than a change in public policy. Throughout the twentieth century, African American history has been entwined with the history of America's prisons. Blacks have been more likely than whites to go to prison, at least since the 1920s. Southern prisons operated quite transparently as instruments of racial domination, using forced labor to farm cotton and build roads. The prison boom, growing quickly in the wake of the civil rights movement, produed a wholly new scale of penal confinement. The basic brute fact of incarceration in the new era of mass imprisonment is that African Americans are eight times more likely to be incarcerated than whites. Incarceration rates climbed to extraordinary levels among young black men, particularly among those with little schooling. The Bureau of Justice Statistics reports that in 2004, over 12 percent of black men aged twenty-five to twenty-nine were behind bars, in prison or jail. Among black men born in the late 1960s who received no more than a high school education, 30 percent had served time in prison by their mid-thirties; 60 percent of high school dropouts had prison records. By the end of the 1990s, criminal justice supervision was pervasive among young black men. This was a historically novel development in American race relations. We need only go back thirty years, to 1970, to find a time when young black men were not routinely incarcerated. The betrayal of the democratic purpose of rehabilitation had diminished the citizenship of African Americans most of all.

How can we understand the fabulous growth in the American penal system and its effects on the poor and minority communities from which prison inmates are drawn and ultimately return? ...

My main arguments rely on two basic insights of the sociology of politics and crime. First, for political sociology, state power flows along the contours of social inequality. From this perspective, the prison boom was a political project that arose partly because of rising crime but also in response to an upheaval in American race relations in the 1960s and the collapse of urban labor markets for unskilled men in the 1970s. The social activism and disorder of the 1960s fueled the anxieties and resentments of working-class whites. These disaffected whites increasingly turned to the Republican Party through the 1970s and 1980s, drawn by a law and order message that drew veiled connections between civil rights activism and violent crime among blacks in inner cities. For these conservative politics, rehabilitation coddled the criminals who had forfeited their rights to fairness and charity. The young black men of poor urban neighborhoods were the main targets of this analysis. Jobless ghettos, residues of urban deindustrialization, lured many young men into the drug trade and left others unemployed, on the street, and exposed to the scrutiny of police. The punitive sentiment unleashed in the 1970s by rising crime and civil rights activism in the 1960s, institutionalized what had become a chronically idle population of young men with little education. Their life path through adulthood was transformed as a result.

Second, for sociologists of crime, the life path through adulthood normalizes young men, so criminal behavior recedes with age. Adolescents are drawn into the society of adults by passing through a sequence of life course stages—completing school, finding a job, getting married, and starting a family. The integrative power of the life course offers a way out of crime for adult offenders. Men involved in crime who can find steady work and a stable marriage also become embedded in a web of social supports and obligations. These social bonds help criminally active men desist from further offending. Men coming out of prison, however, have little access to the steady jobs that usually build work histories and wages. Employers are reluctant to hire job seekers with prison records, and former inmates are generally poorly prepared for the routines of steady employment. Prison also disrupts families. By 2000, over a million black children—9 percent of those under eighteen—had a father in prison or jail. In around half of all cases, these fathers were living with their children at the time they were incarcerated. The forced separation of men from their families also takes a toll on conjugal bonds. For women with men in prison, married life is threatened by the strains of visitation and the temptations of free men who can help support a household. Few couples survive a term of imprisonment. Unmarried men stigmatized by a prison time can also pay a price. Serving time signals a man's unreliability and a prison record can be as repellent to prospective marriage partners as it is to employers.

A common logic underlies the negative effects of incarceration on a former inmate's job prospects and family life. Although the normal life course is integrative, incarceration is disintegrative, diverting young men from the life stages that mark a man's gradual inclusion in adult society.

The employment problems and disrupted family life of former inmates suggest that incarceration may be a self-defeating strategy for crime control. Although incarceration surely prevents those who are locked up from committing crime in society, inmates are ultimately released with few resources to lead productive lives. Without great hopes for job security or a good marriage, crime remains an inviting alternative. Skeptics will counter that through the 1990s, when incarceration rates reached their highest levels, crime rates fell to their lowest levels since the 1960s. Correlation, however, is not causation. There were many forces operating at the end of the 1990s to drive down crime rates. My empirical analysis shows that fully 90 percent of the decrease in serious crime from 1993 to 2001 would have happened even without the run-up in the incarceration rates. The prison boom contributed a little to the decline in crime through the 1990s, but this gain in public safety was purchased at a cost to the economic well-being and family life of poor minority communities.

Even more important than the effects of the prison boom on crime are its effects on American inequality. The repudiation of rehabilitation and the embrace of retribution produced a collective experience for young black men that is wholly different from the rest of American society. No other group, as a group, routinely contends with long terms of forced confinement and bears the stigma of official criminality in all subsequent spheres of social life, as citizens, workers, and spouses. This is a profound social exclusion that significantly rolls back the gains to citizenship hard won by the civil rights movement. The new marginality of the mass-imprisonment generation can be seen not only in the diminished rates of employment and marriage of former prisoners. Incarceration also erases prison and jail inmates from our conventional measures of economic status. So marginal have these men become, that the most disadvantaged among them are hidden from statistics on wages and employment. The economic situation of young black men—measured by wage and employment rates—appeared to improve through the economic expansion of the 1990s, but this appearance was wholly an artifact of rising incarceration rates.

... I see little evidence that growth in the penal population is related to either rising crime, or that increased incarceration among young disadvantaged men is associated with increased offending.... Incarceration rates grew most in states that elected Republican governors and adopted punitive regimes of criminal sentencing. Analyzing rates of prison admission for black and white men at different levels of education shows that class inequalities in imprisonment increased as the economic status of less-educated men decreased.

... I find that young black men obtained no benefit—either in employment or relative wages—from the record-breaking economic growth in the late 1990s. The invisible inequality that burgeoned through the boom times of the 1990s challenges the claim that robust growth by itself, without the supports of social policy, could bring opportunity to the most disadvantaged....

Although the prison boom undermined economic opportunity and split up families, it cannot explain all the unemployment and female-headed households that underpin much of America's racial inequality. Unemployment and broken homes are as much a cause of imprisonment as a consequence. The disadvantaged men who go to prison would still risk unemployment and marital instability even if they weren't incarcerated. Instead, the prison boom helps us understand how racial inequality in America was sustained, despite great optimism for the social progress of African Americans. From this perspective, the prison boom is not the main cause of inequality between blacks and whites in America, but it did foreclose upward mobility and deflate hopes for racial equality. Perhaps more than adding to inequality between blacks and whites, the prison boom has driven a wedge into the black community, where those without college education are now traveling a path of unique disadvantage that increasingly separates them from college-educated blacks.

The prison boom opened a new chapter in American race relations, but the story of race and class inequalities sustained by political institutions is an old one. The punitive turn in criminal justice disappointed the promise of the civil rights movement and its burdens fell heavily on disadvantaged African Americans. By cleaving off poor black communities from the mainstream, the prison boom left America more divided. Incarceration rates are now so high that the stigma of criminality brands not only individuals, but an entire generation of young black men with little schooling. Tocqueville and Beaumont might be surprised that the American prison had failed so completely to realize the promise of its democratic origins. Although the growth in imprisonment was propelled by racial and class division, the penal system has emerged as a novel institution in a uniquely American system of social inequality.

## DISCUSSION QUESTIONS

1. What impact do you think the high rate of imprisonment has on African American families? What would you suggest in terms of social policies that would address this issue?

2. Why does Western think that the high rate of imprisonment stalls the economic health of the nation?

# 50

## Debunking the Myth of Immigrant Criminality:

### Imprisonment Among First- and Second-Generation Young Men

RUBÉN G. RUMBAUT, ROBERTO G. GONZALES, GOLNAZ KOMAIE, AND CHARLIE V. MORGAN

*Myths abound about immigrants, one of which is that they are likely to engage in crime. Based on rigorous analysis of national data, Rumbaut and his colleagues show that this is untrue and that crime is far more likely among the second generation, those born in the United States, than among the first generation.*

The past few decades have seen the confluence of two eras in the United States: an era of mass immigration and an era of mass imprisonment. A great deal has been said and written about each, reinforcing age-old popular stereo-types about immigration and crime (a Google search for "immigration + crime" immediately returns 57.2 million hits). But rarely are carefully researched connections made between the two, based on rigorous evidence.

The new era of mass immigration, accelerating since the 1970s and coming chiefly from Latin America and Asia, has transformed the ethnic and racial composition of the U.S. population and the communities where they settle. Today, nearly 70 million persons are of foreign birth or parentage (that is, first or second generation)—about 23 percent of all Americans, including 76 percent of all "Hispanics" and 90 percent of all "Asians" (two pan-ethnic categories officially constructed during this period that lump together dozens of diverse nationalities)—composing an "immigrant-stock" population with a young age structure. This population is growing rapidly in an otherwise aging society as a result both of sustained migration and the higher fertility of immigrant women.

The Mexican-origin population dwarfs all others in both the first and second generations; it already accounts for 27 percent of the country's total immigrant-stock population. The first generation of Mexican immigrants now totals more than 10 million persons—much larger than the next sizable immigrant groups (the Filipinos,

SOURCE: Migration Information Source, June 1, 2006: Washington, DC: Migration Policy Institute
www.migrationinformation.org.

Chinese, Indians, and Vietnamese, with more than 1 million each, followed by Cubans, Koreans, Salvadorans and Dominicans, with less than 1 million each).

Indeed, the Mexican total is larger than all other immigrant groups from Latin America combined, and of all Asian countries combined. Except for the rapidly dwindling remnants of the "old second generation" of Europeans and Canadians, U.S.-born children of immigrants today are still very young—in fact, they mostly consist *of* children, with median ages ranging from 9 to 15 for almost all the Latin American and Asian-origin groups—a telling marker of the recency of the immigration of their parents.

Immigrants and their children are heavily concentrated in metropolitan areas, are predominantly nonwhite, speak languages other than English, reflect an extraordinarily wide range of national origins and class, religious, and cultural backgrounds, and arrive with a mix of legal statuses, socioeconomic skills, and resources. By far, the most and the least educated adults in the United States today are immigrants. Their incorporation has coincided with a period of economic restructuring and rising inequality, during which the returns to education have sharply increased.

The era of mass immigration has also coincided with an era of mass imprisonment in the United States, which has further transformed paths to adulthood among young men with low levels of education. Indeed, the U.S. incarceration rate has become the highest of any country in the world. In California alone, there are more people imprisoned than in any other country in the world except China.

The number of adults incarcerated in federal or state prisons or local jails in the United States skyrocketed during this period, quadrupling from just over 500,000 in 1980 to 2.2 million in 2005, according to the Department of Justice. Two-thirds of those are in federal or state prisons and one-third in local jails; the vast majority are young men between 18 and 39. An estimated 80 percent of them either violated drug or alcohol laws, were high at the time they committed their crimes, stole property to buy drugs, or had a history of drug and alcohol abuse and addiction—or some combination of those characteristics. Adding those on parole or probation to the incarcerated population, nearly 7 million adults are currently under correctional supervision, 3.2 percent of all US adults 18 or older.

The official statistics are not kept by nativity or generation, but they show that imprisonment rates vary widely by gender (93 percent of inmates in federal and state prisons are men, although women are now being imprisoned faster than men); by racial categories (there were 4,834 black male prisoners per 100,000 black males in the United States, compared to 1,778 Hispanic males per 100,000, and 681 white males per 100,000, although since 1985 Hispanics have been the fastest group being imprisoned); and by level of education (those incarcerated are overwhelmingly high school dropouts).

Among some racial minorities, becoming a prisoner has become a modal life event in early adulthood: As sociologists Becky Pettit and Bruce Western have noted, a black male high school dropout born in the late 1960s had a nearly 60 percent chance of serving time in prison by the end of the 1990s, and recent birth cohorts of black men are more likely to have prison records than military records or bachelor's degrees.

Today's children of immigrants—both the first (foreign-born) and second (US-born with at least one foreign-born parent) generations—confront a complex set of circumstances that shape their incorporation. Many are progressing exceptionally well, as evidenced by a variety of educational and socioeconomic indicators. For a smaller but not insignificant segment of this population, there is a strong pull from the streets, where violence and gangs make up a large part of the realities of central cities. By the time these children of immigrants reach adulthood, the impediments and opportunities faced as adolescents solidify.

For those with troubled pasts, the transition to adulthood can be an especially rough process. Those who lack adequate education, requisite job skills, and family safety nets, are hard put to find steady work and a stable source of income. Moreover, for some, a pattern of delinquency during adolescence signals deeper future involvements in the adult criminal justice system....

## BACKGROUND: THE CONFLATION OF "IMMIGRANT" AND "CRIME"

In the absence of rigorous empirical research, myths and stereotypes about immigrants and crime often provide the underpinnings for public policies and practices, are amplified and diffused by the media, and shape public opinion and political behavior. Periods of increased immigration have historically been accompanied by nativist alarms and pervasive pejorative stereotypes of newcomers, particularly during economic downturns or national crises (such as the "war on terror" of the post-9/11 period), and when the immigrants have arrived *en masse* and differed substantially from the natives in such cultural markers as religion, language, phenotype, and region of origin.

In the past, such were the prevailing perceptions that variously met the Catholic Irish in the mid-19th century; later the Chinese, the Jews, and the Italians; and, more recently, Cuban Marielitos, Colombians, and others. Popular movies like *The Godfather* and *Scarface,* and television series from *The Untouchables* to *Miami Vice* and *The Sopranos,* project the enduring concern with the presence of foreign criminal elements.

The present period is no exception. California's Proposition 187, which was passed with 59 percent of the statewide vote in 1994 (but challenged as unconstitutional and overturned by a federal court), asserted in its opening lines that "the people of California ... have suffered and are suffering economic hardship [and] personal injury and damage caused by the criminal conduct of illegal aliens in this state."

In 2000, the General Social Survey interviewed a nationally representative sample of adults with a newly developed module to measure attitudes and perceptions toward immigration in a "multi-ethnic United States." Asked whether "more immigrants cause higher crime rates," 25 percent said "very likely" and another 48 percent "somewhat likely"—that is, about three-fourths (73 percent) believed that immigration was causally related to more crime. That was a much higher proportion than the 60 percent who believed that "more immigrants were [somewhat or very] likely to cause Americans to lose jobs," or the 56 percent who thought that "more immigrants were [somewhat or very] likely to make it harder to keep the country united."...

## FOREIGN-BORN VS. NATIVE-BORN MEN: WHO ARE MORE LIKELY TO BE INCARCERATED?

Inasmuch as conventional theories of crime and incarceration predict higher rates for young adult males from ethnic minority groups with lower educational attainment—characteristics which describe a much greater proportion of the foreign-born population than of the native born—it follows that immigrants would be expected to have higher incarceration rates than natives. And immigrant Mexican men—who comprise fully a third of all immigrant men between 18 and 39, and who have the lowest levels of education—would be expected to have the highest rates.

Data from the 5 percent Public Use Microsample (PUMS) of the 2000 census were used to measure the institutionalization rates of immigrants and natives, focusing on males 18 to 39, most of whom are in correctional facilities. Of the 45.2 million males age 18 to 39, three percent were in federal or state prisons or local jails at the time of the 2000 census—a total of over 1.3 million, in line with official prison statistics at that time.

Surprisingly, at least from the vantage of conventional wisdom, the data show the above hypotheses to be unfounded. In fact, the incarceration rate of the U.S. born (3.51 percent) was four times the rate of the foreign born (0.86 percent). The foreign-born rate was half the 1.71 percent rate for non–Hispanic white natives, and 13 times less than the 11.6 percent incarceration rate for native black men....

*The advantage for immigrants vis-à-vis natives applies to every ethnic group without exception.* Almost all of the Asian immigrant groups have lower incarceration rates than the Latin American groups (the exception involves foreign-born Laotians and Cambodians, whose rate of 0.92 percent is still well below that for non-Hispanic white natives).

Tellingly, among the foreign born, the highest incarceration rate by far (4.5 percent) was observed among island-born Puerto Ricans, who are not immigrants as such since they are U.S. citizens by birth and can travel to the mainland as natives. If the island-born Puerto Ricans were excluded from the foreign-born totals, the national incarceration rate for the foreign born would drop to 0.68 percent.

Of particular interest is the finding that the lowest incarceration rates among Latin American immigrants are seen for the least educated groups: Salvadorans and Guatemalans (0.52 percent), and Mexicans (0.70 percent). These are precisely the groups most stigmatized as "illegals" in the public perception and outcry about immigration.

### Second Generation

Incarceration rates increase significantly for all U.S.-born coethnics without exception. That is most notable for Mexicans, whose incarceration rate increases more than eightfold to 5.9 percent among the U.S. born; for Vietnamese (from 0.46 to 5.6 percent among the U.S. born); and for the Laotians and Cambodians (from 0.92 percent to 7.26 percent, the highest of any group except for native blacks). Almost all of the U.S. born among those of Latin American and Asian origin can be assumed to consist of second-generation persons, with the exception of Mexicans and Puerto Ricans, whose numbers may include a sizable number (around 25 percent) of third-generation individuals. (Since 1980, when the questions on parents' country of birth were dropped, the decennial census has not permitted the precise identification of second vs. third or higher generations.)

Thus, while incarceration rates are found to be extraordinarily low among immigrants, they are also seen to rise rapidly by the second generation. Except for the Chinese and Filipinos, the rates of all U.S.-born Latin American and Asian groups exceed that of the referent group of non–Hispanic white natives.

## Education and Incarceration Rates

For all ethnic groups, as expected, the risk of imprisonment is highest for men who are high school dropouts (6.91 percent) compared to those who are high school graduates (2.0 percent). However, the differentials in the risk of incarceration by education are observed principally among native-born men, and not immigrants. Among the U.S. born, 9.76 percent of all male dropouts 18 to 39 were in jail or prison in 2000, compared to 2.23 percent among those who had graduated from high school.

But among the foreign born, the incarceration gap by education was much narrower: Only 1.31 percent of immigrant men who were high school dropouts were incarcerated, compared to 0.57 percent of those with at least a high school diploma.

The advantage for immigrants held when broken down by education for every ethnic group. Indeed, nativity emerges in these data as a stronger predictor of incarceration than education. As noted, native-born high school graduates have a higher rate of incarceration than foreign-born, non–high school graduates (2.2 percent to 1.3 percent)....

Among U.S.-born men who had not finished high school, the highest incarceration rate by far was seen among non-Hispanic blacks, an astonishing 22.25 percent of whom were imprisoned at the time of the 2000 census; that rate was triple the 7.64 percent among foreign-born black dropouts.

Other high rates among U.S.-born high school dropouts were observed among the Vietnamese (over 16 percent), followed by Colombians (over 12 percent), Cubans and Puerto Ricans (over 11 percent), Mexicans (10 percent), and Laotians and Cambodians (over 9 percent). Again, almost all these can be assumed to consist of second-generation persons, as well as the large majority of Mexicans and Puerto Ricans.

## Length of Time in the United States and Incarceration Rates

The data examined thus far suggest that the process of "Americanization" leads to downward mobility and greater risks of involvement with the criminal justice system among a small but significant segment of this population. Therefore, the question of what happens to immigrant men over time in the United States was explored.

*For every group without exception, the longer immigrants had resided in the United States, the higher were their incarceration rates.* Here again, the rates of incarceration for island-born Puerto Rican are significantly higher—regardless of how long they have lived on the U.S. mainland—than the rates for all the immigrant groups, underscoring their unique status....

In contrast, foreign-born Mexican men 18 to 39, by far the largest group (at over 3 million), have a lower incarceration rate than many other ethnic and racial groups—even after they have lived in the United States for over 15 years. Thus, the Mexican incarceration story in particular can be very misleading when the data conflate the foreign born and the native born, as official statistics on "Latinos" or "Hispanics" routinely do....

## DISCUSSION: CONFIRMATORY RESULTS FROM OTHER STUDIES, NOW AND THEN

... The finding that incarceration rates are much lower among immigrant men than the national norm, despite their lower levels of education and greater poverty, but increase significantly over time in the United States for those who arrived as children and especially among the second generation, suggests that the process of "Americanization" can lead to downward mobility and greater risk of involvement with the criminal justice system for a significant minority of this population....

In a sense, these systematic findings should not come as news, for they are not new—merely forgotten and overruled by popular myth. In the first three decades of the 20th century, during another era of mass immigration, three major government commissions came to much the same conclusions.

The Industrial Commission of 1901, the [Dillingham] Immigration Commission of 1911, and the [Wickersham] National Commission on Law Observance and Enforcement of 1931, each sought to measure how immigration resulted in increases in crime. Instead, each found lower levels of criminal involvement among the foreign born but higher levels among their native-born counterparts, noting that a disproportionate number of the incarcerated had foreign-born parents. If there was an "immigrant crime problem," it was not found among the immigrants but among their U.S.-born sons, who had a different frame of reference than their parents and faced an entirely different set of circumstances.

## CONCLUSIONS AND IMPLICATIONS

Because many immigrants, especially labor migrants from Mexico and Central America and refugees from Southeast Asia, are young men who have arrived with very low levels of education, conventional wisdom—both in the form of nativist stereotype as well as standard criminological theory—tends to associate them with high rates of crime and incarceration. The unauthorized entry and visa overstays of many, framed as an assault against the "rule of law" by pundits and politicians (most notoriously by a House of Representatives bill, passed in December 2005, which would make felons of all "illegal" immigrants and criminalize those who assist them), reinforces the stereotypical association of immigration and criminality in much public discourse. This association flourishes in a post-9/11 climate of fear and ignorance where "terrorism" and "losing control of our borders" are often mentioned in the same breath, if without any evidence to back them up.

But correlation is not causation. In fact, immigrants have the lowest rates of imprisonment for criminal convictions in American society. Both the national and local-level findings presented here turn conventional wisdom on its head and present a challenge to criminological theory as well as to sociological perspectives on "straight-line assimilation."

For every ethnic group without exception, the census data show an *increase* in rates of criminal incarceration among young men from the foreign-born to the U.S.-born generations, and over time in the United States among the foreign born–exactly the opposite of what is typically assumed both by standard theorias and by public opinion on immigration and crime.

Paradoxically, incarceration rates are lowest among immigrant young men, even among the least educated and the least acculturated among then, but they increase sharply among the U.S.-born and acculturated second generation, especially among the least educated–evidence of downward assimilation that parallels patterns observed for marginalized native minorities.

What is more, these patterns have now been observed consistently over the last three decennial censuses, a period that spans precisely the eras of mass immigration and mass imprisonment–and they recall similar findings reported by three major commissions during the first three decades of the 20th century, a previous era of mass migration and crime concerns.

Nativity emerges in this analysis as a stronger predictor of incarceration than education. When immigration and generational status are taken into account, the association between (lower) education and (higher) crime and incarceration rates is complicated in ways not anticipated by canonical perspectives.

It is in the context of the study of immigrant groups and generational cohorts that such paradoxes are revealed, further underscoring the importance of connecting the research literatures on immigration and on crime and imprisonment, which have largely ignored each other—to the impoverishment of both and to the enrichment of popular prejudice.

## REFERENCE

Pettit, Becky, and Bruce Western. 2004. "Mass Imprisonment and the Life Course." *American Sociological Review* 69:151–169.

## DISCUSSION QUESTIONS

1. What evidence do Rumbaut and his colleagues find regarding public belief in the criminality of immigrants?

2. Why would "Americanization" of immigrants lead to higher rates of involvement with the criminal justice system?

# 51

# The Uneven Scales of Capital Justice

CHRISTINA SWARNS

*Despite constitutional rights to equal protection under the law, the facts show that race, especially when combined with class, has a significant impact on the likelihood of receiving capital punishment (that is, the death penalty).*

In 1972, the U.S. Supreme Court declared the death penalty unconstitutional. The Court found that because the capital-punishment laws gave sentencers virtually unbridled discretion in deciding whether or not to impose a death sentence, "The death sentence [was] disproportionately carried out on the poor, the Negro, and the members of unpopular groups."

In 1976, the Court reviewed the revised death-penalty statutes—which are in place today—and concluded that they sufficiently restricted sentencer discretion such that race and class would no longer play a pivotal role in the life-or-death calculus. In the 28 years since the reinstatement of the death penalty, however, it has become apparent that the Court was wrong. Race and class remain critical factors in the decision of who lives and who dies.

Both race and poverty corrupt the administration of the death penalty. Race severely disadvantages the black jurors, black defendants, and black victims within the capital-punishment system. Black defendants are more likely to be executed than white defendants. Those who commit crimes against black victims are punished less severely than those who commit crimes against white victims. And black potential jurors are often denied the opportunity to serve on death-penalty juries. As far as the death penalty is concerned, therefore, blackness is a proxy for worthlessness.

Poverty is a similar—and often additional—handicap. Because the lawyers provided to indigent defendants charged with capital crimes are so uniformly undertrained and undercompensated, the 90 percent of capitally charged defendants who lack the resources to retain a private attorney are virtually guaranteed a death sentence. Together, therefore, race and class function as an elephant on death's side of the sentencing scale.

When and how does race infect the death-penalty system? The fundamental lesson of the Supreme Court's 1972 decision to strike down the death penalty is that discretion, if left unchecked, will be exercised in such a manner that arbitrary and irrelevant factors like race will enter into the sentencing decision. That

SOURCE: *The American Prospect*, June 18, 2004. Reprinted by permission.

conclusion remains true today. The points at which discretion is exercised are the gateways through which racial bias continues to enter into the sentencing calculation.

Who has the most unfettered discretion? Chief prosecutors, who are overwhelmingly white, make some of the most critical decisions vis-à-vis the death penalty. Because their decisions go unchecked, prosecutors have arguably the greatest unilateral influence over the administration of the death penalty.

Do prosecutors exercise their discretion along racial lines? Unquestionably yes. Prosecutors bring more defendants of color into the death-penalty system than they do white defendants. For example, a 2000 study by the U.S. Department of Justice reveals that between 1995 and 2000, 72 percent of the cases that the attorney general approved for death-penalty prosecution involved defendants of color. During that time, statistics show that there were relatively equal numbers of black and white homicide perpetrators.

Prosecutors also give more white defendants than black defendants the chance to avoid a death sentence. Specifically, prosecutors enter into plea bargains—deals that allow capitally charged defendants to receive a lesser sentence in exchange for an admission of guilt—with white defendants far more often than they do with defendants of color. Indeed, the Justice Department study found that white defendants were almost twice as likely as black defendants to enter into such plea agreements.

Further, prosecutors assess cases differently depending upon the race of the victim. Thus, the Department of Justice found that between 1995 and 2000, U.S. attorneys were almost twice as likely to seek the death penalty for black defendants accused of killing nonblack victims than for black defendants accused of killing black victims.

And, finally, prosecutors regularly exclude black potential jurors from service in capital cases. For example, a 2003 study of jury selection in Philadelphia capital cases, conducted by the Pennsylvania Supreme Court Commission on Race and Gender Bias in the Justice System, revealed that prosecutors used peremptory challenges—the power to exclude potential jurors for any reason aside from race or gender—to remove 51 percent of black potential jurors while excluding only 26 percent of nonblack potential jurors. Such bias has a long history: From 1963 to 1976, one Texas prosecutor's office instructed its lawyers to exclude all people of color from service on juries by distributing a memo containing the following language: "Do not take Jews, Negroes, Dagos, Mexicans or a member of any minority race on a jury, no matter how rich or how well educated." This extraordinary exercise of discretion harms black capital defendants because statistics reveal that juries containing few or no blacks are more likely to sentence black defendants to death.

Such blatant prosecutorial discretion has significantly contributed to the creation of a system that is visibly permeated with racial bias. Black defendants are sentenced to death and executed at disproportionate rates. For example, in Philadelphia, African American defendants are approximately four times more likely to be sentenced to death than similarly situated white defendants. And nationwide, crimes against white victims are punished more severely than crimes against black victims. Thus, although 46.7 percent of all homicide victims are black, only 13.9 percent of the victims of executed defendants are black. In some jurisdictions, all of the defendants on death row have white victims; in other jurisdictions, having a white victim exponentially increases a criminal defendant's likelihood of being sentenced to death. It is beyond dispute, therefore, that race remains a central factor in the administration of the death penalty.

Socioeconomic status also plays an inappropriate yet extremely influential role in the determination of who receives the death penalty. The vast majority of the people who are sentenced to death and executed in the United States come from a background of poverty. Indeed, as noted by the Supreme Court in 1972, "One searches our chronicles in vain for the execution of any member of the affluent strata of this society. The Leopolds and Loebs are given prison terms, not sentenced to death."

The primary reason for this economic disparity is that the poor are systematically denied access to well-trained and adequately funded lawyers. "Capital defense is now a highly specialized field requiring practitioners to successfully negotiate minefield upon minefield of exacting and arcane death-penalty law," according to the Pennsylvania commission. "Any misstep along the way can literally mean death for the client." It is therefore critical that lawyers appointed to represent poor defendants facing death possess the requisite compensation, training, and skill to mount a meaningful challenge to the government's case.

Unfortunately, few if any of the defendants on death row are provided with lawyers possessing the requisite skills and resources. Instead, poorly trained and underfunded court-appointed lawyers who provide abysmal legal assistance typically represent those death-sentenced prisoners. Tales of the pathetic lawyering provided by appointed counsel to their capitally charged clients are legion. Perhaps the most famous example is that of Calvin Burdine, whose court-appointed lawyer slept through significant portions of his trial. Another example is the case of Vinson Washington, whose court-appointed lawyer suggested to the defense psychiatrist that Vinson "epitomized the banality of evil." Death-sentenced defendants are so frequently provided with poor representation that, in 2001, Supreme Court Justice Ruth Bader Ginsberg commented that she had never seen a death-penalty defendant come before the Supreme Court in search of an eve-of-execution stay "in which the defendant was well-represented at trial."

One reason that appointed counsel perform so poorly is that they are grossly undercompensated. In some cases, capital-defense attorneys have been paid as little as $5 an hour. Not surprisingly, these paltry rates of compensation have yielded an equally paltry quality of representation. As was succinctly noted by the 5th U.S. Circuit Court of Appeals in its review of the quality of representation provided by a court-appointed lawyer to a capitally charged defendant in Texas: "The state paid defense counsel $11.84 per hour. Unfortunately, the justice system got only what it paid for."

Lawyers appointed to handle capital trials also often lack the expertise necessary to appropriately defend capitally charged defendants. Many states fail to provide appointed counsel with the training necessary to handle these complex cases, and many fail to impose minimum qualifications for lawyers handling capital cases. As a result, capital defendants have been represented by lawyers with absolutely no experience in criminal, much less capital, law. Although the American Bar Association has promulgated standards for the representation of indigent defendants charged with capital offenses, and although those guidelines have been endorsed by the Supreme Court, no death-penalty jurisdiction has implemented a system that meets these requirements. Thus, lawyers without meaningful training or expertise in the area of capital punishment continue to represent defendants facing death.

Because race and class continue to play a powerful role in the administration of the death penalty, it is clear that the current system is as broken today as it was in 1972. As the Supreme Court explained at the time, "A law that stated that anyone making more than $50,000 would be exempt from the death penalty would plainly fall, as would a law that in terms said that blacks, those who never went beyond the fifth grade in school, those who made less than $3,000 a year, or those who were unpopular or unstable should be the only people executed. A law which in the overall view reaches that result in practice has no more sanctity than a law which in terms provides the same."

Because the current death-penalty law, while neutral on its face, is applied in such a manner that people of color and the poor are disproportionately condemned to die, the law is legally and morally invalid.

## DISCUSSION QUESTIONS

1. Why did the Supreme Court declare the death penalty unconstitutional in 1972?

2. What does discretion mean and why is it important in the criminal justice system? How does it influence who ends up on a jury?

3. Why does Swarns look at both race and social class in exploring the experiences of individuals in the criminal justice system?

# 52

# The Mark of a Criminal Record

DEVAH PAGER

*What are the consequences of the very high rate of incarceration of young Black men? In what is known as an audit study—a study in this case where pairs of Black and White men are identically matched in their social characteristics and compared in employment outcomes following incarceration—Pager shows the barrier that a criminal record creates for subsequent employment, showing the influence of race in the likelihood of former prisoners finding jobs.*

While stratification researchers typically focus on schools, labor markets, and the family as primary institutions affecting inequality, a new institution has emerged as central to the sorting and stratifying of young and disadvantaged men: the criminal justice system. With over 2 million individuals currently incarcerated, and over half a million prisoners released each year, the large and growing numbers of men being processed through the criminal justice system raises important questions about the consequences of this massive institutional intervention.

This article focuses on the consequences of incarceration for the employment outcomes of black and white men. While previous survey research has demonstrated a strong association between incarceration and employment, there remains little understanding of the mechanisms by which these outcomes are produced. In the present study, I adopt an experimental audit approach to formally test the degree to which a criminal record affects subsequent employment opportunities. By using matched pairs of individuals to apply for real entry-level jobs, it becomes possible to directly measure the extent to which a criminal record—in the absence of other disqualifying characteristics—serves as a barrier to employment among equally qualified applicants. Further, by varying the race of the tester pairs, we can assess the ways in which the effects of race and criminal record interact to produce new forms of labor market inequalities.

## TRENDS IN INCARCERATION

Over the past three decades, the number of prison inmates in the United States has increased by more than 600%, leaving it the country with the highest

SOURCE: *American Journal of Sociology* 108 (March 2003): 937–975. Reprinted by permission of University of Chicago Press.

incarceration rate in the world (Bureau of Justice Statistics 2002a; Barclay, Tavares, and Siddique 2001). During this time, incarceration has changed from a punishment reserved primarily for the most heinous offenders to one extended to a much greater range of crimes and a much larger segment of the population. Recent trends in crime policy have led to the imposition of harsher sentences for a wider range of offenses, thus casting an ever-widening net of penal intervention.

While the recent "tough on crime" policies may be effective in getting criminals off the streets, little provision has been made for when they get back out. Of the nearly 2 million individuals currently incarcerated, roughly 95% will be released, with more than half a million being released each year (Slevin 2000). According to one estimate, there are currently over 12 million ex-felons in the United States, representing roughly 8% of the working-age population (Uggen, Thompson, and Manza 2000). Of those recently released, nearly two-thirds will be charged with new crimes and over 40% will return to prison within three years (Bureau of Justice Statistics 2000). Certainly some of these outcomes are the result of desolate opportunities or deeply ingrained dispositions, grown out of broken families, poor neighborhoods, and little social control (Sampson and Laub 1993; Wilson 1997). But net of these contributing factors, there is evidence that experience with the criminal justice system in itself has adverse consequences for subsequent opportunities. In particular, incarceration is associated with limited future employment opportunities and earnings potential (Freeman 1987; Western 2002), which themselves are among the strongest predictors of recidivism (Shover 1996; Sampson and Laub 1993; Uggen 2000).

The expansion of the prison population has been particularly consequential for blacks. The incarceration rate for young black men in the year 2000 was nearly 10%, compared to just over 1% for white men in the same age group (Bureau of Justice Statistics 2001). Young black men today have a 28% likelihood of incarceration during their lifetime (Bureau of Justice Statistics 1997), a figure that rises above 50% among young black high school dropouts (Pettit and Western 2001). These vast numbers of inmates translate into a large and increasing population of black ex-offenders returning to communities and searching for work. The barriers these men face in reaching economic self-sufficiency are compounded by the stigma of minority status and criminal record. The consequences of such trends for widening racial disparities are potentially profound (see Western and Pettit 1999; Freeman and Holzer 1986)....

## RESEARCH QUESTIONS

There are three primary questions I seek to address with the present study. First, in discussing the main effect of a criminal record, we need to ask whether and to what extent employers use information about criminal histories to make hiring decisions. Implicit in the criticism of survey research in this area is the assumption that the signal of a criminal record is not a determining factor. Rather, employers use information about the interactional styles of applicants, or other observed characteristics—which may be correlated with criminal records—and this explains the differential outcomes we observe. In this view, a criminal record does not represent a meaningful signal to employers on its own. This study formally tests the degree to which employers use information about criminal histories in the absence of corroborating evidence. It is essential that we conclusively document this effect before making larger claims about the aggregate consequences of incarceration.

Second, this study investigates the extent to which race continues to serve as a major barrier to employment. While race has undoubtedly played a central role in shaping the employment opportunities of African-Americans over the past century, recent arguments have questioned the continuing significance of race, arguing instead that other factors—such as spatial location, soft skills, social capital, or cognitive ability—can explain most or all of the contemporary racial differentials we observe (Wilson 1987; Moss and Tilly 1996; Loury 1977; Neal and Johnson 1996). This study provides a comparison of the experiences of equally qualified black and white applicants, allowing us to assess the extent to which direct racial discrimination persists in employment interactions.

The third objective of this study is to assess whether the effect of a criminal record differs for black and white applicants. Most research investigating the differential impact of incarceration on blacks has focused on the differential rates of incarceration and how those rates translate into widening racial disparities. In addition to disparities in the rate of incarceration, however, it is also important to consider possible racial differences in the effects of incarceration. Almost none of the existing literature to date has explored this issue, and the theoretical arguments remain divided as to what we might expect.

On one hand, there is reason to believe that the signal of a criminal record should be less consequential for blacks. Research on racial stereotypes tells us that Americans hold strong and persistent negative stereotypes about blacks, with one of the most readily invoked contemporary stereotypes relating to perceptions of violent and criminal dispositions (Smith 1991; Sniderman and Piazza 1993; Devine and Elliott 1995). If it is the case that employers view all blacks as potential criminals, they are likely to differentiate less among those with official criminal records and those without. Actual confirmation of criminal involvement then will provide only redundant information, while evidence against it will be discounted. In this case, the outcomes for all blacks should be worse, with less differentiation between those with criminal records and those without.

On the other hand, the effect of a criminal record may be worse for blacks if employers, already wary of black applicants, are more hesitant when it comes to taking risks on blacks with proven criminal tendencies. The literature on racial stereotypes also tells us that stereotypes are most likely to be activated and reinforced when a target matches on more than one dimension of the stereotype (Quillian and Pager 2002; Darley and Gross 1983; Fiske and Neuberg 1990). While employers may have learned to keep their racial attributions in check through years of heightened sensitivity around employment discrimination, when combined with knowledge of a criminal history, negative attributions are likely to intensify.

A third possibility, of course, is that a criminal record affects black and white applicants equally. The results of this audit study will help to adjudicate between these competing predictions.

## THE AUDIT METHODOLOGY

… The basic design of an employment audit involves sending matched pairs of individuals (called testers) to apply for real job openings in order to see whether employers respond differently to applicants on the basis of selected characteristics. The appeal of the audit methodology lies in its ability to combine experimental methods with real-life contexts. This combination allows for greater generalizability than a lab experiment and a better grasp of the causal mechanisms than what we can normally obtain from observational data. The audit methodology is particularly valuable for those with an interest in discrimination. Typically, researchers are forced to infer discrimination indirectly, often attributing the residual from a statistical model—which is essentially all that is not directly explained—to discrimination. This convention is rather unsatisfying to researchers who seek empirical documentation for important social processes. The audit methodology therefore provides a valuable tool for this research….

## STUDY DESIGN

The basic design of this study involves the use of four male auditors (also called testers), two blacks and two whites. The testers were paired by race; that is, unlike in the original Urban Institute audit studies, the two black testers formed one team, and the two white testers formed the second team. The testers were 23-year-old college students from Milwaukee who were matched on the basis of physical appearance and general style of self-presentation. Objective characteristics that were not already identical between pairs—such as educational attainment and work experience—were made similar for the purpose of the applications. Within each team, one auditor was randomly assigned a "criminal record" for the first week; the pair then rotated which member presented himself as the ex-offender for each successive week of employment searches, such that each tester served in the criminal record condition for an equal number of cases. By varying which member of the pair presented himself as having a criminal record, unobserved differences within the pairs of applicants were effectively controlled. No significant differences were found for the outcomes of individual testers or by month of testing.

Job openings for entry-level positions (defined as jobs requiring no previous experience and no education greater than high school) were identified from the Sunday classified advertisement section of the *Milwaukee Journal Sentinel*. In addition, a supplemental sample was drawn from *Jobnet,* a state-sponsored web site for employment listings, which was developed in connection with the W-2 Welfare-to-Work initiatives.

The audit pairs were randomly assigned 15 job openings each week. The white pair and the black pair were assigned separate sets of jobs, with the same-race testers applying to the same jobs. One member of the pair applied first, with the second applying one day later (randomly varying whether the ex-offender was first or second). A total of 350 employers were audited during the course of this study: 150 by the white pair and 200 by the black pair. Additional tests were performed by the black pair because black testers received fewer callbacks on average, and there were thus fewer data points with which to draw comparisons. A larger sample size enables me to calculate more precise estimates of the effects under investigation.

Immediately following the completion of each job application, testers filled out a six-page response form that coded relevant information from the test. Important variables included type of occupation, metropolitan status, wage, size of establishment, and race and sex of employer. Additionally, testers wrote narratives describing the overall interaction and any comments made by employers (or included on applications) specifically related to race or criminal records.

One key feature of this audit study is that it focuses only on the first stage of the employment process. Testers visited employers, filled out applications, and proceeded as far as they could during the course of one visit. If testers were asked to interview on the spot, they did so, but they did not return to the employer for a second visit. The primary dependent variable, then, is the proportion of applications that elicited callbacks from employers. Individual voicemail boxes were set up for each tester to record employer responses. If a tester was offered the job on the spot, this was also coded as a positive response....

## THE EFFECT OF A CRIMINAL RECORD FOR WHITES

I begin with an analysis of the effect of a criminal record among whites. White noncriminals can serve as our baseline in the following comparisons, representing the presumptively nonstigmatized group relative to blacks and those with criminal records. Given that all testers presented roughly identical credentials, the differences experienced among groups of testers can be attributed fully to the effects of race or criminal status.

Figure 1 shows the percentage of applications submitted by white testers that elicited callbacks from employers, by criminal status. As illustrated below, there is a large and significant effect of a criminal record, with 34% of whites without criminal records receiving callbacks, relative to only 17% of whites with criminal records. A criminal record thereby reduces the likelihood of a callback by 50%....

2The results here demonstrate that criminal records close doors in employment situations. Many employers seem to use the information as a screening mechanism, without attempting to probe deeper into the possible context or complexities of the situation. As we can see here, in 50% of cases, employers were unwilling to consider equally qualified applicants on the basis of their criminal record.

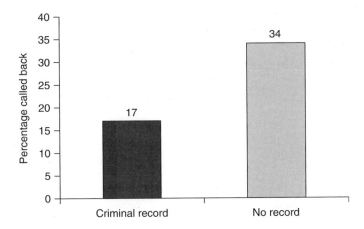

**FIGURE 1**  The effect of a criminal record on white job applicants.

Of course, this trend is not true among all employers, in all situations. There were, in fact, some employers who seemed to prefer workers who had been recently released from prison. One owner told a white tester in the criminal record condition that he "like[d] hiring people who ha[d] just come out of prison because they tend to be more motivated, and are more likely to be hard workers [not wanting to return to prison]."…

## THE EFFECT OF RACE

Figure 2 presents the percentage of callbacks received for both categories of black testers relative to those for whites. The effect of race in these findings is strikingly large. Among blacks without criminal records, only 14% received callbacks, relative to 34% of white noncriminals…. In fact, even whites with criminal records received more favorable treatment (17%) than blacks without criminal records (14%). The rank ordering of groups in this graph is painfully revealing of employer preferences: race continues to play a dominant role in shaping employment opportunities, equal to or greater than the impact of a criminal record….

## RACIAL DIFFERENCES IN THE EFFECTS OF A CRIMINAL RECORD

The final question this study sought to answer was the degree to which the effect of a criminal record differs depending on the race of the applicant. Based on the results presented in Figure 2, the effect of a criminal record appears more pronounced for blacks than it is for whites. While this interaction term is not statistically significant, the magnitude of the difference is nontrivial. While the ratio of callbacks for nonoffenders

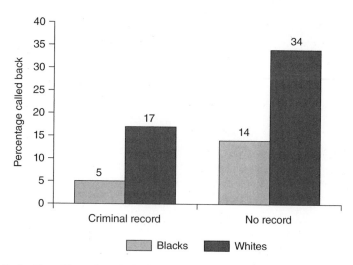

**F I G U R E  2**   The effect of a criminal record for black and white job applicants.

relative to ex-offenders for whites is 2:1, this same ratio for blacks is nearly 3:1. The effect of a criminal record is thus 40% larger for blacks than for whites.

This evidence is suggestive of the way in which associations between race and crime affect interpersonal evaluations. Employers, already reluctant to hire blacks, appear even more wary of blacks with proven criminal involvement. Despite the fact that these testers were bright articulate college students with effective styles of self-presentation, the cursory review of entry-level applicants leaves little room for these qualities to be noticed. Instead, the employment barriers of minority status and criminal record are compounded, intensifying the stigma toward this group.

The salience of employers' sensitivity toward criminal involvement among blacks was highlighted in several interactions documented by testers. On three separate occasions, for example, black testers were asked in person (before submitting their applications) whether they had a prior criminal history. None of the white testers were asked about their criminal histories up front.

The strong association between race and crime in the minds of employers provides some indication that the "true effect" of a criminal record for blacks may be even larger than what is measured here. If, for example, the outcomes for black testers *without* criminal records were deflated in part because employers feared that they may nevertheless have criminal tendencies, then the contrast between blacks with and without criminal records would be suppressed....

## DISCUSSION

There is serious disagreement among academics, policy makers, and practitioners over the extent to which contact with the criminal justice system in itself leads to harmful consequences for employment. The present study takes a strong stand in this debate by offering direct evidence of the causal relationship between a criminal record and employment outcomes.... The finding that ex-offenders are only one-half to one-third as likely as nonoffenders to be considered by employers suggests that a criminal record indeed presents a major barrier to employment. With over 2 million people currently behind bars and over 12 million people with prior felony convictions, the consequences for labor market inequalities are potentially profound.

Second, the persistent effect of race on employment opportunities is painfully clear in these results. Blacks are less than half as likely to receive consideration by employers, relative to their white counterparts, and black nonoffenders fall behind even whites with prior felony convictions. The powerful effects of race thus continue to direct employment decisions in ways that contribute to persisting racial inequality. In light of these findings, current public opinion seems largely misinformed. According to a recent survey of residents in Los Angeles, Boston, Detroit, and Atlanta, researchers found that just over a quarter of whites believe there to be "a lot" of discrimination against blacks, compared to nearly two-thirds of black respondents (Kluegel and Bobo 2001). Over the past decade, affirmative action has come under attack across the country based on the argument that direct racial discrimination is no longer a major barrier to opportunity. According to this study, however, employers, at least in Milwaukee, continue to use race as a major factor in their hiring decisions. When we combine the effects of race and criminal record, the problem grows more intense. Not only are blacks much more likely to be incarcerated than whites; based on the findings presented here, they may also be more strongly affected by the impact of a criminal record. Previous estimates of the aggregate consequences of incarceration may therefore underestimate the impact on racial disparities.

Finally, in terms of policy implications, this research has troubling conclusions. In our frenzy of locking people up, our "crime control" policies may in fact exacerbate the very conditions that lead to crime in the first place. Research consistently shows that finding quality steady employment is one of the strongest predictors of desistance from crime (Shover 1996; Sampson and Laub 1993; Uggen 2000). The fact that a criminal record severely limits employment opportunities—particularly among blacks—suggests that these individuals are left with few viable alternatives.

As more and more young men enter the labor force from prison, it becomes increasingly important to consider the impact of incarceration on the job prospects of those coming out. No longer a peripheral institution, the criminal justice system has become a dominant presence in the lives of young disadvantaged men, playing a key role in the sorting and stratifying of labor market opportunities. This article represents an initial attempt to specify one of the important mechanisms by which incarceration leads to poor employment outcomes. Future research is needed to expand this emphasis to other mechanisms (e.g., the transformative effects of prison on human and social capital), as well as to include other social domains affected by incarceration (e.g., housing, family formation, political participation, etc.); in this way, we can move toward a more complete understanding of the collateral consequences of incarceration for social inequality.

At this point in history, it is impossible to tell whether the massive presence of incarceration in today's stratification system represents a unique anomaly of the late 20th century or part of a larger movement toward a system of stratification based on the official certification of individual character and competence. Whether this process of negative credentialing will continue to form the basis of emerging social cleavages remains to be seen....

# REFERENCES

Barclay, Gordon, Cynthia Tavares, and Arsalaan Siddique. 2001. "International Comparisons of Criminal Justice Statistics, 1999." U.K. Home Office for Statistical Research. London.

Bureau of Justice Statistics. 1997. *Lifetime Likelihood of Going to State or Federal Prison*, by Thomas P. Bonczar and Allen J. Beck. Special report, March. Washington, D.C.

———. 2001. *Prisoners in 2000*, by Allen J. Beck and Paige M. Harrison. August. Bulletin. Washington, D.C.: NCJ 188207.

———. 2002a. *Sourcebook of Criminal Justice Statistics*. Last accessed March 1, 2003. Available http://www.albany.edu/sourcebook/

Darley, J. M., and P. H. Gross. 1983. "A Hypothesis-Confirming Bias in Labeling Effects." *Journal of Personality and Social Psychology* 44:20–33.

Fiske, Susan, and Steven Neuberg. 1990. "A Continuum of Impression Formation, from Category-Based to Individuating Processes." Pp. 1–63 in *Advances in Experimental Social Psychology*, vol. 23. Edited by Mark Zanna. New York: Academic Press.

Freeman, Richard B. 1987. "The Relation of Criminal Activity to Black Youth Employment." *Review of Black Political Economy* 16 (1–2): 99–107.

Freeman, Richard B., and Harry J. Holzer, eds. 1986. *The Black Youth Employment Crisis*. Chicago: University of Chicago Press for National Bureau of Economic Research.

Kluegel, James, and Lawrence Bobo. 2001. "Perceived Group Discrimination and Policy Attitudes: The Sources and Consequences of the Race and Gender Gaps." Pp. 163–216 in *Urban Inequality: Evidence from Four Cities*, edited by Alice O'Connor, Chris Tilly, and Lawrence D. Bobo. New York: Russell Sage Foundation.

Loury, Glenn C. 1977. "A Dynamic Theory of Racial Income Differences." Pp. 153–86 in *Women, Minorities, and Employment Discrimination*, edited by P. A. Wallace and A. M. La Mond. Lexington, Mass.: Heath.

Moss, Philip, and Chris Tilly. 1996. "'Soft Sills' and Race: An Investigation of Black Men's Employment Problems." *Work and Occupations* 23(3): 256–76.

Neal, Derek, and William Johnson. 1996. "The Role of Premarket Factors in Black-White Wage Differences." *Journal of Political Economy* 104(5): 869–95.

Pettit, Becky, and Bruce Western. 2001. "Inequality in Lifetime Risks of Imprisonment." Paper presented at the annual meetings of the American Sociological Association. Anaheim, August.

Quillian, Lincoln, and Devah Pager. 2002. "Black Neighbors, Higher Crime? The Role of Racial Stereotypes in Evaluations of Neighborhood Crime." *American Journal of Sociology* 107(3): 717–67.

Sampson, Robert J., and John H. Laub. 1993. *Crime in the Making: Pathways and Turning Points through Life*. Cambridge, Mass.: Harvard University Press.

Shover, Neil. 1996. *Great Pretenders: Pursuits and Careers of Presistent Thieves*. Boulder, Colo.: Westview.

Slevin, Peter. 2000. "Life after Prison: Lack of Services Has High Price." *Washington Post*, April 24.

Smith, Tom W. 1991. *What Americans Say about Jews*. New York: American Jewish Committee.

Sniderman, Paul M., and Thomas Piazza. 1993. *The Scar of Race*. Cambridge, Mass.: Harvard University Press.

Uggen, Christopher, Melissa Thompson, and Jeff Manza. 2000. "Crime, Class, and Reintegration: The Socioeconomic, Familial, and Civic Lives of Offenders." Paper presented at the American Society of Criminology meetings, San Francisco, November 18.

Western, Bruce. 2002. "The Impact of Incarceration on Wage Mobility and Inequality." *American Sociological Review* 67(4): 526–46.

Western, Bruce, and Becky Pettit. 1999. "Black–White Earnings Inquality, Employment Rates, and Incarceration." Working Paper no. 150. New York: Russell Sage Foundation.

Wilson, William Julius. 1987. *The Truly Disadvantaged: The Inner City, the Underclass, and Public Policy*. Chicago: University of Chicago Press.

———. 1997. *When Work Disappears: The World of the New Urban Poor*. New York: Vintage Books.

## DISCUSSION QUESTIONS

1. What effect does Pager find of the impact of having a criminal record on one's employment status? What is the effect of race on this relationship?

2. Audit studies reveal the impact of race in a variety of settings. In what other context might you do an audit study to conduct research on the impact of race? See if you can design such a study, being careful to identify all of the things you would have to consider in setting it up.

# Student Exercises

1. What are the conditions under which a person might commit a crime? Are some groups more likely to be apprehended for crimes than others? If so, who and why?

2. The death penalty is very controversial in the United States, with supporters and critics. What are the practices with regard to the death penalty in your state? How does an understanding of racial bias affect debates about the death penalty?

# PART IV

# Building a Just Society

# Section XIII

## Moving Forward:

### Analysis and Social Action

### Elizabeth Higginbotham
### and Margaret L. Andersen

R acial hierarchies, like all social hierarchies, change over time, but how do they change? They can change because the political economy needs different groups of workers with particular skills, which can result in dominant groups changing their views of others. For example, African Americans were barred from industrial work in one era but then were recruited for those same jobs when European immigration was curtailed. Access to industrial work not only changed their opportunities but brought them into the cities in great numbers and gave them the resources to create new representations of themselves, as in the flowering of the Harlem Renaissance (Marks 1989).

More often, however, change occurs because of actions taken by oppressed people. Dominant groups use their economic and political power to develop ideologies that make the current racial order appear natural. Disadvantaged groups in turn mobilize not only to challenge those ideologies but perhaps to change laws and create a more inclusive and just society. Such actions show us that, although systems of inequality may be established by those with the most power, they are not just quietly accepted by subordinates. Rather, these arrangements are contested by oppressed groups and their allies. For example, notions of racial and cultural inferiority justified slavery, the mistreatment of American Indians, and the denial of rights to Chinese immigrants. Members of these groups had different interpretations of their situations. For example, many slaves rejected the system, escaping to form Maroon colonies in the South, and some even engaged in armed rebellions. American Indians knew that European settlers had guns, while they had only bows and arrows; they also knew that the laws of the land gave advantages to White people, who had economic and political power. Still, they continued to contest and challenge these views and are still doing so now. Likewise, denied the opportunity to become citizens, Chinese Americans used the state and federal court systems to fight the erosion of their rights as human beings (Takaki 1993). Among all oppressed racial and ethnic groups, there is a long history of active resistance, even as the elites of the nation use the power of government to control those who do not conform.

Racism is a tool for exclusion, but it also can be used to facilitate building community. People who are denied participation in the mainstream society build their own communities, and segregation itself can end up fostering community building. Segregated communities can be places where people question the controlling images and negative stereotypes imposed by the dominant group to justify inequality. Within minority communities there have long been alternative explanations for their lack of resources and opportunities. Such messages contradict explanations from the ruling system of power that typically define people of color as somehow less able. Alternative explanations identify the structural sources of inequality and the power differentials, and can motivate oppressed people and those who support their goals to demand wider opportunities.

Michael Omi and Howard Winant ("On Racial Formation," in Section I) present the concept of **racial projects**, defined as organized efforts to distribute social and economic resources along racial lines. One form of racial project is when people with power exclude others, as in the case of early immigration legislation. But racial projects are not just used by those with the most power in the society. Racial projects can also be organized to redistribute resources with equality in mind, such as when disadvantaged people challenge interpretations of themselves as undeserving. Racial projects waged by disadvantaged groups are often undertaken to build and sustain communities so that they can survive and raise the next generation. With these actions, people challenge dominant, controlling images with new ones that better represent themselves (Omi and Winant 1994; Takaki 1993). Sharing new images with the majority groups and others in the society challenges the ideological justification for their unequal treatment.

Section XIII looks at the various ways that knowledge and social action are used to end injustice. The goal is social change so that more people can enjoy the fruits of a more equitable society. This section opens with Thomas F. Pettigrew ("Post-Racism? Putting Obama's Victory in Perspective"), who analyzes the many factors that came together to elect our first African American president in 2008. Pettigrew makes it clear that racism has not vanished from the nation, but many factors created this promising outcome. Yet he also notes that groups have to continue to mobilize for racial justice.

Part of this progressive change is recognizing that people benefit from work and educational environments that are diverse and supportive of workers. Frank

Dobbin, Alexandra Kalev, and Erin Kelly ("Diversity Management in Corporate America") study different corporate strategies designed to develop more diversity in corporate management to identity those interventions that are most effective. Their work can help you think about how to build workplaces that are positive for all who work there and are successful in promoting new inclusion.

The privileges of many people are often invisible to us. Often people saddled with disadvantages can see them, while those with privileges cannot. Thus, changing the perspectives of all parties is important in working for social change. How do you work for change? The Southern Poverty Law Center ("Ways to Fight Hate") offers some actions that individual people can take. This important center has long been active in challenging injustices. They have also been pioneers in educating people and monitoring hate crimes. Here, they offer ten suggestions of actions that anyone can take to have an immediate impact on those around them.

## REFERENCES

Marks, Carole. 1989. *Farewell—We're Good and Gone: The Great Black Migration.* Bloomington: Indiana University Press.

Omi, Michael, and Howard Winant. 1994. *Racial Formation in the United States.* New York: Routledge.

Takaki, Ronald. 1993. *A Different Mirror: A History of Multicultural America.* Boston: Little, Brown and Company.

# FACE THE FACTS:

## Percent of Each Group in Management Occupations

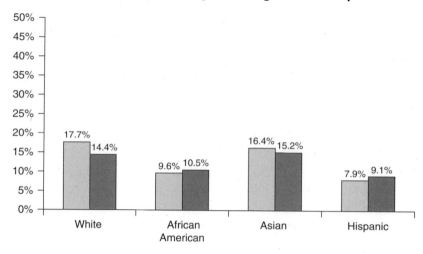

## Percent of Each Group in Professional Occupations

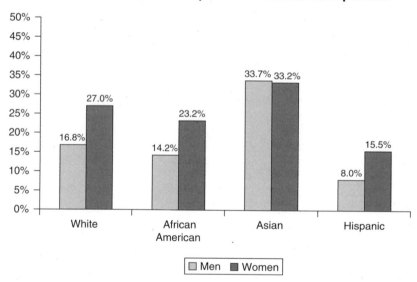

SOURCE: U.S. Department of Labor, 2010. *Employment and Earnings.* www.dol.gov.

***Think about it:*** These data reflect very broad categories that would include very different occupations. For example, "management" includes managers of both very small as well as very large businesses. And "professions" includes a wide range of things, including teachers, social workers, engineers, physicians, and so forth. Given that, what questions would you want to ask about the diversity within management and the professions? How might a more or less diverse management affect the experiences of women and people of color?

# 53

# Post-Racism?

## Putting President Obama's Victory in Perspective

THOMAS F. PETTIGREW

*How did Barack Obama win the 2008 election, and what does that victory mean about race relations in the United States? Pettigrew argues that we are not beyond race, but clearly in a new phase of race relations in this country. He identifies the many forces that made the election of an African American man possible, but also how the government has to address many persistent problems in society to achieve real social justice.*

## INTRODUCTION

President Barack Obama arrived upon the public stage almost two decades ago when he was elected the first African American editor of the prestigious *Harvard Law Review*. Publishers soon offered him book contracts. In his engaging and candid volume, *Dreams From My Father*, Obama modestly attributes this sudden attention to "… America's hunger for any optimistic sign from the racial front—a morsel of proof that, after all, some progress has been made." (Obama 1995, p. xviii)

Since Obama won the presidency, this national hunger for racial optimism is overflowing with self-congratulation. The codeword for this phenomenon has become "post-racism"; the claim is that we are now entering a new era in America in which race has substantially lost its special significance….

Clearly, the momentous occasion of Obama's presidential triumph needs to be placed in perspective. Toward this end, six points deserve special emphasis. First, it took "a perfect storm" for Obama to win. Second, many White bigots actually voted for Obama. Third, two logical fallacies underlie the post-racism contention. Fourth, racist attitudes and actions repeatedly erupted during both the primary and final campaigns. Fifth, in many southern states, the White vote for Obama significantly shrunk from the 2000 and 2004 Democratic Party totals, and the elderly White Democratic vote throughout

SOURCE: *Du Bois Review*, 6:2 (2009): 279–292 © 2009 W. E. B. Du Bois Institute for African and African American Research.

the nation similarly declined. Finally, the increases in both voter registration and turnout of young, minority, and independent voters were critical. Let's examine each of these inter-connected phenomena, because each undercuts the sweeping claim of post-racism.

## THE PERFECT STORM

It took four decades of steady Black political progress to prepare the scene for the 2008 presidential race. The Voting Rights Act of 1965 was without question the most effective civil rights legislation of the twentieth century. Just prior to the act, only about one hundred Black Americans held elected office in the entire United States; today almost 10,000 Blacks occupy elected offices of all kinds....

Yet even with this long preparation, Obama's decisive victory in 2008 required "a perfect storm." Not to take away anything from his obvious intelligence, cool charisma, oratorical skill, and well-organized campaign, Obama has also needed some political good fortune throughout his career. When running for the U.S. Senate in Illinois in 2004, he trailed in the polls in both the primary and final races. But his major opponent in the Democratic Party primary, Blair Hull, had to drop out of the race due to a marital scandal. Then, in the final race, his major Republican foe, Jack Ryan, also had to drop out of the race due—once again—to a marital scandal.

Next, Obama had the good fortune of being selected to give the keynote speech at the 2004 Democratic Party Convention, and this provided him with a national audience to demonstrate his remarkable speech-making ability. The national buzz about him as a future presidential aspirant started with that memorable address.

In the 2008 presidential campaign, he revealed surprising toughness, confidence, and the ability to assemble and manage a dedicated and able staff. His caucus success in Iowa was an absolutely necessary start to his campaign. This victory in the sixth Whitest state in the nation (89% non-Hispanic White in 2007) convinced many voters, especially skeptical Blacks, that he was indeed a serious candidate who, unlike previous Black presidential candidates, could garner White votes....

Moreover, 2008 offered the Democratic Party a golden opportunity to revive itself after eight long years of George W. Bush's disastrous administration—six years of which the Republican Party controlled all three branches of government....

Yet even all these promising elements might well not have been enough for Obama's victory. The final and possibly necessary component of the perfect storm was the sudden and severe turndown in the economy. Indeed, 63% of voters in an exit-poll listed the economy as most important issue facing the nation compared with only 10% listing the Iraq war (Pew Research Center 2008). With McCain widely perceived as incapable of handling the crisis and

Obama coolly following the advice of leading economists, the final piece of the necessary conditions were in place for a victory of the first African American president....

## MANY BIGOTS VOTED FOR OBAMA

A fundamental mistake of the post-racism advocates is their assumption that a White vote for a Black candidate for president is proof that the person is not prejudiced and would not racially discriminate in other situations. Given the gravity of the nation's economic crisis, this assumption was especially mistaken in 2008. Prejudiced people, like all of us, act in diverse, even conflicting ways in different situations. Hence, one survey found that a third of White Democrats ascribed one or more negative adjectives—violent, boastful, lazy, irresponsible— to African Americans, yet 58% of them still supported Obama. And while Nebraska voters gave one of its electoral votes to Obama, they also passed by 58% to 42% a ban on race- and gender-based affirmative action. Previously voters in California, Michigan and Washington State had passed similar referenda, yet all three provided Obama wide winning margins in 2008....

## TWO LOGICAL FALLACIES

The post-racism thesis suffers from two logical and interrelated fallacies: *the ecological fallacy* and *the constant turnout composition fallacy*. The *ecological fallacy* draws conclusions about individuals from macro-level data alone (Pettigrew 1996). It is a fallacy because macro-units are too broad to determine individual data, and individuals have unique properties that cannot be directly inferred from just macro data. This mistake is often seen in statements made about individual voters from aggregate voting results alone. For example, television's talking heads often are confused by the fact that rich states tend to support Democrats while poor states tend to support Republicans—yet rich individual voters tend to vote Republican while the poor overwhelmingly vote Democratic....

The *constant turnout composition fallacy* involves the assumption that the presidential electorate in 2008 was essentially the same as it was in 2004—thus, easy comparisons can be made between the two elections without allowing for these changes.... In 2004, both political parties achieved historically large turnouts. Four years later, the Republican turnout fell with many on the extreme right of the party failing to vote. But the skillful Obama campaign orchestrated a record number of Democrats coming to the polls—especially among the young, minorities, political independents and the most liberal wing of the Party. So, rather than simply representing a massive change in the nation's racial attitudes, the 2008 vote for Obama reflected, at least in part, a shift in the electorate from generally prejudiced right-wingers to generally less prejudiced moderates and left-wingers....

## RACIST ATTITUDES AND ACTIONS
## DURING THE CAMPAIGN

One might think from reading the post-racism advocates that the 2008 presidential campaign unfolded with scant signs of the nation's traditional racism. This impression is far from the truth. Expressions of racist sentiments in many forms, from violence and campaign tactics to verbal slips, were ever present.

Racial violence escalated and erupted in the North as well as the South. Cross-burnings, threats, intimidation, and racist graffiti proliferated across the nation but typically did not receive nationwide publicity. In New York, a Black teenager was assaulted by bat-wielding White men shouting, "Obama." School children on a bus in Idaho chanted "assassinate Obama"; and, on election night, a Black church in Massachusetts was burned in a suspicious fire. The Secret Service reported that there had been more threats on Obama's life than for any previous president-elect. Ironically, some leaders of violent White supremacist groups actually advocated voting for Obama. They reasoned that a Black president would ignite their movements, swell their ranks, and lead to more violence against African Americans....

One can always shrug off these horrific occurrences as the work of a mere lunatic fringe. But unadulterated racism also arose in elite circles in the campaign....

## ELDERLY AND SOUTHERN WHITES RESISTED

Elderly and Southern Whites demonstrated that racism is alive and well in these sizeable segments of the American population. Largely overlooked because they do not fit with the optimistic post-racism view, many of these Americans revealed in exit-polls that they had voted against Obama largely on racial grounds.... Whites older than sixty-four years were the only age group to vote more Republican than in 2004 (Pew Research Center 2008). And these data emerge despite the fact that four years had passed with many highly prejudiced elderly Whites having died and many of the elderly now facing economic problems.

In four Southern states—Alabama, Arkansas, Louisiana, and Mississippi—Whites voted less for Obama than they had in 2004 for the regionally unpopular John Kerry....

To hail the end of racism when these large population groups still strongly resist racial change is obviously premature. But change is evident. The oldest cohort of White Americans is dying off, and more tolerant cohorts are replacing them. Likewise, two processes are changing the South. New residents, Hispanics as well as Northerners, have come into the South; and the region's modest desegregation has produced a more liberal young cohort of White Southerners. Obama surprised many observers by winning in Florida, North Carolina, and Virginia—and these three states reflect these processes. All three are among the seven states that have received the largest percentages of newcomers in recent years, and they have witnessed somewhat more racial desegregation than most other Southern states.

## FOUR CRITICAL GROUPS: BLACKS, YOUTH, LATINOS, AND INDEPENDENTS

Diverging from previous Democratic Party presidential races, the Obama campaign adopted a bold, fifty-state strategy. It focused on so-called red states as well as blue states, with particular attention to Colorado, Indiana, Florida, New Mexico, North Carolina, and Virginia—all of which Obama won even though they had cast their electoral votes for Bush in previous elections. This strategy required special attention to increasing both the registration and turnout of African Americans, voters younger than thirty, Latinos, and political independents. All these efforts succeeded.

We have noted the success of the turnout efforts. Young, minority, and independent voters all increased their registrations and turnout while Republicans suffered a decline in turnout. Young voters have been historically difficult to register and vote; had they voted in their true proportion of the population, both Al Gore and Kerry would have easily defeated Bush. Yet the Obama campaign achieved record registration and turnout for White as well as Black youth....

Record minority registration and turnout were also key ingredients of Obama's triumph. Record total turnouts occurred in thirty-seven states and the District of Columbia in spite of lower Republican turnout....

Latinos joined Blacks in these trends. In Colorado, Florida, and New Mexico, sharp increases in Latino registrations, turnout, and support were critical for Obama's surprising wins in these previously Republican states....

Finally, 52% of political independents and 60% of moderates told exit pollsters they had just voted for Obama (Pew Research Center 2008a). This represented a 3% shift of independents and 6% shift of moderates from their voting four years earlier and was not generally predicted....

Obama's special appeal for new, young, and Latino voters will have long-term benefits for the Democratic Party. Karl Mannheim held that the political events of early life decisively shape each age cohort's political orientation: "... the older generation cling[s] to the re-orientation that had been the drama of their youth" (1952, p. 301)....

## WHAT CAN WE NOW EXPECT IN AMERICAN RACE RELATIONS?

Although Obama's presidential election does not mean the end of American racism, it obviously signifies a significant transformation in American race relations.

It is difficult for White Americans to appreciate how much it means for African Americans and other racial minorities as well. Even Black Europeans rejoiced.... African Americans have been in North America for four centuries. And they have been loyal to the United States throughout its history—building the nation and dying in its wars despite constant discrimination and second-class citizenship. Obama's victory, made possible by significant White support, belatedly rewards the faith Black America has somehow managed to maintain in their nation.

There are many other positive consequences of the 2008 election. Racial prejudice, especially among the young, is likely to recede—not precipitously but steadily. Americans are going to see almost everyday in their media for the next four to eight years an able, likable African American in charge of their nation. We know from research that such vicarious contact with high-status figures, even from television viewing, can erode prejudice and ease the anxiety that often accompanies interracial contact....

This positive attitude trend is helpful, of course. It makes later positive change more possible, but it does not directly address the core of African American distress today. The basic problems are structural—rampant housing segregation, poverty, job discrimination, poor education, massive imprisonment—and the list goes on. Most of these structural barriers are direct legacies of two centuries of slavery and another century of legal segregation. Thus, these barriers are deeply embedded in American society and have proven extremely difficult to correct.

The American presidency is powerful, but it has its definite limits. For the most part, at least early in his term, Obama will only be able to work on the margins of these issues. For example, he can weed out of government those who are adamant opponents of racial change appointed by the previous administration—even some members of the U.S. Civil Rights Commission. But he will be severely restricted by the economic crisis that overshadows everything, by a reactionary majority on the U.S. Supreme Court that has overturned *Brown v. The Board of Education* and other important High Court racial decisions without saying so directly (Pettigrew 2004), and by increasingly southern-based Republican minorities in both houses of Congress.

In short, there will undoubtedly be a new and positive tone to the Federal Government's treatment of minorities. But the needed fundamental structural remedies in American race relations will largely have to await a likely Obama second-term. And that, too, depends on Obama becoming a transformative president in the mold of Lincoln and Roosevelt.

## A FINAL WORD

Obama's decisive victory marks a momentous milestone in the history of America's most persistent domestic problem. I do not wish to diminish that fact in any way. Nevertheless, we cannot yet unfurl the "mission accomplished" banner. The post-racism era in American society has yet to arrive—even if we have taken a giant step forward. We are no longer "two nations," but neither are we a one, non-racialized nation.

Frank Rich sums it up best: "Obama doesn't transcend race. He isn't post-race. He is the latest chapter in the ever-unfurling American racial saga. It is an astonishing chapter.... But we are a people as practical as we are dreamy. We'll soon remember that the country is in a deep ditch, and we turned to the [B]lack guy not only because we hoped he could lift us up but because he looked like the strongest leader to dig us out" (2008).

## REFERENCES

Mannheim, K. (1952). The Problem of Generations. In K. Mannheim (Ed.), *Essays in the Sociology of Knowledge*, pp. 276–320. London: Routledge & Kegan Paul.

Obama, B. (1995). *Dreams from My Father: A Story of Race and Inheritance*. New York: Three River Press.

Pettigrew, T. F. (1996). *How to Think Like a Social Scientist*. New York: Harper Collins.

Pettigrew, T. F. (2004). Justice Deferred: A Half Century after *Brown v Board of Education*. *American Psychologist*, 59(6): 521–529.

Pew Research Center (2008). Inside Obama's Sweeping Victory. Retrieved January 19, 2009 at http://pewresearch.org/pubs/1023/exit-poll-analysis-2008.

Rich, F. (2008). Guess Who is Coming to Dinner. *New York Times*. Retrieved November 10, 2008 at http://www.nytimes.com/2008/11/02/opinion/02rich.

## DISCUSSION QUESTIONS

1. What does the fact that Obama's election required "a perfect storm" suggest about the state of race relations in the United States?

2. How did young people behave in this election, and what does that suggest about their future?

3. What does Pettigrew suggest is required for continued movement towards social justice?

# 54

## Diversity Management in Corporate America

FRANK DOBBIN, ALEXANDRA KALEV, AND ERIN KELLY

*Laws change and new doors of employment open, but as women and racial minorities move into management positions, how do they fare? Many are locked out of the social networks necessary for advancement. We need interventions to help these new groups succeed in corporations, but what strategies are successful? These researchers investigated different initiatives by corporations, finding that accountability is key to effectiveness in this arena.*

Do America's costly diversity-management programs work? Some do and some don't. The best idea is to assign clear responsibility for change.... In the decades since Congress passed the Civil Rights Act of 1964, firms have experimented with dozens of diversity measures. Consultants have been pushing diversity training, diversity performance evaluations for managers, affinity networks, mentoring programs, diversity councils, and diversity managers, to name a few. But some experts question whether diversity programs are counterproductive, raising the hackles of white men who, after all, still do most of the hiring and firing....

... Until recently, no one had looked systematically at large numbers of companies to assess which kinds of programs work best, on average. Our research shows that certain programs do increase diversity in management jobs—the best test of whether a program works—but that others do little or nothing.

The good news is that companies that give diversity councils, or diversity managers, responsibility for getting more women and minorities into good jobs typically see significant increases in the diversity of managers. So do companies that create formal mentoring programs. Much less effective are diversity training sessions, diversity performance evaluations for managers, and affinity groups for women and minorities. There is no magic bullet for the problem of inequality. Programs that work in one firm may not work in another. Programs that fail on average may be just what the Acme Rocket Co. needs. But we are beginning to understand what works in general and what does not....

SOURCE: *Contexts*, Vol. 6, Number 4: 21–27.

# FROM AFFIRMATIVE ACTION TO
# DIVERSITY MANAGEMENT

Until the 1960s, companies in both the North and the South practiced discrimination openly. King David Holmes had grown up in Connecticut and went off to college in the 1940s. When he returned with his degree at the end of the decade, he went to the local brass mill where family members had worked: "I came back, went in the employment office, said 'I want a job;' I filled out my form—'College graduate.'" Holmes wanted a job in sales, for which any college graduate should have been eligible. The personnel man made no bones about it: they were not hiring blacks in sales: "Oh, that's reserved," he said. They hired Holmes to make sheet metal in the old north mill, a job that required no reading or writing.

In his first year as president, John F. Kennedy decreed that companies wanting to do business with the federal government would have to take affirmative action to end discrimination of that sort, expanding orders from Roosevelt, Truman, and Eisenhower that applied to military contractors. The year after Kennedy's assassination, Lyndon Johnson signed the Civil Rights Act of 1964, outlawing discrimination in education, housing, public accommodations, and employment.

In principle, Kennedy's affirmative action order and the Civil Rights Act made discrimination illegal in the private sector. In practice, most employers did not think the order would much affect them. Most thought they would have to take down "No Negroes" signs, and that would be the end of it.

Over time, however, Congress and the courts strengthened equal opportunity legislation. By the mid-1970s, the federal government was getting thousands of complaints a year, and personnel experts were setting up equal opportunity programs in the hope of fending off complaints and providing a defense in lawsuits. Because the law did not specify how employers should combat discrimination, firms experimented. Early on, they installed formal job tests, job ladders, and annual performance evaluations—measures designed to take the guesswork (and bias) out of hiring and promotion. The Supreme Court vetted many of these programs but overturned others, such as job tests that had been used as a subterfuge to exclude blacks from the workforce.

By the time of Ronald Reagan's election in 1980, most big employers had hired equal opportunity managers, if not entire departments, and were in the process of creating race-relations workshops, special recruitment systems, and a host of programs designed to improve opportunities for women and minorities. After Reagan's campaign vow to put the brakes on federal regulation, equal opportunity managers began to rebrand themselves as diversity managers. New programs emphasized business preparedness for an increase of women and minorities in the workforce. The forward-thinking executive would plan for change not because it was the cautious thing to do given the likelihood of lawsuits, but because it was the rational thing to do given growing numbers of African-Americans, Latinos, Asians, and women in the workforce.

Employers have invested in three broad approaches to increasing diversity: (1) changing the attitudes and behavior of managers; (2) improving the social ties of women and minorities; and (3) assigning responsibility for diversity to special managers and taskforces. We set out to discover which of these approaches works best.

## THE STUDY

Texaco and Coca-Cola spent more than $7 million a year on diversity programs created under their consent decrees. Even Fortune 500 companies not under consent decrees can spend millions of dollars a year on trainers alone, without accounting for lost work time or travel expenses. Yet it is not easy to determine whether these programs actually improve opportunity for women and minorities.

Our challenge was to get accurate data on workforce diversity and on diversity programs for enough companies, over a long enough period of time, so that we could use sophisticated statistical techniques to isolate the effects of, say, diversity training on the percentage of black women in management. Our first break came when we obtained access to the EEOC's EEO-1 reports under a federal program requiring strict confidentiality. Every private firm with 100 or more employees must submit a form each year detailing workforce race, gender, and ethnicity in nine broad job categories. Those data are in good shape back to 1971....

We surveyed 829 of the companies covered in the EEOC's massive data file, putting together a life history of employment practices at each. We asked companies whether they had used dozens of different programs, and when those programs had been in place. Our research question was simple: If a company adopts a particular diversity program, what effect does it have on the share of women and minorities in management? To answer the question, we conducted statistical analyses on the 829 firms over 31 years.... Our analyses considered more than 60 workplace characteristics that might affect diversity so as to isolate the effects of diversity taskforces, training, and so on.

## CHANGING MANAGERS' BEHAVIOR

One source of gender and racial inequality in the workplace is stereotyping and bias among managers who make hiring and promotion decisions. Research shows that educating people about members of other groups may reduce stereotyping. Such training first appeared in race-relations workshops conducted for government agencies and large federal contractors in the early 1970s. An industry of diversity trainers emerged in the 1980s to argue that people were unaware of their own racial, ethnic, and gender biases and that sensitivity training would help them to overcome stereotypes....

Some diversity experts dismissed training, arguing that attitudes are difficult to alter but that behavior can be changed with feedback. Instead they supported performance evaluations offering feedback on managers' diversity efforts. Laboratory experiments support this idea, showing that subjects who are told that their deci-

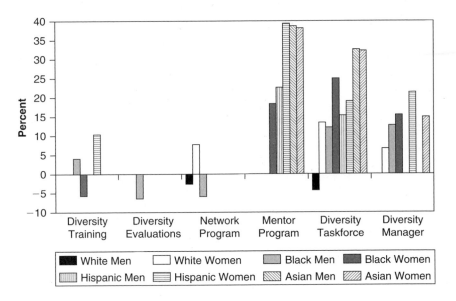

**FIGURE 1**

sions will be reviewed are less likely to use stereotypes in assigning people to jobs. General Electric first created equal opportunity performance evaluations in 1969, arguing that the key to achieving equal opportunity is a "measurement system with rewards and penalties designed to produce behavioral changes in managers." ...

The current enthusiasm for training and evaluations has made them widely popular. Four in ten of the firms we surveyed offered training, and one in five had diversity evaluations. But these programs did not, on average, increase management diversity.

Our findings show shows the percentage change in the proportion of managers from each group as a consequence of each program—holding the effects of other factors constant.... African-American men held about 3 percent of management jobs in 2002, and so a 25 percent increase does not translate into big numerical gains. For white women, the numerical changes are more dramatic. But for all groups, an increase is an increase. Where the effect is not statistically significant, we show it in the figure as zero.

Training had small negative effects for African-American women and small positive effects for Hispanic women. Evaluations had only small negative effects on black men. In very large workplaces, with 1,500 or more workers, diversity training has a small positive effect on other groups, but nothing like the effects of the mentoring programs and taskforces discussed below. Evaluations were no more effective in large workplaces.

Some psychological research supports our finding that training may be ineffective. Laboratory experiments and field studies show that it is difficult to train away stereotypes, and that white men often respond negatively to training—particularly if they are concerned about their own careers. If training cannot eliminate stereotypes, and if it can elicit backlash, perhaps it is not surprising that, on average, it does not revolutionize the workplace....

Overall, companies that try to change managers' behavior through training and evaluations have not seen much change. That is disappointing, because training is the single most popular program and, by most accounts, the most costly, and because many companies have put their money on diversity evaluations in recent years.

## CREATING SOCIAL CONNECTIONS

Another way to view the problem of inequality in management jobs is from the supply side. Do women and minorities have the social resources needed to succeed? Many firms have well-educated white women and minorities in their ranks who fail to move up, and some sociologists suggest this is because they lack the kinds of social connections that ambitious white men develop easily with coworkers and bosses. Strong social networks and mentors have long been thought crucial to career success. Mark Granovetter's 1974 classic book, *Getting a Job,* showed that people often find jobs through their social networks.... The idea that women and minorities have trouble moving up because they lack good network contacts and mentoring relationships has caught on among diversity experts.

Consultants argued that formal programs could help women and minorities develop social networks at work. Company affinity networks became popular.... By 2001 Texaco had established seven employee networks, and in 2004 Coca-Cola set up a "networking for success" program. Under these affinity networks, people gather regularly to hear speakers and talk about their experiences.

Many companies also put in formal mentoring programs that match aspiring managers with seasoned higher-ups for regular career-advising sessions. These programs can target women and minorities but are often open to all employees....

When it comes to improving the position of women and minorities in management, network programs do little on average, but mentor programs show strong positive effects. In the average workplace, network programs lead to slight increases for white women and decreases for black men. Perhaps that is not surprising, as studies of these network programs suggest that while they sometimes give people a place to share their experiences, they often bring together people on the lowest rungs of the corporate ladder. They may not put people in touch with what they need to know, or whom they need to know, to move up.

Mentor programs, by contrast, appear to help women and minorities. They show positive effects for black women, Latinos, and Asians. Moreover, in industries with many highly educated workers who are eligible for management jobs, they also help white women and black men. Mentor programs put aspiring managers in contact with people who can help them move up, both by offering advice and by finding them jobs. This strategy appears to work.

## MAKING SOMEONE RESPONSIBLE

... The best way to bring about change, management theorists argue, is to make new programs the responsibility of a person or a committee. Instead of just sending around a memo asking managers to practice equal opportunity, military contractors, fearful of losing their contracts in the 1960s, put someone in charge of affirmative action programs. That person's sole job was to make sure the firm hired, and kept, women and minorities. Now that is the job of the diversity manager, or chief diversity officer. Recently firms have also put in taskforces, or diversity councils, comprising managers from different departments and charged with finding ways to increase diversity. The accounting giant Deloitte and Touche set up taskforces responsible for analyzing the gender gap, recommending remedial steps, and establishing systems for monitoring results and holding managers responsible....

Our analyses show that making a person or a committee responsible for diversity is very effective. Companies that establish taskforces typically see small decreases in the number of white men in management, and large increases for every other group (white men dominate management to begin with, so a small percentage decrease for white men can make space for big percentage gains for the other groups). Firms that put in diversity managers see increases for all groups of women, and for black men.

In interviews, executives tell us that diversity managers and taskforces are effective because they identify specific problems and remedies. If the taskforce sees that the company has not been recruiting African-American engineers, it will suggest sending recruiters to historically black colleges. If a company has trouble retaining women, the diversity manager may talk to women at risk of leaving and try to work out arrangements that will keep them on the job. Managers and taskforces feel accountable for change, and they monitor quarterly employment data to see if their efforts are paying off. If not, it is back to the drawing board to sketch new diversity strategies. Taskforces may be so widely effective, some tell us, because they cause managers from different departments to "buy into" the goal of diversity.

## LESSONS FOR EXECUTIVES

What are the most cost-effective strategies for increasing diversity? ... Many companies were not investing in the strategies that have proven most effective. Three of the four most popular programs—diversity training, evaluations, and network programs—have no positive effects in the average workplace. The two least popular initiatives, mentoring and diversity managers, were among the most effective.

On average, programs designed to reduce bias among managers responsible for hiring and promotion have not worked. Neither diversity training to extinguish stereotypes, nor diversity performance evaluations to provide feedback and oversight to people making hiring and promotion decisions, have accomplished much. This is not surprising in the light of research showing that stereotypes are difficult to extinguish.

There are two caveats about training. First, it does show small positive effects in the largest of workplaces, although diversity councils, diversity managers, and mentoring programs are significantly more effective. Second, optional (not mandatory) training programs and those that focus on cultural awareness (not the threat of the law) can have positive effects. In firms where training is mandatory or emphasizes the threat of lawsuits, training actually has negative effects on management diversity. Psychological studies showing backlash in response to diversity training suggest a reason: managers respond negatively when they feel that someone is pointing a finger at them. Most managers are still white and male, so forced training focusing on the law may backfire. Unfortunately, among firms with training, about three-quarters make it mandatory and about three-quarters cover the dangers posed by lawsuits.

Our findings suggest that firms should look into how they can make training and evaluations more effective. Some experts suggest that training should focus on hiring and promotion routines that can quash subjectivity and bias. Others suggest that diversity performance evaluations based on objective indicators (minority hiring and retention, for instance) and tied to significant incentives (attractive bonuses, or promotions) would work. But few companies follow that model.

One piece of good news is that mentor programs appear to be quite effective. Such programs can provide women and minorities with career advice and vital connections to higher-ups. While mentoring can be costly—involving release time for mentors and mentees, travel to meetings, and training for both groups— the expense generally pays off. One reason mentoring may not elicit backlash, as training often does, is that many companies make it available to men and women and to majority and minority workers. The theory is that women and minorities less often find mentors on their own (if they did, there would be less of a problem to start with), and thus benefit more from formal programs.

Another piece of good news is that companies that assign responsibility for diversity to a diversity manager, or to a taskforce made up of managers from different departments, typically see significant gains in diversity. Management experts have long argued that if a firm wants to achieve a new goal, it must make someone responsible for that goal. To hire a diversity manager or appoint a taskforce or council is to make someone responsible. Both managers and taskforces scrutinize workforce data to see if their efforts are paying off, and both propose specific solutions to the company's problems in finding, hiring, keeping, and promoting women and minorities. Taskforces have the added advantage of eliciting buy-in—they focus the attention of department heads from across the firm who sit together with a collective mission.

Many executives bemoan the fact that their companies saw increases in diversity in the 1970s but have been at a standstill ever since. Innovative strategies come and go, but diversity in management jobs does not seem to budge....

If companies regrouped and put their time and energy into programs known to work elsewhere, they would likely see small but steady increases in the representation of women and minorities in management jobs. Programs that assign responsibility for change and that connect promising management talent with mentors (another, less formal way to assign responsibility), seem to hold the best hope. Managers might also spend more time assessing programs that don't seem to be working and trying to figure out how to make them effective.

## DISCUSSION QUESTIONS

1.  Why do corporations develop plans and policies to enhance diversity?
2.  Why is "diversity training" problematic for many managers?
3.  What types of strategies demonstrate the most success in increasing the presence of women and racial minorities?

# 55

# Ten Ways to Fight Hate:
## A Community Response Guide

### SOUTHERN POVERTY LAW CENTER

*Many social problems are centuries old, even if they have changed in their scope. What can an individual do to make a difference? The Southern Poverty Law Center has long worked for social justice, and they offer ten suggestions to fight hate and make a difference in your own community.*

## 1 ACT

Do something. In the face of hatred, apathy will be interpreted as acceptance— by the perpetrators, the public and, worse, the victims. Decent people must take action; if we don't, hate persists.

## 2 UNITE

Call a friend or co-worker. Organize allies from churches, schools, clubs and other civic groups. Create a diverse coalition. Include children, police and the media. Gather ideas from everyone, and get everyone involved.

## 3 SUPPORT THE VICTIMS

Hate-crime victims are especially vulnerable, fearful and alone. If you're a victim, report every incident—in detail—and ask for help. If you learn about a hate-crime victim in your community, show support. Let victims know you care. Surround them with comfort and protection.

## 4 DO YOUR HOMEWORK

An informed campaign improves its effectiveness. Determine if a hate group is involved, and research its symbols and agenda. Understand the difference between a hate crime and a bias incident.

SOURCE: Southern Poverty Law Center, . Montgomery, Alabama, 2010. www.splc.org

## 5 CREATE AN ALTERNATIVE

Do not attend a hate rally. Find another outlet for anger and frustration and for people's desire to do something. Hold a unity rally or parade to draw media attention away from hate.

## 6 SPEAK UP

Hate must be exposed and denounced. Help news organizations achieve balance and depth. Do not debate hate-group members in conflict-driven forums. Instead, speak up in ways that draw attention away from hate, toward unity.

## 7 LOBBY LEADERS

Elected officials and other community leaders can be important allies in the fight against hate. But some must overcome reluctance—and others, their own biases— before they're able to take a stand.

## 8 LOOK LONG RANGE

Promote tolerance and address bias before another hate crime can occur. Expand your community's comfort zones so you can learn and live together.

## 9 TEACH TOLERANCE

Bias is learned early, usually at home. Schools can offer lessons of tolerance and acceptance. Sponsor an "I Have a Dream" contest. Reach out to young people who may be susceptible to hate-group propaganda and prejudice.

## 10 DIG DEEPER

Look inside yourself for prejudices and stereotypes. Build your own cultural competency, then keep working to expose discrimination wherever it happens—in housing, employment, education and more.

## DISCUSSION QUESTIONS

1. Can you think about a time when you took action in the face of an injustice? How did you feel about the action that you took?

2. Why do you think the Southern Poverty Law Center suggests that people organize rather than act alone?

# Student Exercises

1.  There is a long history of people working for social change. We can look to the past, but also to current events. Investigate your campus and/or your community for a person who is involved in working for social change. The person can be involved in any number of causes: the environmental justice movement, fair housing, housing the homeless, health care, political reform, or educational reform. Find out some details about the mission and objective of the organization that this person works with via the Internet or media. Arrange an interview to talk with the person about his or her activities. What can you learn from the person that might be missing in the media? Can you see how this person might have a perspective on issues that is different from the dominant viewpoint of the institution she or he wants to change?

2.  We all want work environments that are supportive and promote growth. Reviewing the types of programs that Dobbin, Kalev, and Kelly researched, think about what type of program you would like to see in your current or future workplace. What might be some of the barriers that you might face as you attempt to move up in a professional or managerial position? How might they be related to your race, gender, and social class? Why is mentoring and social networking important for all people in the workplace? Can you see the difference that intervention can make in the workplace?

# Index